05-WEM-1COa

ESTIMATION OF MERCURY BIOACCUMULATION POTENTIAL FROM WASTEWATER TREATMENT PLANTS IN RECEIVING WATERS: PHASE II

by:

J. David Dean
ArcTellus

Robert P. Mason
University of Connecticut

2009

WERF

The Water Environment Research Foundation, a not-for-profit organization, funds and manages water quality research for its subscribers through a diverse public-private partnership between municipal utilities, corporations, academia, industry, and the federal government. WERF subscribers include municipal and regional water and wastewater utilities, industrial corporations, environmental engineering firms, and others that share a commitment to cost-effective water quality solutions. WERF is dedicated to advancing science and technology addressing water quality issues as they impact water resources, the atmosphere, the lands, and quality of life.

For more information, contact:
Water Environment Research Foundation
635 Slaters Lane, Suite 300
Alexandria, VA 22314-1177
Tel: (703) 684-2470
Fax: (703) 299-0742
www.werf.org
werf@werf.org

This report was co-published by the following organization. For non-subscriber sales information, contact:

IWA Publishing
Alliance House, 12 Caxton Street
London SW1H 0QS, United Kingdom
Tel: +44 (0) 20 7654 5500
Fax: +44 (0) 20 7654 5555
www.iwapublishing.com
publications@iwap.co.uk

Library of Congress Catalog Card Number: 2008942452

IWAP ISBN: 978-1-84339-356-6/1-84339-356-5

ArcTellus, University of Connecticut

The research on which this report is based was developed, in part, by the United States Environmental Protection Agency (EPA) through Cooperative Agreement No. CR83155901-1 with the Water Environment Research Foundation (WERF). However, the views expressed in this document are solely those of ArcTellus and the University of Connecticut and neither EPA nor WERF endorses any products or commercial services mentioned in this publication. This report is a publication of WERF, not EPA. Funds awarded under the Cooperative Agreement cited above were not used for editorial services, reproduction, printing, or distribution.

ACKNOWLEDGMENTS

Report Preparation

Co-Principal Investigators:

J. David Dean
ArcTellus

Dr. Robert P. Mason
University of Connecticut

Project Team:

Patricia McGovern Engineers

Frontier Geosciences, Inc.

Internal Peer Review Team:

Dr. Helen Hsu-Kim
Duke University

Tim Tuominen
Western Lake Superior Sanitation District

Dr. James Wiener
University of Wisconsin, LaCrosse, WI

Project Subcommittee

Michele Pla
Bay Area Clean Water Agencies (BACWA)

Dr. P.L. Fan, Ph.D.
American Dental Association

Keith Linn
Northeast Ohio Regional Sewer District

Richard Looker
San Francisco Bay Regional Water Control Board

Dr. Michael S. Connor
PSC Chair, San Francisco Estuary Institute

David W. Tucker
City of San Jose, CA

Vicki Fry
Sacramento Regional County Sanitation District

Water Environment Research Foundation Staff

Director of Research:	Daniel M. Woltering, Ph.D.
Program Manager:	Jane M. Casteline, M.S.

ABSTRACT AND BENEFITS

Abstract:

This project was initiated in response to the establishment of mercury TMDLs around the country and issues raised by this process, specifically concerning the issue of mercury bioavailability. While many TMDLs recognize that point sources constitute a small fraction of the mercury load to a waterbody, a question has been raised concerning the *relative* bioavailability of mercury coming from various sources. For instance, is the mercury discharged from a wastewater treatment plant more or less bioavailable than mercury in precipitation, mercury in urban stormwater, or mercury in sediments? This project seeks to address this question by developing a reliable definition and approach to estimating bioavailability, by profiling various sources of mercury in a watershed with regard to the species of mercury present and by profiling those factors or conditions in either the effluent or the receiving water that enhance or mitigate the bioavailability of those forms.

The report consists of two parts. Phase I is a background document for evaluating the bioavailability of mercury in wastewater effluents and receiving waters and establishes relevant project objectives. Phase II is a guidance document for wastewater treatment professionals interested in assessing the bioavailability of mercury in their wastewater, comparing it to other sources, and assessing changes in bioavailability in their effluent when it is mixed in a receiving waterbody.

It is important to note that this project deals mainly with the estimation of bioavailable species of mercury and how they are affected by wastewater treatment, although the acquisition of data to support bioaccumulation studies is addressed in the Phase II report. It is important to note that bioavailability does not necessarily equate to bioaccumulation. However, provided that bioavailable mercury species are present in an aquatic ecosystem, bioaccumulation will occur. The degree to which this happens will depend upon the biological species and the dynamics of the food chain present. Bioaccumulation and consumption must occur in order for humans or wildlife to be at risk.

The project concludes that, based on available data and bioavailability as defined in this report, wastewater effluent is one of the lowest among the sources evaluated with respect to mercury bioavailability due to its typically low levels of methylmercury. Due to their typically low levels of suspended solids and low levels of mercury in solids, wastewater treatment plants employing post-secondary treatment should not contribute appreciably to local sediment mercury burdens.

Benefits:

- Provides a better understanding of bioavailable mercury and the effects of wastewater treatment on bioavailable mercury species.
- Includes results of a seven-plant survey of reactive- and methylmercury in treated wastewater.
- Provides a guidance document for wastewater treatment professionals on how to assess and manage risk vis-a-vis mercury discharges, in light of their contribution to overall mercury loads in various types of receiving waters.

Keywords: Mercury, bioavailability, bioaccumulation, wastewater effluent, wastewater treatment, receiving waters, TMDLs.

TABLE OF CONTENTS

LIST OF TABLES

LIST OF FIGURES

LIST OF ACRONYMS

ADA	American Dental Association
AVS	Acid volatile sulfides
BAF	Bioaccumulation factor
B-Hg	Bioavailable mercury
%B-Hg	Percent bioavailable mercury normalized to total mercury (also relative bioavailable Hg)
BNR	Biological nutrient removal
BOD	Biological oxygen demand
CBDA	California Bay-Delta Authority
CCC	Criterion continuous concentration
CDF	Cumulative distribution function
CMC	Criterion maximum concentration
COD	Chemical oxygen demand
CV	Coefficient of variation
CVAFS	Cold vapor atomic fluorescence spectroscopy
D-Hg	Dissolved mercury
D-MeHg	Dissolved methylmercury
DO	Dissolved oxygen
DOC	Dissolved organic carbon
DOE	Department of Energy
DOM	Dissolved organic matter
EDTA	Ethylene diamine tetra acetic acid
EMC	Event mean concentration
EPRI	Electric Power Research Institute
fB-Hg	Fraction of bioavailable mercury normalized to total mercury
fMeHg	Fraction of methylmercury normalized to total mercury
fR-Hg	Fraction of reactive mercury normalized to total mercury
$f_{U(x)}$	Fractional uncertainty in the variable 'x'
GAC	Granular activated carbon
Hg	Mercury

Hg^0	Hg species in the nonionic inorganic form
Hg^{2+}	Hg species in the inorganic (2+) valence state
Hg^{II}	Hg species in the ionic inorganic (II) form
I-Hg	Inorganic mercury
IB-Hg	Inorganic bioavailable mercury
ICP-MS	Inductively-coupled plasma mass spectrometry
ID-Hg	Inorganic dissolved mercury
INR-Hg	Inorganic non-reactive mercury
K_D	Partition coefficient
k_{demeth}	First-order demethylation constant
k_{meth}	First-order methylation constant
LA	Load allocation
LC	Lethal concentration
LIS	Long Island Sound
$logK_D$	base 10 logarithm of the partition coefficient
LOI	Loss on ignition
MDL	Maximum detection limit
MeHg	Methylmercury (equivalent to CH_3Hg^+ or CH_3Hg^{II})
%MeHg	Percent MeHg normalized to total mercury (also relative MeHg)
MGD	Million gallons per day
MINEQL	A chemical equilibrium model
MINTEQ	A chemical equilibrium model
MOS	Margin of safety
MRL	Maximum reporting limit
NACWA	National Association of Clean Water Agencies
NOM	Natural organic matter
NOAA	National Oceanographic and Atmospheric Administration
NPDES	National Pollutant Discharge Elimination System
OM	Organic matter
ORP	Oxidation-reduction potential
P-Hg	Particulate phase mercury
P-MeHg	Particulate methylmercury
POC	Particulate organic carbon

POM	Particulate organic matter
Q	flow rate
R-Hg	Reactive mercury
%R-Hg	Percent reactive mercury normalized to total mercury (also relative reactive Hg)
RO	Reverse Osmosis
RPD	Relative percent difference
SQRT	Square root
SRB	Sulfate reducing bacteria
SRM	Standard reference material
T-Hg	Total mercury
TMDL	Total maximum daily load
TOC	Total organic carbon
TSS	Total suspended solids
U.S. EPA	United States Environmental Protection Agency
U.S. FDA	United States Food and Drug Administration
USGS	United States Geological Survey
UV	Ultraviolet
U(x)	Uncertainty in the variable 'x'
VSS	Volatile suspended solids
WLA	Waste load allocation
WQT	Water quality target
WWTP	Wastewater treatment plant
YOY	Young of the year
ZID	Zone of initial dilution

EXECUTIVE SUMMARY

Mercury levels in fish tissue and the associated human health concern have led regulatory agencies to issue fish consumption advisories for over 2,000 waterbodies in the United States. The existence of fish consumption advisories in a waterbody have in turn been used as a basis for placing these waterbodies on the states' lists of impaired waters, the 303(d) lists. The sources of mercury are known to be diverse, as it is a naturally occurring element, although it is generally accepted by the scientific community that the emissions of mercury have increased several fold during the industrial period. Therefore, attempts to regulate mercury discharges to waterbodies have focused on point source dischargers and air emissions.

The TMDL process has been used to estimate mercury loadings to waterbodies, both to estimate the acceptable load that would result in the attainment of regulatory levels in fish tissue and to estimate the current load of mercury to the waterbody that is resulting in nonattainment. In the case of mercury, a particularly difficult aspect of this exercise arises from the fact that it must be transformed from inorganic mercury that is typically deposited or discharged to waterbodies (primarily divalent mercury) to methylmercury, the form that is bioaccumulated in fish. Overall, only a fraction of the total Hg is in a bioavailable form in any aquatic system.

Currently, there are no accepted techniques available to "trace" mercury from its emitted origin to its appearance in fish tissue. However, one approach toward tracing the bioaccumulative fraction is an assessment of the bioavailability of mercury from various mercury sources once it reaches a waterbody. It is reasonable that a source contributing mercury which is more difficult to transform into a bioaccumulative form should be of less concern than one discharging a similar amount of mercury which is more easily transformed. However, this assessment is more difficult than simply measuring who is discharging inorganic versus methylmercury. There are a number of species of mercury, other than methylmercury, that may be considered "bioavailable", especially if not only the instantaneous bioavailability but the overall potential bioavailability is to be assessed. Some of the forms of mercury may be bioaccumulated in their own right and some may be precursors to methylmercury formation. The presence or absence of these species are determined by the biogeochemistry of effluent waters and receiving waterbody types (i.e. fresh, brackish, marine), individual waterbody characteristics (i.e. pH, dissolved organic carbon (DOC) levels, dissolved oxygen (DO) levels, etc.), the ecosystem biological activity, and the aquatic foodwebs present to move bioaccumulative mercury up the food chain.

The current project was designed to address these issues and ascertain the potential for defining the bioavailability of Hg in sources and receiving waters. This project was divided into two phases and this document reports the results of Phase II. The main objective of the second phase of the project was to develop a guidance document that would assist wastewater treatment professionals to assess the bioavailability of mercury in their effluents when mixed with receiving waters. Receiving waters include fresh, brackish (estuarine), and marine systems. The methodology was to include the ability to perform both near-field and far-field assessments, in order to assess the instantaneous bioavailability upon mixing and the resultant longer term steady state situation. It was also directed that the methodology consider the uncertainty in the results and in making decisions about when to collect additional data to improve the assessments.

The guidance document divides the assessment in two tiers. The first is a screening level assessment and the guidance provides data and tools to complete it. The second tier is a detailed level assessment. In this case, the document describes the circumstances under which a more detailed assessment may be needed and provides guidance on how to collect the information required for a detailed assessment of bioavailability, both in terms of chemical and biological assessment of bioavailability.

Screening Assessment

Chapter 2.0 of this document presents a screening methodology for estimating the bioavailability of mercury in effluents when mixed with receiving waters. It is based on the concepts developed in the Phase I report, which defined bioavailable Hg as the sum of inorganic bioavailable mercury and methylmercury in an effluent or aquatic system. The methodology implements the computational approach developed in Phase I using a spreadsheet tool (herein after the "Bioavailability Tool" or the "Tool"). Using this tool, bioavailable mercury can be estimated for various effluent or effluent/receiving water mixtures, using several approaches. If the user has reactive mercury measurements as a surrogate for inorganic bioavailable mercury, then the tool will calculate bioavailable mercury from methyl and reactive mercury measurements. If the user does not have reactive mercury measurements, then the tool will estimate bioavailable inorganic mercury using the techniques identified in the Phase I report.

For receiving water and effluent mixtures, the tool produces both "near-field" and "far-field" bioavailability estimates. The near-field estimates are the concentration estimates resulting from initial mixing, assuming that net methylation reactions have not had time to occur. For the far-field estimates, the tool uses estimates of the strength of the receiving water methylating environment to estimate net methylation once these reactions have had sufficient time to be exerted (i.e., approaching steady-state conditions). The strength of the methylating environment is used to identify the ratio k_{meth}/k_{demeth}, where k_{meth} is a first order reaction constant for methylation of bioavailable inorganic mercury, which incorporates a number of processes in this instance, and k_{demeth} is a first-order demethylation reaction constant for methyl mercury. The steady-state computations assume that total bioavailable mercury (the sum of bioavailable inorganic mercury and methyl mercury) in the receiving water is constant following initial mixing and that the relative distribution between the two forms is governed by the selected reaction constants.

In freshwaters, there are two ways in which the strength of the methylating environment can be assessed. The first is by using the percent methyl/total ratio in the receiving water (%MeHg). The second depends on the values of other water quality parameters (i.e. pH, dissolved organic carbon) if methylmercury measurements are unavailable or inadequate in number. In estuarine waters, the strength of the methylating environment depends on the salinity level in the vicinity of the discharge. Methylation is stronger in waters with lower salinity and lower in waters with higher salinity. For marine waters, the methylating environment is weaker than in either fresh or estuarine waters. In fresh waters, the ratios of k_{meth}/k_{demeth} are estimated to vary between 0.2 and 16, depending of the relative strength of the methylating environment. In estuarine waters, the ratio is estimated to be 0.2 in waters with salinity less than 15 parts per thousand and 0.14 in waters with salinities greater than 15 parts per thousand. In marine waters the ratio is assumed to be a constant, 0.02. These ratios are incorporated in the Tool as tables in the spreadsheets and can be changed by the user if information more relevant to the aquatic system of interest is available.

The ratios of reaction constants for fresh, estuarine, and marine waters were developed on the assumption that at steady-state, the ratio of the reaction constants is equivalent to the ratio of methyl and inorganic bioavailable mercury in the water column. The researcher's approach, based on an examination of the literature, assumes that demethylation in the water column is primarily a photo-catalyzed process and that the demethylation constant can be reliably estimated for receiving waters. Once this is done, the ratio of reaction constants can be selected based on the methylation potential of the receiving water into which the effluent is mixed, and an estimate of the degree of methylation of reactive mercury in an effluent and receiving water mixture can be made.

The analysis performed to link methylation potential to ambient conditions in freshwaters, either in terms of %MeHg or to other water quality variables is presented in Section 2.4 of this report. It builds on data collected in national surveys by the USGS. Recent examination of databases by the USGS and the resultant production of so-called sensitivity maps, served as a basis for our examination of the potential for methylation within a system and its relationship to water quality variables. The validity of our approach to estimating k_{demeth}, k_{meth}, and for using the bioavailability ratio as a diagnostic variable, was assessed using data from the USGS (Brumbaugh et al., 2001; Krabbenhoft et al., 1999) for over 100 different rivers where a variety of water quality parameters were measured, including Hg and MeHg. While MeHg measurements were available directly from the data, inorganic bioavailable mercury (IB-Hg) was calculated for each river system using the relationship between DOC and IB-Hg developed in the Phase I report. The ratios of methylmercury and inorganic bioavailable mercury were then calculated. The result indicated a linear correlation between %MeHg and the ratio of MeHg/IB-Hg. In addition, the results of the data analysis indicated a relationship between pH, DOC, and the ratio of MeHg/IB-Hg. The initial parameterization of the resulting methodology was tested using three additional data sets, with favorable results.

To perform a similar assessment of estuarine and coastal waters, four data sets were utilized including two from Long Island Sound and the Hudson River, one from the Chesapeake Bay and the mid-Atlantic shelf, and one from San Francisco Bay. For the estuarine and coastal waters, the results of water column methylation/demethylation studies suggest that the water column demethylation rate is a factor of 10 lower than that for freshwater systems for comparable conditions of DOC and the relative number of binding sites per gram of DOC is approximately a factor of 10 lower for estuarine/coastal waters compared to that of freshwaters. The calculation of inorganic bioavailable mercury was adjusted to reflect this.

The results of these calculations showed the %MeHg is typically lower than that for freshwaters. However, as the DOC values for the various ecosystems are relatively low, the MeHg/IB-Hg ratios are also lower than those for freshwaters. This ratio provides an indication of the net potential for conversion of bioavailable inorganic Hg into MeHg, and therefore is a measure of the potential for overall MeHg production and transport within ecosystems, which is independent of total Hg concentration. The values of the ratio are reasonable in that they suggest that the lowest relative MeHg is offshore, and for the estuaries, is lowest for the Hudson River, and is similar for Long Island Sound, the Chesapeake Bay and San Francisco Bay. A plot of the k_{meth} determined versus the ratio MeHg/IB-Hg showed that the relationship falls on the same line as that of freshwater ecosystems but that the estuarine and coastal data is at the lower end of the range.

The Bioavailability Tool

In performing a first-tier or screening level assessment, the user is aided by a Bioavailability Tool which has been developed and is provided with the report. The Tool performs bioavailability computations in a spreadsheet environment. It is designed to address questions that should be of common interest to most users. The spreadsheet allows the user to assess the bioavailability of mercury in an effluent and compare it to another effluent or to a nonpoint source. It also allows the effluent to be mixed with fresh, brackish, and marine receiving waters. It predicts near-field concentrations of bioavailable mercury species (methylmercury and reactive mercury). It also uses one of several user-selected methods to assess the fate of methylmercury and reactive mercury in the far-field, once methylation/ demethylation reactions poised to occur when the effluent is mixed with the receiving water have occurred. Inputs can, for the most part, be developed with information in the Phase I and II reports. However, measurement of key parameters and water quality variables is strongly recommended for accurate assessment. This report provides extensive examples on the use of the Tool and guidance for parameter estimation.

Detailed Assessment

This report also provides guidance for making more detailed, or tier-two, assessments. In doing so, it deals with the quantification of uncertainty in the desired results and provides a means for the user to assess the uncertainty in results given the uncertainty in the inputs using error propagation theory. The guidance provides an approach to assessing whether additional data is needed based on the desired uncertainty in the result of an initial assessment using available data.

If additional data is needed or desired, the document provides guidance on the collection of chemical and water quality data and biological data. Guidance is provided for collecting water quality data for determining relevant concentrations, including mercury, in riverine, estuarine and marine systems. A focus of the guidance is on spatially locating samples for performing both near-field and far-field bioavailability assessments. As an aid to this assessment, a simple methylation/demethylation model is presented which allows the user to assess the percent completion of these reactions given the travel time from the discharge.

The primary focus of the biological assessment is the decision of which species to collect. In defining the best bioindicator species to answer the questions posed under this project, the researchers suggest it would be an organism that accumulates either inorganic Hg or MeHg or both from its environment (from water or pelagic organisms/particles, preferably) in such a way that there is a significant and relatively simple relationship between the concentration of the chemical in the organism and that in the surrounding waters. Controlled studies show that many organisms can fit such criteria. In addition, the advantage of a bioindicator is that the accumulation is continuous and the average tissue concentration after exposure is related to the average bioavailable concentration in the water rather than the total concentration, or some short-term fluctuation in concentration. Thus, such organisms should be extremely useful in any attempt to ascertain the bioavailable concentration of Hg species in a particular environment. Finally, from a practical point of view, the best organisms should be abundant, hardy, cosmopolitan in the environment, easy to catch and handle, but responsive to changes in Hg and MeHg on a relatively short timescale (months).

Table ES-1 provides a summary of conclusions about the relative usefulness of the various species and approaches to examine Hg bioavailability and bioaccumulation in effluents and receiving waters. Details of each approach for recommended organisms are provided in the document.

Table ES-1. Bioindicator Organism Summary Table.

Organism	Approach	Complexity	Recommendation
Phytoplankton, Microorganism	Uptake measurements with water from field. Lab-based study	Medium	Perhaps
Seston	Field collection of in situ organisms	Low	No
Bioreporters	Lab examination of bioavailability	High	No
Zooplankton	Collection of in situ organisms	Low	Yes
Filter Feeders	Collection of in situ organisms	Low	Yes
	Transplantation of caged organisms	Medium	Yes
Benthic infauna	Collection of in situ organisms	Low	No
Prey Fish	Collection of in situ organisms	Low	Yes
	Transplantation of caged organisms	Medium	Yes
Predator Fish	Collection of in situ organisms	Low	No
Mammals, birds	Collection of organisms or blood, feathers or other tissue	Medium	No

Reactive Mercury Assessment in Wastewater Effluents

As part of this project, a study was conducted to examine reactive mercury in the effluents of seven wastewater treatment plants, as this is a potentially useful measure which has not been previously investigated. The objectives of this study were to assess the bioavailability of mercury in the effluents of these seven plants. As the definition of bioavailable mercury includes not only methylmercury, the form that readily bioaccumulates in the tissues of aquatic organisms, but also inorganic bioavailable mercury; consisting of those species which may become methylated, principally by organisms at the bottom of aquatic foodchains (i.e. sulfate reducing bacteria), the principal objective was to measure methylmercury (MeHg) and an operationally defined surrogate for inorganic bioavailable mercury, reactive mercury (R-Hg). Since the definition of bioavailability in this report includes both adsorbed and dissolved MeHg, total MeHg was measured in the samples. By definition, R-Hg is measured in the dissolved fraction. In addition to these two species, both unfiltered and filtered total mercury was measured as well; the former to enable us to calculate %MeHg and %R-Hg as a percentage of the total mercury in the sample. Dissolved mercury (D-Hg), was measured so that partition coefficients for total mercury could be calculated. For one round of sampling, dissolved MeHg was measured so that partition coefficients could be measured for MeHg for each plant. These measurements were required so that a correction could be applied to D-Hg, in order to calculate inorganic dissolved mercury ID-Hg, as there is no direct way of measuring ID-Hg.

While MeHg and total mercury (T-Hg) measurements have been made in a number of WWTPs, this study represents the first to measure bioavailable mercury, in terms of MeHg and R-Hg.

Sampling was conducted weekly for 10 weeks beginning June 26, 2008 and ending August 27, 2008. Samples were collected by the participating WWTPs and submitted for analysis to Frontier Geosciences, Inc. in Seattle, WA. All samples were collected by WWTP employees or contractors using Method 1669. Field blanks were collected with each set of samples. During the course of the study, one field duplicate was collected from each of the seven participating WWTPs. Six of the seven facilities collected field measurements typically consisting of temperature, pH, conductivity, DO, and turbidity.

The primary objectives of the data analysis were to examine the relationships among the various species of mercury measured, to assess the level of bioavailable mercury species, and to examine the relationships between mercury species and other conventional water quality parameters. The speciation of the mercury present in these effluents is summarized in Figure ES-1.

Unfiltered methylmercury as a percent of total unfiltered mercury ranged from 2.0% (Facility B) to 19% (Facility E) and averaged 5.1%. Reactive mercury as a percent of total unfiltered ranged from 3.1% (Facility D) to 34% (Facility G) and averaged 15.7%. There did not appear to be a strong relationship between %MeHg and %R-Hg. Bioavailable mercury, the sum of these two species, ranged from 5.3% (Facility D) to 38% (Facility E). The average total bioavailable mercury at the seven facilities was 21%. This percentage corresponds closely to the percentage of bioavailable mercury anticipated by the analysis in Phase I, Section 5.0, which suggested that bioavailable Hg in wastewater effluents should on the order of 20% or less.

Given the levels of mercury observed in the suspended solids from the seven advanced wastewater treatment plants sampled in the reactive mercury study (average of 0.46 mg/kg), and the low suspended sediment concentrations (average 5.1 mg/L), it appears that near-field sediment impacts caused by deposition of suspended solids would be minimal.

In general, no significant relationships were apparent among the mercury species and the conventional parameters measured; with the exception of weak relationship observed between chloride levels and %R-Hg and a weak relationship observed between DO levels in the effluents and %MeHg.

Perhaps the most significant finding of this research was the confirmation of the theorized relationship between the fraction of reactive to dissolved mercury (expressed as a percent) with DOC. This relationship was parameterized as a declining exponential function in

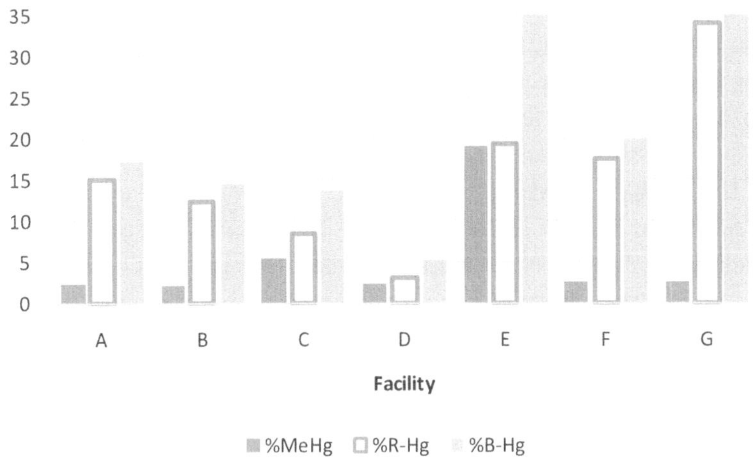

Figure ES-1. Relationships Among Bioavailable Mercury Species in Seven Wastewater Effluents.

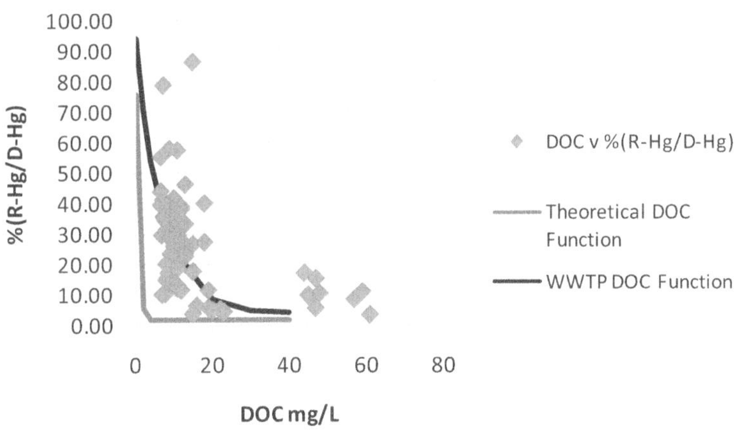

Figure ES-2. Relationship Between DOC and Reactive Mercury as a Percent of Dissolved Inorganic Mercury.

Phase I, Chapter 5.0, based on speciation model results. Using the measurements in this study, this plot is reconstructed as Figure ES-2.

From this figure it is apparent that the levels of reactive mercury in these seven wastewater effluents are higher than would be predicted based on the DOC characteristics and binding constants reported for natural waters. Therefore a new function was developed based on the observed data for this study for wastewater effluents and was incorporated into the Bioavailability Tool. While there is an apparent relationship between DOC and reactive mercury, it by no means explains the major proportion of the variance between the two parameters. This suggests that there are other factors that may be of equal or greater importance than DOC in predicting reactive mercury levels in wastewater effluents.

Values of the binding constant, logK', for Hg by DOC were determined over the range of DOC concentrations for each facility. The average value determined was logK' = 21.9. This value compares well with values estimated by other researchers for the sewage treatment plant effluent (logK' = 23). The calculated binding constants are similar for all facilities - the variability is 0.3 log units. The results suggest that there is overall little difference between treatment facilities in the capacity of the DOC to bind Hg and that the relatively fraction of reactive Hg is driven primarily by the amount of DOC in solution. Overall, the consistency of these results with the work of other researchers, the relationship between fR-Hg and DOC, and their theoretical basis, suggest that the measurement of R-Hg provides a suitable measure of the fraction of Hg in the dissolved fraction that is bioavailable or reactive on a kinetically fast timescale. Thus, it seems reasonable to use the R-Hg measurement as a surrogate for bioavailability of the inorganic Hg pool in WWTPs and in their receiving waters as well.

Given the results of this focused sampling program, the researchers concluded that the concepts that form the basis of the approach in this report are reasonable for examining the issue of bioavailability of mercury in complex environmental situations.

Overall, the information and calculations contained in this report build on the information presented in Phase I and support the main objective of this phase of the project, which was to develop a guidance document to assist wastewater treatment professionals in assessing the bioavailability of mercury in their effluents when mixed with fresh, estuarine or marine receiving waters. This has been achieved through a detailed examination of the literature and is based on sound scientific principles and relationships. It provides procedures for assessing the bioavailability of mercury in a particular source, and to assess the impact of mixing this source with the receiving water on the resultant mercury level and the bioavailability of that mercury. The report provides guidance and input at two levels: a preliminary (screening) level that can be done with relatively little information on mercury speciation and chemistry, and a detailed assessment level which requires a more thorough collection and analysis of data. It is expected that an initial screening in a particular instance could be carried out with existing data and then the results of this approach would inform the decision of whether a more detailed examination is necessary. The information gained from such a study should be able to answer the questions concerning the amount of mercury being input to a receiving water, which is often the critical question being asked by the regulatory process, such as a TMDL, as well as the relative likelihood of the mercury from the source being bioaccumulated into the food chain compared to the mercury already within the receiving waters and the overall ecosystem. If a detailed study is the decision, then this report provides recommendations for the level of effort and focus needed to successfully estimate bioavailability and biological uptake.

CHAPTER 1.0

INTRODUCTION

Mercury levels in fish tissue are considered by regulatory agencies to be a widespread human health concern and fish consumption advisories have been issued for over 2,000 waterbodies in the United States. The existence of fish consumption advisories in a waterbody have in turn been used as a sufficient basis for the listing of these waterbodies on the states' lists of impaired waters, the 303(d) lists. The sources of mercury are known to be diverse, as it is a naturally occurring element, although it is generally accepted by the scientific community that the emissions of mercury have increased several fold during the industrial period. Therefore, attempts to regulate mercury discharges to waterbodies have focused on point source dischargers and air emissions.

The TMDL process has been used to attempt to quantify mercury loadings to waterbodies, both to estimate the acceptable load that would result in the attainment of acceptable levels in fish tissue, and to estimate the current load of mercury to the waterbody that is resulting in nonattainment. In the case of mercury, a particularly difficult aspect of this exercise arises from the fact that it must be transformed from the form that is typically deposited or discharged to waterbodies (inorganic, primarily divalent mercury) to the form that is bioaccumulated in fish (methylmercury). Currently, there are no accepted techniques available to "trace" mercury from its emitted origin to its appearance in fish tissue[1]. Therefore, the question of "whose mercury is it?" remains unanswered.

The movement toward an assessment of the bioavailability of various mercury sources is a step in this direction and is a rational one. It stands to reason that a source which may be contributing mercury that is more difficult to transform into a bioaccumulative form should receive less attention than one discharging mercury that is more easily transformed. However, this assessment is more difficult than simply measuring who is discharging inorganic versus methylmercury. There are a number of species of mercury, other than methylmercury, that may be considered "bioavailable". Some of these may be bioaccumulated in their own right and some may be precursors to methylmercury formation. The presence or absence of these species are determined by the biogeochemistry of effluent waters and receiving waterbody types (i.e. fresh, brackish, marine), individual waterbody characteristics (i.e. pH, dissolved organic carbon (DOC) levels, dissolved oxygen (DO) levels, etc.) and the aquatic foodwebs present to move bioaccumulative mercury up the food chain.

[1] Although stable isotopes of mercury have been used in scientific studies such as METTALICUS (Harris et al., 2007) to attempt to trace the movement of mercury in watersheds and aquatic foodwebs under experimental conditions, there are still no distinctive isotopic "signatures" for the various mercury sources that would allow such a technique to be of widespread utility in a regulatory context.

1.1 Purpose of Research

1.1.1 Background

The motivation for this study grew out of the San Francisco Bay mercury Total Maximum Daily Load (TMDL). As part of that process it was estimated that the wastewater treatment plants (WWTPs) discharging into the Bay accounted for only a small percentage of the estimated total mercury entering the Bay on an annual basis (about 2.5%). However, questions were raised concerning the bioavailability of mercury from the WWTPs versus other sources of mercury including stormwater, mining sources, contaminated sediments and air deposition, to name a few. The WWTPs were thus obligated to perform a study to make an assessment of this type. WERF became aware of this research challenge and made the decision to host the project. Additional funding was provided by the Bay Area Clean Water Agencies (BACWA).

This was a prudent course of action for several reasons:

- WERF sponsorship lent a high level of objectivity and legitimacy to the study and study results,
- WERF sponsorship elevated the study to a nationwide effort, as many other WWTPs in various parts of the country will inevitably face similar challenges, and
- The project geography and the types of receiving water environments were expanded so that the study would be useful to a wider population of the WERF membership.

1.1.2 Objectives

1.1.2.1 Phase 1

This project was divided into two phases. The objectives of this first phase of the project were as follows:

- Develop a working definition of bioavailable mercury (Chapter 2.0),
- Identify factors that enhance or mitigate production of bioavailable mercury in aquatic environments (Chapter 3.0),
- Profile and rank sources of mercury with respect to bioavailable Hg, enhancers, and mitigators (Chapters 4.0 and 5.0),
- Apply the ranking procedure in a "proof-of-concept" fashion (Chapter 5.0),
- Assess the state of the science with respect to those processes that are being used, and may potentially be used, by WWTPs to treat and remove mercury from their discharges (Appendix A), and
- Assess the potential risks to WWTP owners and operators vis-à-vis the mercury TMDL process (Appendix B).

The development of a working definition of bioavailable mercury was critical to the success of this project. It was important that the definition of bioavailable mercury not only included methylmercury, the form bioaccumulated by higher trophic-level organisms, but also the forms that are bioavailable to organisms that methylate mercury. The definition in turn must not only encompass the mercury species that are immediately bioavailable but also those *potentially* bioavailable within a reasonable time-scale. Only at this level is the definition broad enough to account for biogeochemical transformations that occur from the discharge point to the end point of methylmercury in fish tissue in various receiving water environments.

It is also essential to identify the major environmental factors that enable the transformations of the species in the pool of labile bioavailable mercury to bioavailable mercury species (enhancers) and those that mitigate or retard these processes (mitigators). These factors will be different in different receiving water environments and in the various mercury sources to a given waterbody. To this end it is also essential that the major mercury sources are identified and their effluents profiled with regard to bioavailable and potentially bioavailable species discharged, along with those substances that potentially enhance or mitigate transformations from the labile to the bioaccumulative pools. In Phase 1, a ranking protocol was developed to enable a "relative bioavailability" scoring for effluents from selected sources.

Once the methodology for ranking various effluents was developed, it was important to use it in one or more "real-world" situations in order to determine its usability, its relevance, and to assess whether the data required to use it are generally available to potential users.

A literature review was performed to assess the effectiveness of currently deployed technology to treat mercury in wastewater effluents. In addition, emerging technologies that might be applicable for mercury treatment in WWTPs were reviewed, both in terms of effectiveness and cost. To make a connection to the assessment of bioavailability, these various processes and technologies were reviewed in terms of their potential effect on mercury bioavailability. Finally, an assessment of mercury TMDLs was provided so the potential risks to wastewater treatment plant owners and operators could be articulated.

1.1.2.2 Phase 2

This document reports on Phase 2 of the project. The report consists of two main chapters, in addition to this introduction, and constitutes a guidance document for WWTPs interested in assessing the level of bioavailable mercury in their effluent. The guidance is two-tiered. The first tier is a screening level assessment tool that builds upon the ranking protocol developed in Phase 1. This tool will be most useful in assessing overall risk relative to other dischargers in a watershed, making the decision to perform a more detailed assessment, and allocation of resources if a more detailed assessment is warranted. The second tier is a set of guidelines and protocols to conduct a detailed-level assessment, including sampling, analysis, data analysis and interpretation, to make a more definitive assessment of bioavailability in WWTP effluents. Such an assessment may be warranted if the WWTP is directly involved in a TMDL or similar regulatory action and is asked to make a demonstration of the degree of mercury bioavailability in its effluent and mixtures of its effluent with the receiving water, taking into account both the characteristics of the effluent and the characteristics of the receiving waterbody.

1.2 How To Use This Document

The purpose of this guidance document is to assist wastewater treatment professionals in assessing the bioaccumulation potential of trace levels of mercury in effluents from their plants in receiving waters. There are several types of studies treatment facility owners or operators may be required to perform in which this issue may arise:

- investigatory studies,
- TMDL studies,
- permit studies, and
- management studies.

Investigatory studies include those in which an assessment is made of whether there is an impact of discharged mercury to the receiving water aquatic ecosystem. If, on the basis of such an assessment, an impact is indicated, then a TMDL or TMDL-like study may be required. In this type of study, the loading of mercury to the waterbody is evaluated as well as the degree to which the ecosystem is impacted. This study may be done for an individual source or it may be done in the context of a larger assessment which includes all upstream mercury sources. In such a case, the impact of the individual sources may be indentified and allocations of reductions in loadings may be assigned to individual dischargers or groups of dischargers. Permitting studies may be required. Such a study may differ from a TMDL-like study in that a water column- based discharge limit may be calculated in order to be imposed on the discharger. A study of this type will likely take the load required to bring fish tissue body burdens of methylmercury to acceptable levels and translate that into a permit limit in terms of total mercury. Finally, management studies may be required to assess the impact of changes or proposed changes in plant operations to the health of the aquatic ecosystem, in terms of observed or anticipated reductions of mercury if fish tissue. All of these studies potentially involve the assessment of bioaccumulation potential of discharged mercury.

This guidance document provides tools and information to potential users in order to address these types of studies. It does not provide *pro forma* approaches or study plans because of the diversity of studies that may be required. It is the intent of this section, however, to guide the user to various portions of the document that contain the tools and information to perform these studies.

There are several options that this document provides the user:

- degree of complexity (screening or detailed);
- type of approach (water chemistry, biological, or both); and
- spatial assessment (near-field or far-field).

For instance, the user may choose to first complete a screening assessment with the information at hand or that can be estimated based on literature sources. If a more certain answer is required, the user may opt to collect more data by doing a detailed assessment. If no biological data is available, the user may opt to make an assessment using available water quality data to assess whether additional data is needed. If the environment to which the effluent is discharged is weak in terms of its methylation potential, then a near-field assessment may be preferred, since the greatest impact would likely be in the vicinity of the discharge.

The screening approach is the subject of Chapter 2.0. The screening focus is on water chemistry. If biological data are to be collected, the user is directed to the detailed methods in Section 3.0. The screening approach is designed to be used with little data; although the assessment will be better the more data is available. Information for using the screening approach outlined in this report can be supplemented by information contained in the Phase I report for this project (in particular, see Chapter 4.0 of the Phase I report). The screening approach uses a set of spreadsheets referred to as the "Bioavailability Tool". This tool is designed to answer a number of questions with regard to the bioaccumulation potential of mercury in a discharged effluent. The Tool provides a means of assessing the bioavailability, bioavailable concentrations, and bioavailable load of mercury in an effluent and comparing these measures to another effluent. The Tool also allows the user to assess how bioavailability measures changes when the effluent is mixed with a receiving waterbody, which includes fresh, estuarine, and marine waterbody types. It is strongly recommended that the user apply the screening approach to become familiar with the available data and to develop a feel for the effects of mixing the effluent and receiving water on the bioavailability of mercury in the discharge. The screening assessment will also help the user understand data deficiencies and will likely help focus any detailed studies, if they are selected to be performed.

Detailed approaches to assessing bioaccumulation potential, which include both water chemistry and biological approaches, are covered in Chapter 3.0. While biological approaches may give a more definitive assessment of impact, water chemistry approaches are more likely to illuminate the linkage between discharged and bioaccumulated mercury. A more robust, but costly, assessment will result from the implementation of both approaches. Section 3.1 discusses preliminary data gathering relevant to both approaches, Section 3.2 deals with issues surrounding the decision to collect additional data, Section 3.3 deals with water chemistry approaches, and Section 3.4 with biological approaches. The user is advised to make this decision based on the acceptable level of certainty (or uncertainty) in the answer that is sought. The decisions of what type of data and how many data to collect can be based on error propagation principles.

The generic steps to be taken in making the data collection decision are:

- Step 1. Decide on the end points of the analysis (this will depend on the kind of study, e.g. investigative, permitting, etc.)
- Step 2. Decide on the acceptable levels of uncertainty in the end point calculations (this is a management/regulatory decision)
- Step 3. Assess the uncertainty in the measured components that contribute to the calculated endpoints (Section 3.2)
- Step 4. Assess the uncertainty in the end point using error propagation (if the end point is calculated rather than measured) (Section 3.2)
- Step 5. Compare the calculated uncertainty with the acceptable level

If additional data collection is desired, the following generic steps are recommended:

- Step 1. Decide which types of data to collect (Water, Sections 3.2, 3.3; Biota, Section 3.4)
- Step 2. Decide how many data to collect (Water, Section 3.2; Biota, Section 3.4)
- Step 3. Decide where in the system to collect data (near-field or far-field) (Water, Section 3.3; Biota, Section 3.4)
- Step 4. Decide when to collect data (Section 3.3)

CHAPTER 2.0

A Screening Approach for Estimating Bioavailability of Mercury in Effluents and Receiving Waters

2.1 Introduction

This chapter presents a screening methodology for estimating the bioavailability of mercury in effluents when mixed with receiving waters. It is based on the concepts developed in the Phase I report, which define bioavailable Hg as the sum of inorganic bioavailable mercury and methylmercury in an effluent or aquatic system. The methodology implements the computational approach developed in Phase I using a spreadsheet tool. Using this tool, bioavailable mercury can be estimated for various effluent or effluent/receiving water mixtures, using several approaches. If the user has reactive mercury measurements as a surrogate for inorganic bioavailable mercury, then the tool will calculate bioavailable mercury from methyl and reactive mercury measurements. If the user does not have reactive mercury measurements, then the tool will estimate bioavailable inorganic mercury using the techniques identified in the Phase I report.

For receiving water and effluent mixtures, the tool produces both "near-field" and "far-field" bioavailability estimates. The near-field estimates are the concentration estimates resulting from initial mixing, assuming that net methylation reactions have not had time to occur. For the far-field estimates, the tool uses estimates of the strength of the receiving water methylating environment to estimate net methylation once these reactions have had time to occur (i.e., under steady-state conditions). The strength of the methylating environment is used to identify the ratio k_{meth}/k_{demeth}, where k_{meth} is a first order reaction constant for methylation of bioavailable inorganic mercury and k_{demeth} is a first-order demethylation reaction constant. The steady-state computations assume that total bioavailable mercury (the sum of bioavailable inorganic mercury and methyl mercury) in the receiving water is conservative and that the relative distribution between the two forms is governed by the selected reaction constants.

There are two ways in which the strength of the methylating environment can be assessed. The first is by using the percent methyl/total ratio in the receiving water. The second depends on the values of other water quality parameters (i.e. pH, dissolved organic carbon) if methylmercury measurements are unavailable.

In this chapter, a user guide to the Bioavailability Tool is presented. In Section 2.2, the basics of the computational approach are discussed and in Section 2.3, the user will find guidance on how to use the tool to answer each of several questions the tool was designed to address. In Section 2.4, the development of the net methylation approach is discussed in detail.

2.2 Screening Approach

This screening methodology was developed to answer some particular questions concerning the bioavailability of mercury in effluents. The questions the approach was designed to answer are:

1) a) How does the bioavailability of mercury in my effluent compare to another effluent or point source?
   b) Is the MeHg content of my effluent the major contributor to bioavailable Hg or is it the inorganic pool?

2) Is my effluent more/less bioavailable than one of the following non-point sources:
   Mining runoff?
   Urban stormwater?
   Non-urban stormwater?
   Precipitation?
   Sediment pore water?

3) How do the concentrations and fluxes of bioavailable mercury in my effluent compare to those of another source?

4) a) To what degree is my effluent affecting the bioavailability of mercury in the receiving waters?
   b) To what degree is the bioavailability of mercury in my effluent (or another) affected when it is mixed with my receiving water?

5) a) What happens to the concentration of bioavailable mercury when it is mixed with my receiving water?
   b) What characteristics of the effluent and receiving waters are contributing or enhancing the bioavailability?

Note that these questions address the issue of bioavailability or the bioavailable concentration of mercury in effluents and/or receiving waters. They do not go as far as discussing bioaccumulation, as this depends on the food web in individual systems. However, the researchers recognize it is important to keep the system biology in mind when interpreting results. In addition, guidance for conducting biological assessments of bioavailability is covered in Chapter 3.0 of this document. Additional detail concerning the approaches to answering each of these questions is provided in the following sections. Practical information on where and how to obtain the information required to implement these approaches is provided in Section 2.3 for screening level assessments and in Chapter 3.0 for more detailed assessments.

2.2.1 Approach to Question #1 - How Does the Bioavailability of Mercury in My Effluent Compare to Another Effluent or Point Source?

This was the fundamental question which established the need for this project; essentially, is the mercury in WWTP effluent more or less bioavailable that that from other sources. In order to answer question #1, the following information is needed:

- MeHg and total Hg (T-Hg) of the effluents; and
- R-Hg; or
- TSS, K_D (for T-Hg) and DOC of the effluents.

For a screening assessment, the information can be obtained either from limited measurements, estimated using values taken from the literature, or derived from a scientific evaluation, but more likely one or both of the first two. Clearly, different types of information will result in a higher or lower level of certainty in the estimation. The screening assessment will provide methods to answer the questions posed above even if some of these parameters are not known.

The procedure for evaluating bioavailability of the effluent is provided in the Phase I report. By implementing this approach, one can directly compare the bioavailability of any two or more effluents. Bioavailable Hg consists of two components; MeHg and inorganic bioavailable Hg (IB-Hg) that may be readily converted to MeHg. The relative importance or contribution of MeHg or inorganic bioavailable mercury can be easily assessed from the calculations.

To briefly review, the procedure is to calculate the sum of the fraction MeHg and fraction IB-Hg for each effluent. If the MeHg and T-Hg are known, the fraction of MeHg is simply the ratio of the two:

$$f(\text{MeHg}) = \text{MeHg}/\underline{\text{T-Hg}} \qquad \text{Equation 2.2.1-1}$$

If reactive mercury (R-Hg), an estimate of bioavailable inorganic mercury is known, the fraction of bioavailable inorganic mercury may be estimated by:

$$f(\underline{\text{IB-Hg}}) = \underline{\text{R-Hg}}/\underline{\text{T-Hg}} \qquad \text{Equation 2.2.1-2}$$

The results are summed to obtain the fraction of bioavailable Hg:

$$f(\text{B-Hg}) = f(\text{MeHg}) + f(\text{IB-Hg}) \qquad \text{Equation 2.2.1-3}$$

If R-Hg is not known, then f(IB-Hg) must be estimated by other means. From the relationships in Chapter 5.0 of the Phase I report the following equation for estimating this quantity can be obtained:

$$\underline{\text{IB-Hg}} = \underline{\text{ID-Hg}} * f(\underline{\text{IDB-Hg}}) \qquad \text{Equation 2.2.1-4}$$

Inorganic dissolved Hg (ID-Hg) can be estimated from T-Hg and the partition coefficient (K_D) for T-Hg. This is because most of the particle bound mercury will tend to be inorganic Hg, as the K_D is about 10x higher for inorganic mercury than for MeHg. Therefore the expression for ID-Hg, in terms of K_D, is:

$$\underline{\text{ID-Hg}}\ (\text{ng/L}) = \underline{\text{T-Hg}}/(1 + 10^{-6} * \text{TSS} * K_D) \qquad \text{Equation 2.2.1-5}$$

where TSS is in mg/L and K_D is essentially dimensionless (units of L/kg).

Other approaches can be used to estimate ID-Hg as well. Once this parameter is known, then the fraction of dissolved bioavailable inorganic mercury is estimated using a relationship that relates the bioavailability to the concentration of dissolved organic carbon (DOC) by:

$$f(\underline{IDB\text{-}Hg}) = [2.0 + 73.5 * \exp(-1.48 * DOC)]/100 \qquad \text{Equation 2.2.1-6}$$

where DOC is the dissolved organic carbon in mg/L[2].

The relationship that was derived for freshwaters between the fraction of dissolved bioavailable inorganic mercury and the concentration of dissolved organic carbon (DOC) is not directly applicable to estuarine and coastal waters. This is because the binding constants used in the formulation of the relationship were derived using terrestrial organic matter of high molecular weight and this is not truly representative of the organic matter found in estuarine and coastal waters. Furthermore, while there is essentially always sufficient organic matter in freshwaters to bind at least some of the Hg, this is not the case for coastal waters, given the lower binding constants and the low organic matter content possible in these waters, as well as the high levels of chloride. As chloride binds strongly to Hg, it can out compete DOC as the dominant binding ligand at low DOC concentrations (Fitzgerald et al. 2007). As noted by Lamborg et al. (2004), at low DOC content, the concentration of DOC at which the Cl complexation dominates is related to the strength of the binding of Hg by the organic ligands, and this binding varies with molecular size, weight and thiol content. Given these complexities, it is difficult to provide a definitive answer concerning the DOC concentration at which the complexation switches from a DOC-dominated system to Cl-dominated complexation. Also, it is possible that a number of weaker binding low molecular weight ligands could provide a low level of complexation even in relatively low DOC waters. Given these considerations, for estuarine and marine waters, the relationship derived earlier for freshwater systems (Equation 2.2.1-6) is modified for estuarine (>10 ppt) and coastal waters to the following:

For DOC > 1.5 mg/L:

$$f(\underline{IB\text{-}Hg}) = [2.0 + 73.5 * \exp(-1.48 * DOC)]/10 \qquad \text{Equation 2.1.1-7}$$

where DOC is the dissolved organic carbon in mg/L.

For DOC < 1.5 mg/L:

$$f(\underline{IB\text{-}Hg}) = 0.95 \qquad \text{Equation 2.1.1-8}$$

Recall that f(IDB-Hg) is the fraction of the *dissolved* inorganic Hg concentration and is the fraction f(IB-Hg) that is added to f(MeHg) to obtain the total bioavailable Hg fraction. So:

---

[2] Note that this relationship was developed for fresh waters. For brackish and marine waters, the fraction of dissolved bioavailable mercury was determined to be a factor of ten higher, as discussed in the following paragraph. In addition, the relationship appears to be different for wastewater effluents. Data collected as part of the reactive mercury study presented in Appendix A of this report show the best fit relationship for the data to be:

$$f(\underline{IDB\text{-}Hg}) = [4 + 90.1 * \exp(-0.147 * DOC)]/100.$$

$$f(\underline{IB\text{-}Hg}) = \underline{IB\text{-}Hg}/\underline{T\text{-}Hg} \qquad \text{Equation 2.2.1-9}$$

2.2.2 Approach to Question #2 - Is my Effluent More/Less Bioavailable than a Non-point Source?

In order to answer question #2, the same information for the effluent is needed. The additional information for the other non-point sources is in the Phase I report (for the screening assessment). Averages and ranges of parameter values for particular source types are given.

2.2.3 Approach to Question #3 - How Do the Concentrations and Fluxes of Bioavailable Mercury in My Effluent Compare to Those of Another Source?

The approach to answering question #3 is straightforward given that concentrations of mercury species such as MeHg, R-Hg and total bioavailable mercury are calculated and effluent flow rates or volumes are provided. The mass flux or mass flux rate is simply the product of the concentration of the species of concern with the volume or flow rate multiplied by the appropriate conversion factors. Since the flow from WWTPs is likely measured in MGD, and, given a mercury species concentration measured in ng/L, the conversion factor is 3785, assuming the output flux is reported in μg/day.

2.2.4 Approach to Question #4 - What Happens to the Bioavailability of Mercury in My Effluent When It Is Mixed with Receiving Water?

The approach to answering question #4 is a primary concern of this study. In order to answer question #4, the above information for the effluent and the receiving water is needed, along with a mixing ratio. The approach to answering this question utilizes both a near-field and far-field analysis as discussed below.

2.2.4.1 Near-Field Analysis

The mixing ratio may be known directly or it may be estimated from flow rates and/or volumes of the effluent (Q_{eff}) and the receiving water (Q_{rw}). For instance, if the mixing ratio is defined as Q_{eff}/Q_{rw} then with a flow of the effluent of 1 MGD and a reference flow of the receiving water of 5 MGD, the mixing ratio would be 1:5 or 0.2. The content of total Hg, MeHg, and other parameters of the effluent (C_{eff}) and receiving water (C_{rw}) are used to determine the concentrations after mixing by using the mixing ratio and initially assuming conservative mixing. The expression for the mixed concentration using the effluent and receiving water flows is:

$$C_m = (C_{eff}*Q_{eff} + C_{rw}*Q_{rw}) / (Q_{eff} + Q_{rw}) \qquad \text{Equation 2.2.4.1-1}$$

Multiplying both the numerator and denominator of the above equation by Q_{eff}/Q_{rw} and rearranging terms, an expression for the mixed concentration can be derived in terms of the mixing ratio, R_m[3]:

[3] Typically, information of this type in terms of dilution ratios or mixing ratios will be available in the permit studies for the WWTP.

$$C_m = C_{eff} * R_m/(1+R_m) + C_{rw}/(1+R_m) \qquad \text{Equation 2.2.4.1-2}$$

For example, using a flow of 1 MGD for the effluent and a receiving water flow of 5 MGD ($R_m = 0.2$), and T-Hg for the effluent of 2 ng/L and T-Hg for the receiving water of 15 ng/L,

$$C_m = 2*(0.2)/(1+0.2) + 15/(1+0.2) = 0.3 + 12.5 = 12.8 \text{ mg/L.}$$

The bioavailable inorganic Hg concentrations of the effluent and the receiving water are determined using the DOC levels particular to each. This is because the reactivity of dissolved mercury in the effluent and the receiving water may be different. The mixed concentration of IB-Hg is then determined using Equation 2.2.4.1-2. Similarly, ID-Hg is determined from the concentrations of T-Hg in the effluent and the receiving water and the K_D particular to each.

In the near-field, the MeHg concentration is simply the mixed concentration of the MeHg in the effluent and receiving waterbody (i.e. conservative mixing), as methylation/demethylation reactions would not have had time to occur and the concentration or fraction of the mixed MeHg would be the concentration or fraction of bioavailable mercury.

Given the levels of mercury observed in the suspended solids from the seven advanced wastewater treatment plants sampled in the reactive mercury study (average of 0.46 mg/kg; see Appendix A), and the low suspended sediment concentrations (average 5.1 mg/L), it appears that near-field sediment impacts caused by deposition of suspended solids would be minimal. The average concentration of 0.46 mg/kg is in the range of concentrations from slightly to moderately impacted sediments (see Tables 4-3 and 4-5, Figure 4-3, Phase I).

2.2.4.2 Far-Field Analysis

For a far-field analysis, additional steps must be taken to account for methylation and demethylation that will occur after initial mixing. The next step is to assess how much of the bioavailable inorganic mercury would be converted to MeHg in the receiving water and how much of the added methylmercury would be demethylated. To do this, an estimate of the methylation potential of the receiving water is needed. The most straightforward way to evaluate the methylation potential is to make use of MeHg and R-Hg of the receiving water, if they are known, which gives the best answer. If R-Hg is not known then we can resort to a screening level assessment for an initial estimate. There are two screening methods for freshwaters, one based on knowing MeHg and T-Hg, and the other based on knowing DOC and pH, if MeHg and/or T-Hg are unknown. If and MeHg and T-Hg are known, then the ratio of MeHg to T-Hg for the receiving water can be determined, assuming this reflects the net methylation potential. The result signifies the *strength* of the methylating environment or equivalently the methylation potential of the receiving water.

For freshwater environments, the procedure predicts whether a strong, moderate, or weak methylating environment exists and interprets that based on the MeHg and T-Hg for the receiving water. Based on the results represented in Section 2.4, the strength of the methylating environment based on %MeHg is interpreted as follows:

- Very strong > 10%
- Strong 5 to 10%
- Moderate 3 to 5%
- Weak 2 to 3%
- Very weak < 2%

This approach provides an answer consistent with the existing status of the receiving water and may not be consistent with the mixed effluent/ambient conditions, unless the MeHg percentage is measured where such mixing has occurred. If the receiving water volume or flow rate dominates the mixing calculations, then this assessment is certainly appropriate. However, if the effluent flow or volume is such that the effluent might impact the methylation potential of the receiving water, then the methylation potential should be reevaluated using the procedure below.

If the MeHg or T-Hg of the receiving water is unknown, an estimate of the strength of the receiving water environment can still be made from DOC and pH. Note that if the T-Hg of the receiving waters is unknown, then the assessment can only be done with percentages (answering questions #1, 2, and 4, and not with actual concentrations (questions #3 and #5).

The strength of the methylating environment is used to identify the ratio k_{meth} / k_{demeth}, where k_{meth} is a first order reaction constant for methylation of bioavailable inorganic mercury and k_{demeth} is a first-order demethylation reaction constant. The computations assume that the sum of methyl and inorganic bioavailable mercury is conservative in the effluent/receiving water mixture and that at steady-state their distribution is in equilibrium according to the ratio k_{meth} / k_{demeth}. Thus,

$$\underline{\text{B-Hg}} = \text{MeHg} + \underline{\text{IB-Hg}} \qquad \text{Equation 2.2.4.2-1}$$

and

$$\text{MeHg} = k_{meth} * \underline{\text{B-Hg}} / (k_{meth} + k_{demeth}) \qquad \text{Equation 2.2.4.2-2}$$

where B-Hg is the total bioavailable Hg concentration (ng/L).

The mixing calculations are performed and the methylation potential of the resulting water is determined using a series of matrices. The matrices for fresh water are shown below (Tables 2-1 and 2-2). These matrices contain values of the ratio k_{meth} / k_{demeth}. The matrix entries are indexed by either the percent MeHg value or the pH and DOC characteristics of the effluent/receiving water mixture.

The matrices are used by the Tool in the following way. With an estimate of either %MeHg or the pH and DOC concentration in the mixed effluent/receiving water, a ratio of k_{meth} /k_{demeth} is selected from the appropriate matrix. Once this ratio is known, the value of the demethylation constant k_{demeth} is determined using the appropriate equation from Section 2.4. The value of the methylation constant, k_{meth} is then determined from the value of k_{demeth} and the methylmercury concentration can be estimated using Equation 2.2.4.2-2 above.

Table 2-1. Matrix of Methylation/Demethylation Constant Ratios Based on %MeHg.

% MeHg	<2%	2 to 3%	3 to 5%	5 to 10%	>10%
k_{meth} / k_{demeth}	0.2	2	4	8	16

Table 2-2 Matrix of Methylation/Demethylation Constant Ratios (k_{meth} /k_{demeth}) Based on DOC and pH.

		DOC	
pH	**< 2 mg/L**	**2 to 7 mg/L**	**> 7 mg/L**
<7.5	0.2	5	16
7.5 to 8	0.2	2	9
> 8	0.2	2	4

Thus, in freshwaters with low DOC and low pH, the methylating environment would, on the basis of the table above, be determined to be weak. The corresponding ratio k_{meth} / k_{demeth} is 0.2, meaning that the methylation rate constant is only one fifth of the demethylation rate constant. In waters with high DOC and low pH, the methylation potential is maximized and the ratio of the k_{meth} / k_{demeth} constants is on the order of 16. Based on this analysis, methylation potential in very strong methylating environments is 60 times greater than in very weak environments.

Note again that if MeHg and R-Hg measurements are available, then these can be used directly to estimate the ratio of the methylation/demethylation constants, and resort to the screening assessment using the matrices is unnecessary. In closing this discussion, note that this approach was presaged in a paper by Hammerschmidt et al. (2006) in which they observed that the molar ratio of MeHg to R-Hg in precipitation was equal a constant value and that it was equivalent to the steady-state ratio of the methylation/demethylation reaction constants using values calculated from prevailing atmospheric conditions.

There are likely to be other variables that influence the extent of net methylation in a particular environment and measures of the microbial activity may be useful in explaining the relationships. For example, there is much evidence that the primary microorganisms responsible for Hg methylation in aquatic systems are sulfate-reducing bacteria (SRB) (Benoit et al., 2003) and thus the extent of methylation could conceivably be linked to their activity, or to other proxies of organic matter decomposition. However, it is also known that not all SRB's methylate Hg and they can even methylate Hg when growing fermentatively (i.e. without reducing sulfate (i.e. using it as an electron acceptor). Thus, as has been noted in many studies, there is often little direct correlation between the activity of the SRBs and the production of MeHg, even though they are the principal methylators, as summarized in Benoit et al. (2003).

The various factors that may influence the ratio are discussed in more detail in the next chapter. For the screening assessment, the variables and ranges given here are based on field results and are empirical rather than determined from a specific assessment of the factors controlling methylation and demethylation in a particular ecosystem. It should be noted that the methylation constant used here is not the methylation constant related to a single process, such as may be measured by a methylation assay. Rather, it is actually a combination of a variety of important processes. As the focus of the calculations and estimation in this study is the water column, and as it is known that methylation in the water column is minimal, the constant is actually a combination of a variety of processes: 1) transport of inorganic bioavailable Hg to the sites of methylation (mostly the surface sediment layers); 2) its subsequent methylation; and 3) the resultant transport of the MeHg produced back into the water column. Thus, this overall constant incorporates the processes occurring in the sediments that influence Hg methylation, be they related to microbial activity or to bioavailability (chemical speciation) and also include the transport of species between the sediments and the water column. Thus, as will be demonstrated in detail below, while the demethylation rate constant can be relatively simply related to water column chemistry and DOC concentration, this is not so for the overall net MeHg production constant, as defined here. Indeed, given the processes that are inherent in the computation of the methylation rate constant as used here, there is often little relationship between MeHg in sediments and MeHg in water, or their relative fraction (%MeHg). Thus, it was concluded that information on %MeHg in sediments was not a useful parameter to examine in the context of our approach.

However, there is sufficient evidence in the literature to suggest that the major variables used here are those that have the dominant impact, or reflect the factors that are associated with the degree of net methylation of inorganic Hg in a particular system. Finally, as noted above, the estimates and approaches discussed here are focused on freshwaters and, as detailed below. Similar approaches have been developed for estuarine and marine waters[4].

2.2.5 Approach to Question #5 - What Happens to the Concentration of Bioavailable Mercury in My Effluent When It Is Mixed with Receiving Water?

Essentially, the same analysis is required to answer question #5 as is required to answer question #4. To answer question #5, the percentages developed from the question #4 analysis can be multiplied by the total Hg concentration to yield bioavailable Hg. With regard to which factors most strongly affect the result, sensitivity analyses can also be performed using the method by perturbing input values and observing the change in various outputs.

2.3 The Bioavailability Tool

A set of EXCEL spreadsheets was developed to implement the computational approach to the questions identified in Section 2.2. The set of spreadsheets is called the "Bioavailability Tool". It has been developed to assist wastewater utility managers and engineers in performing the computations required to assess the bioavailability of mercury in their effluents and in their effluents when mixed with receiving waters.

When the user opens the Bioavailability Tool, it will open to a tab called FW_I_O Sheet. The FW prefix stands for "Fresh Water", the "I" and "O" for input and output. All of the inputs to and outputs from the Tool are on this worksheet.

[4] Note that the values of the ratios vary over several orders of magnitude. When implementing the screening level assessments, it is sufficient to round the %MeHg data to the nearest percent, DOC to the nearest integer value, and pH to the nearest 0.5 unit. For cases where the result is near the boundary of two matrix entries, the analyst may wish to evaluate the result based on both nearest neighbors. If there is concern about the level of uncertainty involved, it may be desirable to do a detailed assessment, rather than a screening level assessment. This is discussed in further detail in Chapter 3.0.

There are eight worksheets altogether; three for the freshwater analysis, three for the estuarine water analysis, and two for the marine water analysis. The first three are the freshwater analysis sheets, in order of appearance, FW_I_O Sheet, FW Computations, and FW_Net Meth Tables. The FW Computations sheet is where the freshwater bioavailability computations are performed. The FW_Net Meth Tables worksheet contains the net methylation tables for supporting the far-field fresh water analysis. Similarly, for the estuarine water analysis, the three worksheets are called EW_I_O Sheet, EW Computations, and EW_Net Meth Tables. For the marine waters analysis, there are only two worksheets, MW_I_O Sheet and MW Computations. There are no net methylation tables for the marine water analysis as the default ratio of the methylation/demethylation constants is a single value. All of the worksheets are locked so that only the user inputs may be changed. However, if the user desires to unlock the worksheets, the password is "WERF".

In addition to supporting the screening-level analysis of bioavailability, the Tool also supports more detailed assessments. The following sections will guide the user through the application of the Bioavailability Tool to answer the five questions posed in the initial section of this chapter. Since the spreadsheets for fresh, estuarine, and marine waters are similar, most of the examples utilize the freshwater sheets, and where differences may exist between those and the estuarine and marine analyses they are pointed out.

2.3.1 Question #1 User Guidance and Examples

2.3.1.1 Description of Inputs and Outputs

The first question is "How does the bioavailability of mercury in my effluent compare to another effluent or point source?" A corollary question is "Is the MeHg content of my effluent the major contributor to bioavailable Hg or is it the inorganic pool?"

In order to answer these questions, look at the left-hand side of the FW_I_O Sheet. Notice that the work sheet is arranged so that all the inputs necessary to answer question #1 are input in this area under the yellow headings "Evaluate Bioavailability of a Single Effluent" (Area "A", Figure 2-1) and further down "Compare Bioavailability of Two Effluents" (Area "B", Figure 2-1). Under the "Effluent #1" heading in the upper left portion of the worksheet, the first two inputs are total mercury and methylmercury in units of ng/L. The next parameters in the column for the first effluent are reactive mercury (ng/L), DOC (mg/L), TSS (mg/L), $\log K_D$ (L/kg), pH, and flow rate or volume (million gallons or million gallons per day (MGD). Notice that to the right of the TSS, $\log K_D$, pH, and flow fields, there is a note that these parameters are optional. If reactive mercury measurements are utilized, it is not necessary to enter DOC, TSS, or $\log K_D$ to answer this question. However, DOC is required if the receiving water analysis is to be done. If a pH for the effluent is not entered and the user desires to use pH and DOC of the effluent/receiving water mixture, the pH of the receiving water will be the default value. If effluent flow or volume is left out, the flux output calculation will not be performed (See Section 2.3.3).

The same descriptions apply to the inputs for effluent #2, which are entered in under the column titled "Effluent #2", in the lower left portion of the worksheet. Also note that if the user ultimately desires to perform a mixing analysis with the receiving water, the effluent to be mixed should be input as Effluent #1 rather than as Effluent #2.

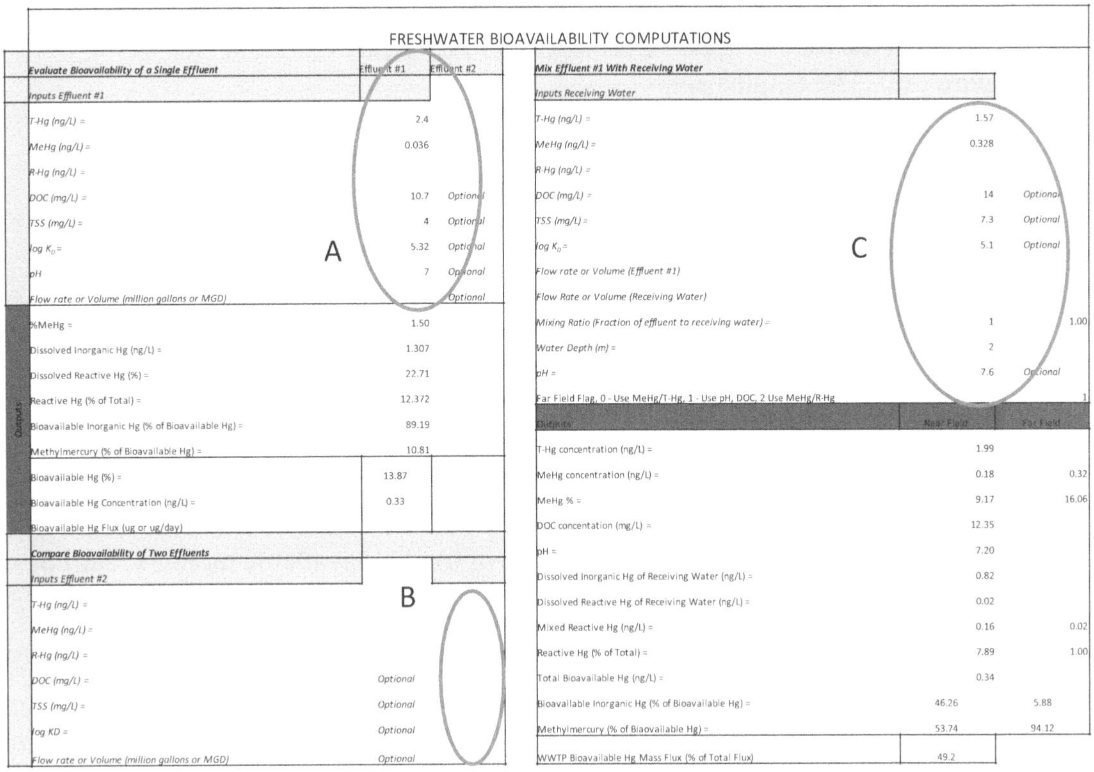

Figure 2-1. Illustration of Input Parameter Fields for the Bioavailability Tool.

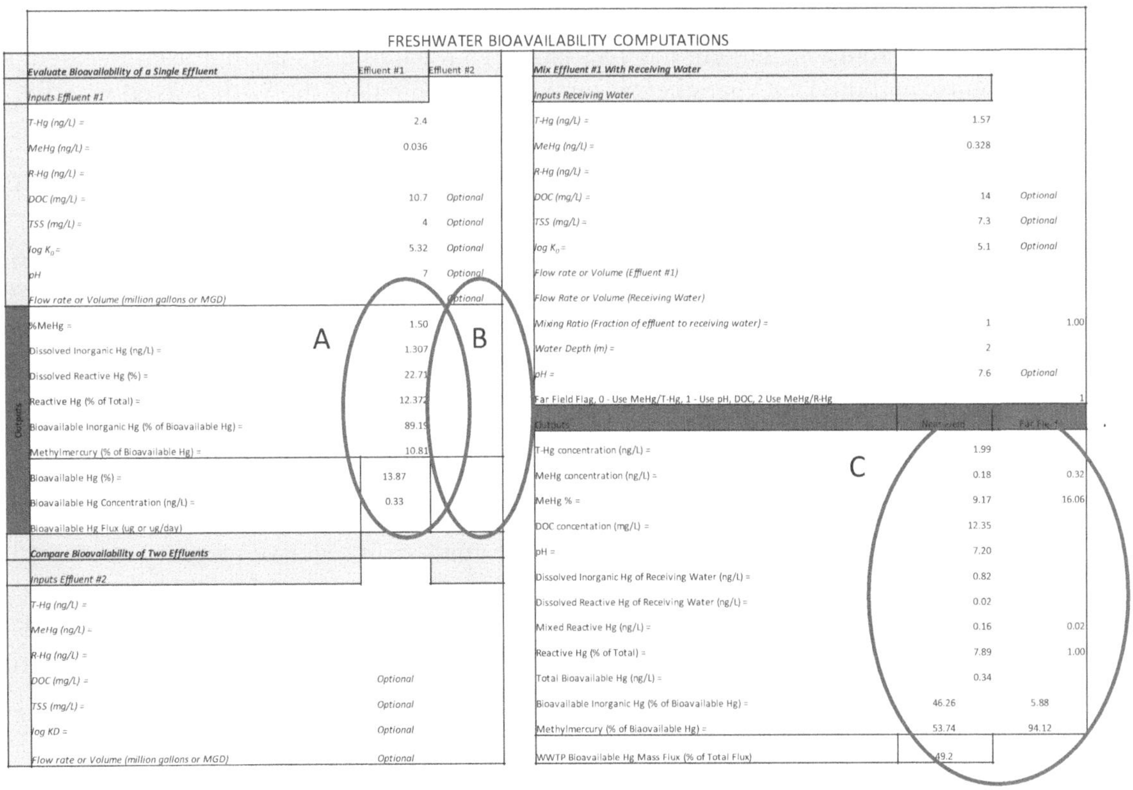

Figure 2-2. Illustration of Output Parameter Fields for the Bioavailability Tool.

The outputs for answering question #1 are found immediately below the inputs in the same column, beneath the red "Outputs" heading in middle of the left-hand side of the worksheet (Area "A", Figure 2-2). The outputs are:

- % MeHg;
- dissolved inorganic Hg (ng/L);
- dissolved reactive Hg (as a % of total dissolved inorganic Hg);
- reactive Hg (% of total mercury);
- bioavailable inorganic Hg (as a % of total bioavailable Hg);
- methylmercury (as a % of total bioavailable Hg);
- bioavailable Hg (total bioavailable mercury, the sum of bioavailable inorganic and methylmercury, as a % of total mercury);
- bioavailable Hg Concentration (concentration of total bioavailable mercury, sum of bioavailable inorganic mercury and methylmercury in ng/L); and
- bioavailable Hg Flux (product of concentration and volume or flow rate in μg or μg/day).

The outputs for effluent #1 are found in the "Effluent #1" column (Area "A", Figure 2-2), and those for effluent #2 are found immediately to the right in the "Effluent #2" column (Area "B", Figure 2-2).

Notice that if a value for Reactive mercury is entered, the "dissolved inorganic Hg" and "dissolved reactive Hg" outputs are blank, as these quantities do not have to be calculated. If a value for Reactive mercury is entered, the TSS and $\log K_D$ parameters are not used to calculate dissolved mercury, and the DOC parameter is not used to calculate the % dissolved reactive Hg, even if they are also provided by the user.

The principal output related to question #1 is bioavailable Hg (%). Bioavailable Hg (%) is the percentage of bioavailable Hg in the effluent and may be used to compare the bioavailability of effluent #1 to a second effluent. This comparison constitutes the answer to question #1, and as such is outlined with a heavy border in the output section. The answer to the corollary question is provided by outputs titled Bioavailable Inorganic Hg (% of bioavailable Hg) and Methylmercury (% of bioavailable Hg). These two outputs should always sum to 100% and show the percentage of bioavailable mercury that is made up of the inorganic and methylated species.

2.3.1.2 Differences between Fresh, Estuarine, and Marine Worksheets

DOC is not needed for the estuarine or marine analysis if reactive mercury is entered as an input parameter. In the freshwater analysis, it has a dual role; to calculate the reactive percentage of dissolved inorganic mercury (if reactive Hg is not input), and to index the net methylation tables for the far-field analysis. In the estuarine and marine analyses, it has no role in determining the ratio of methylation/demethylation constants.

2.3.1.3 Guidance for Input Parameter Estimation

Guidance is provided in this section for providing values of input parameters required to address question #1. Because some of this guidance will be redundant for later questions, the reader will be referred back to this section frequently in this discussion. Also note that this guidance leans toward a screening level assessment. Guidance for obtaining values for these parameters for a more detailed assessment can be found in Chapter 3.0 of this report.

Total Mercury (T-Hg) - Total mercury values should be obtained for the user's effluent by direct measurement. While any number of measured values can provide a starting point, the researchers recommend that a minimum number of samples be collected in order to assure a moderate level of accuracy. Using statistical procedures and knowing something about the variability of the measurement, the analyst can select a number of samples that will produce an acceptable level of uncertainty. Assuming the parameter is normally distributed[5] with a mean 'x_m' and a standard deviation 's', the number of required samples 'n' to produce an estimate of the mean within a tolerance level 'd' is

$$N = [z_{\alpha/2}\ s / d]^2 \qquad \text{Equation 2.3.1.3-1}$$

where $z_{\alpha/2}$ is the standard normal deviate at probability level α (for a two sided test) and the other parameters are as defined above.

So, for instance, if 90% confidence that the sample results will be within 20% of the true population mean (of say 10 ng/L) is desired and the standard deviation of the data is expected to be 5 ng/L, then

$z_{(\alpha/2 = 0.05)} = 1.645$

$s = 5$

$d = 0.2(10) = 2$, and

$N = [1.645(5)/2]^2 = 17.$

Therefore, 17 sample measurements would be required. Data taken from nine WWTPs as part of the NACWA study (NACWA, 2007) indicates that the coefficient of variation ($CV = 100\ s/x_m$) for T-Hg measurements from most plants ranged from about 25% to 50%, although one was as high as 113%. Other studies (Parmer et al., 2005; NEORSD, 2006; Tuominen, 2007 personal communication) have reported lower CVs in the range of 6% to 16.5%. The CV for T-Hg in the reactive mercury study reported in Appendix A of this report is 65%. Unless the plant has specific knowledge about the variability of T-Hg in its effluent, the researchers recommend using a CV of 50%. Thus, the number of samples calculated above (i.e. 17) is a reasonable number of samples to collect for this parameter.

Seasonal considerations are important when sampling for total mercury in WWTP effluents. Since T-Hg has an affinity for particles, it will tend to be elevated when suspended solids are elevated. In wastewater treatment, elevated solids may coincide with poorer settling associated with colder weather. Samples collected during the same season should be reasonably similar as long as the treatment process remains the same. If worst case values are desired, the

---

[5] This assumption is reasonable for WWTP effluent where the objective is to produce an effluent with a consistent quality. Variability in the effluent in chemical constituents of the effluent is caused by random fluctuations in the processes that produce that effluent. This will not necessarily be so for measurements from other sources or in receiving waters where changes in flow and/or chemical loadings can cause dramatic changes in the levels of chemical constituents in water. Most often, the concentrations of water quality constituents from such sources and in receiving waters will be log-normally distributed. Procedures for estimating number of samples for water quality constituents which may not be normally distributed are dealt with in Chapter 3.0 of this report.

winter season would be a good time for sampling with samples collected roughly each week over a four to five month period. If a representative year-round value is desired, the samples should be collected every three weeks over an annual cycle. Care should be taken not to collect samples when significant events (such as sewer cleanouts) or plant upsets are occurring, as these will tend to bias the values upward. Typically, field duplicates are obtained (one for every 10 samples collected) for quality assurance purposes.

Methylmercury (MeHg) - Methylmercury values should be collected from the user's own effluent if at all possible, coincident with samples for total mercury. CVs associated with MeHg measurements appear to about the same as for total mercury, with one plant exhibiting a CV of 45% (Parmer et al., 2005). In the reactive mercury study presented in this report, the CV for unfiltered methylmercury was 59%, about the same as for unfiltered total mercury. The researchers strongly recommend that MeHg and T-Hg measurements be taken at the same time so that the MeHg percentage can be calculated. Therefore, the CV recommended to calculate sample size would be the same as for T-Hg or 50%.

Recently, the California Regional Water Quality Control Board, San Francisco Bay Region, began requiring WWTPs discharging within that region to sample for MeHg (Pla 2008, personal communication). Thus far, the results show that, of the 38 plants sampled, MeHg was detected on average 84% of the time. The mean number of samples over all plants was 8.5; the least sampled plant had two samples collected, and the most sampled plant had 40 samples. The mean MeHg detected was 0.33 ng/L (the lowest mean at any one plant was 0.04 ng/L, the highest mean at any one plant was 1.21 ng/L). The average standard deviation over all the plants was 0.26 ng/L and the average of the computed CVs over all the plants was 66%. Thus a CV of 50% appears to be a reasonable first estimate when planning a sampling program.

Seasonality also may affect MeHg levels measured in the effluent of WWTPs. As discussed in the Phase I report, it is possible that higher temperatures in summer may be associated with slight elevation of MeHg in WWTP effluent. Therefore, if worst case levels are of interest, weekly sampling over a four to five month period in late summer or early fall is recommended. If a year-round average is desired, sampling every three weeks over an annual cycle is recommended.

If measurement of MeHg in one's effluent is not possible, the use of %MeHg information in the Phase I report would be appropriate to make a first approximation of bioavailability related to MeHg in the effluent.

Reactive Mercury (R-Hg) - Reactive mercury values should be collected from the analyst's own effluent if possible, coincident with samples for T-Hg and MeHg. Collecting and analyzing samples for R-Hg directly (rather than relying on a computation to estimate bioavailable inorganic mercury) will result in a more precise (less uncertain) answer. Since there was not a great deal of data for R-Hg from WWTPs, the researchers recommend using CVs from dissolved T-Hg measurements to estimate the variability of this measurement[6]. CVs for dissolved T-Hg in the NACWA study ranged from about 12% to about 70%, not unlike the range in CVs found for unfiltered T-Hg. The average CV for filtered total mercury in the reactive mercury study presented in this report was 51% and the average CV for reactive mercury was 54%. Therefore, they recommend the same number of samples and sampling plan for R-Hg as

[6] The researchers make this recommendation on the basis that the analysis for R-Hg is very similar to that for filtered T-Hg.

for T-Hg and MeHg. In terms of seasonal effects on R-Hg, there is not enough experience with this measurement to give an informed recommendation.

Dissolved Organic Carbon (DOC) - DOC is not often measured in the effluents of WWTPs. However, some recent studies provide useful information. The NEORSD mercury study of the Cuyahoga River (NEORSD, 2006) reported DOC levels in its plant effluent over roughly an annual cycle (n=4). The mean DOC level in this effluent was 12.4 mg/L and the CV was 9.5%. Based on a CV of 10%, the number of samples required to be within 20% of the true value of the population mean would be only 1. However, to allow for the calculation of basic statistics, a minimum of three samples are recommended. In the reactive mercury study reported in Appendix A of this report, mean DOC levels ranged from 7.3 mg/L to 18.5 mg/L, with the exception of one plant, which received effluent from two paper mills and had a mean effluent DOC concentration of 46.9 mg/L. The average CV for the DOC data was 16.7%.

Total Suspended Solids (TSS) - There are comparatively more measurements of TSS available from WWTPs as this is commonly measured and reported as part of NPDES compliance. TSS normally decreases with increasing levels of treatment and our reporting here is limited to those WWTPs for which the researchers had mercury samples. Note that in Chapter 4.0 of Phase I, TSS in the effluent of plants employing secondary treatment was reported to be on the order of 10 to 30 mg/L, while plants employing advanced treatment typically discharged effluent containing 1 to 5 mg/L TSS. The NACWA study (NACWA, 2007) reports TSS in the effluent of nine WWTPs. These nine WWTPs had mean TSS levels ranging from 2.9 mg/L to 15.3 mg/L. CVs ranged from 28% to 70%. The WLSSD plant (Tuominen, 2007 personal communication) had an average TSS concentration in its effluent of 4.6 mg/L with a CV of 47.5% (n=30). The Southerly WWTP (NEORSD, 2006) reported an average TSS level of 3.2 mg/L in its discharge with a CV of 56%. The average CV for TSS reported in the reactive mercury study presented in this report was 32%. Since one would not expect the CV for TSS to be greater than that for T-Hg, the researchers recommend a CV of 50% when planning measurements for TSS, which would result in the same number of samples as that for T-Hg.

Distribution Coefficient (K_D) - The distribution coefficient is normally reported as a $\log_{10}$ value because of its magnitude. Several recent studies have reported these values for total mercury. In Chapter 3.0 of Phase I, $\log K_D$ values for two WWTPs of 5.16 and 5.10 with standard deviations of 0.35 and 0.18, respectively (CVs of 6.8% and 3.5%, respectively) were reported. K_Ds were calculated by the authors of this report from the data in the NACWA mercury study (NACWA, 2007). LogK_Ds ranged from 5.05 to 5.8 with CVs ranging from 3.2% to 8%. Thirty measurements of filtered and unfiltered T-Hg and corresponding TSS measurements were used to calculated $\log K_D$s for the WLSSD effluent, yielding values which had a mean of 5.35 with a standard deviation of 0.26 (CV = 4.9%). Data for the Southerly WWTP (NEORSD, 2006) provides data to calculate K_Ds as well. These data (n=5) yield a mean $\log K_D$ of 5.37 with a standard deviation of 0.37 (CV = 6.9%). Finally, Parmer et al. (2005) report a mean $\log K_D$ for T-Hg in the final effluent of the Sacramento Regional County Sanitation District WWTP of 5.1 with a standard deviation of 0.2 (CV = 3.9%). The mean $\log K_D$ values reported in the reactive mercury study presented in this report ranged from 4.7 to 5.5. The average CV for the $\log K_D$ data from this study was 7.7%.

Based on these data, the CV for $\log K_D$ measured at a WWTP is likely to have a maximum value of about 10%. Notice however, that this is the CV for the log of the K_D value, not for the K_D value itself, which is used in calculations. To convert the CV from log space to arithmetic space, first calculate the standard deviation of the $\log K_D$ (or use the value if it is known). Assuming a value of 10% for the CV and a value of 5.2 for $\log K_D$, a typical standard deviation for $\log K_D$ of 0.51 is obtained. This standard deviation can be used to calculate the CV for the K_D data using Chow's derivation (Chow, 1954) according to the following:

$$CV = [e^{(2.3*VAR(\log(x))} - 1]^{1/2} \qquad \text{Equation 2.3.1.3-2}$$

where VAR(log(x)) is the variance of the values of $\log K_D$ (the square of the standard deviation). The factor 2.3 is to correct for the base 10 logarithm rather than the natural logarithm in the calculation.

Carrying out the calculation, a value for CV of the actual K_D values of about 90% is obtained.[7] This is reasonable given that the CVs calculated for actual K_D values from the NACWA data which range from 41% to 92%. The average CV for KD in the reactive mercury study reported herein was 82%. The implication of this for the estimation of bioavailable Hg is that, because the CV for the calculated K_D value is high, more samples must be collected to assure the desired accuracy and precision.

Obviously, K_D cannot be measured directly. In order to calculate a value for it one must have either unfiltered and filtered Hg measurements in the water column, or a T-Hg measurement on filtered solids (T-Hg(solids)) and a filtered T-Hg water column measurement. If the former is available, the K_D is calculated by:

$$K_D = 10^6\ [\underline{\text{T-Hg}}(\text{unfiltered}) - \underline{\text{T-Hg}}(\text{filtered})] / [\text{TSS} * \underline{\text{T-Hg}}(\text{filtered})] \qquad \text{Equation 2.3.1.3-3}$$

where T-Hg is in ng/L and TSS is in mg/L. If the solids concentration is available, the expression is:

$$K_D = 10^6\ [\underline{\text{T-Hg}}(\text{solids})] / [\underline{\text{T-Hg}}(\text{filtered})] \qquad \text{Equation 2.3.1.3-4}$$

pH - The pH of the WWTP effluent should be well known as it is often required to measure for compliance with the NPDES permit. Since the pH is often adjusted to the final effluent value by chemical addition, the pH of the effluent should vary only by a small amount.

Flow or Volume - Although flow is not essential to the bioavailability or bioavailable concentration calculations, it is useful to have it order to compare the flux of mercury from one source versus another. Every WWTP will have flow data and additional data is not necessary to collect as part of a separate sampling program. The best use of the data is to use flow or volume data collected from the same days as the mercury samples. If long term averages are desired, it is

---

[7] This elevated CV is not unexpected since the K_D value results from a calculation involving the division of several measured values, all of which exhibit a somewhat high degree of variability themselves. Further discussion of uncertainty in these parameters and the propagation of error in bioavailability calculations can be found in Chapter 3.0.

advisable to calculate means of the chemical data and multiply them by long term flow averages or percentiles.

2.3.1.4 Examples

Example 1, Estimation of Bioavailability Using MeHg and R-Hg. This example uses data from the reactive mercury study performed as a part of this study (see Appendix A). The data for this example are from Facility C. The example demonstrates the use of the Bioavailability Tool to estimate the bioavailable mercury in the effluent of this WWTP. T-Hg, MeHg, and R-Hg were measured and are entered in the upper left-hand corner of the spreadsheet in the Effluent #1 input section. These values are 3.83 ng/L, 0.230 ng/L and 0.333 ng/L, respectively, and represent the mean values of ten samples taken for these parameters from this effluent. The outputs are shown under the Effluent #1 column in the middle of the left hand side of the spreadsheet (Figure 2-3). The computation shows that 6.0% of the mercury is in the MeHg form and that 8.7% is in the reactive form. Of the total bioavailable mercury (B-Hg) of 0.56 ng/L, about 59% is in the reactive form and 41% is in the methylated form. B-Hg makes up about 15% of the average total mercury in the effluent.

Example 2, Comparison of the Bioavailability of Two Effluents (Using MeHg and R-Hg Data). In this example, the bioavailability of two effluents are compared, the first being that discussed in Example 1. The second effluent is that of Facility D in Appendix A. T-Hg, MeHg, and R-Hg were measured and are entered in the lower left-hand corner of the spreadsheet in the Effluent #2 input section. These values are 13.4 ng/L, 0.271 ng/L and 0.342 ng/L, respectively, and represent the mean values of ten samples taken for these parameters from this effluent. The outputs are shown under the Effluent #2 column in the middle of the left hand side of the spreadsheet, just to the right of the outputs for Effluent #1 (Figure 2-4). The computation shows that 2% of the mercury is in the MeHg form versus 6% for Effluent #1 and that 2.6% is in the reactive form versus about 8.7% for Effluent #1. Of the total bioavailable mercury (B-Hg) of 0.61 ng/L, about 56% is in the reactive form and 44% is in the methylated form. Compared to Effluent #1, B-Hg is slightly higher, R-Hg as a % is slightly lower, and MeHg as a percent is slightly higher. B-Hg makes up about 4.6% of the total Hg in this effluent versus 14.7% of the average total mercury in the effluent. Thus, even though the average total mercury in the effluent is higher in the effluent of Facility D (Effluent #2) than in the effluent of Facility C (Effluent #2), the concentrations of bioavailable mercury (B-Hg) are similar and therefore the bioavailability (the sum of MeHg and R-Hg, expressed as percents of T-Hg) of the mercury in the effluent of Facility D is substantially lower.

2.3.2 Question #2 User Guidance and Examples

This analysis answers the question "Is my effluent more/less bioavailable than one of the following non-point sources?"

- *Mining runoff*
- *Urban stormwater*
- *Non-urban stormwater*
- *Precipitation*
- *Sediment pore water.*

The inputs and outputs are identical to those required to answer the first question.

FRESHWATER BIOAVAILABILITY COMPUTATIONS

Evaluate Bioavailability of a Single Effluent	Effluent #1	Effluent #2
Inputs Effluent #1		
T-Hg (ng/L) =	3.83	
MeHg (ng/L) =	0.23	
R-Hg (ng/L) =	0.333	
DOC (mg/L) =		
TSS (mg/L) =		Optional
log K_D =		Optional
pH		Optional
Flow rate or Volume (million gallons or MGD)		Optional
Outputs		
%MeHg =	6.01	
Dissolved Inorganic Hg (ng/L) =		
Dissolved Reactive Hg (%) =		
Reactive Hg (% of Total) =	8.695	
Bioavailable Inorganic Hg (% of Bioavailable Hg) =	59.15	
Methylmercury (% of Bioavailable Hg) =	40.85	
Bioavailable Hg (%) =	14.70	
Bioavailable Hg Concentration (ng/L) =	0.56	
Bioavailable Hg Flux (ug or ug/day)		
Compare Bioavailability of Two Effluents		
Inputs Effluent #2		
T-Hg (ng/L) =		
MeHg (ng/L) =		
R-Hg (ng/L) =		
DOC (mg/L) =		
TSS (mg/L) =	Optional	
log KD =	Optional	
Flow rate or Volume (million gallons or MGD)	Optional	

Mix Effluent #1 With Receiving Water		
Inputs Receiving Water		
T-Hg (ng/L) =		
MeHg (ng/L) =		
R-Hg (ng/L) =		
DOC (mg/L) =		
TSS (mg/L) =		Optional
log K_D =		Optional
Flow rate or Volume (Effluent #1)		
Flow Rate or Volume (Receiving Water)		
Mixing Ratio (Fraction of effluent to receiving water) =		0.00
Water Depth (m) =		
pH =		Optional
Far Field Flag, 0 - Use MeHg/T-Hg, 1 - Use pH, DOC, 2 Use MeHg/R-Hg		0
Outputs	Near Field	Far Field
T-Hg concentration (ng/L) =		
MeHg concentration (ng/L) =		
MeHg % =		
DOC concentation (mg/L) =		
pH =		
Dissolved Inorganic Hg of Receiving Water (ng/L) =		
Dissolved Reactive Hg of Receiving Water (ng/L) =		
Mixed Reactive Hg (ng/L) =		
Reactive Hg (% of Total) =		
Total Bioavailable Hg (ng/L) =		
Bioavailable Inorganic Hg (% of Bioavailable Hg) =		
Methylmercury (% of Biaovailable Hg) =		
WWTP Bioavailable Hg Mass Flux (% of Total Flux)		

Figure 2-3. Bioavailability Tool Inputs and Outputs for Example 1.

FRESHWATER BIOAVAILABILITY COMPUTATIONS

Evaluate Bioavailability of a Single Effluent	Effluent #1	Effluent #2
Inputs Effluent #1		
T-Hg (ng/L) =	3.83	
MeHg (ng/L) =	0.23	
R-Hg (ng/L) =	0.333	
DOC (mg/L) =		
TSS (mg/L) =		Optional
log K_D =		Optional
pH		Optional
Flow rate or Volume (million gallons or MGD)		Optional
Outputs		
%MeHg =	6.01	2.02
Dissolved Inorganic Hg (ng/L) =		
Dissolved Reactive Hg (%) =		
Reactive Hg (% of Total) =	8.695	2.552
Bioavailable Inorganic Hg (% of Bioavailable Hg) =	59.15	55.79
Methylmercury (% of Bioavailable Hg) =	40.85	44.21
Bioavailable Hg (%) =	14.70	4.57
Bioavailable Hg Concentration (ng/L) =	0.56	0.61
Bioavailable Hg Flux (ug or ug/day)		
Compare Bioavailability of Two Effluents		
Inputs Effluent #2		
T-Hg (ng/L) =		13.4
MeHg (ng/L) =		0.271
R-Hg (ng/L) =		0.342
DOC (mg/L) =		
TSS (mg/L) =	Optional	
log KD =	Optional	
Flow rate or Volume (million gallons or MGD)	Optional	

Mix Effluent #1 With Receiving Water		
Inputs Receiving Water		
T-Hg (ng/L) =		
MeHg (ng/L) =		
R-Hg (ng/L) =		
DOC (mg/L) =		
TSS (mg/L) =		Optional
log K_D =		Optional
Flow rate or Volume (Effluent #1)		
Flow Rate or Volume (Receiving Water)		
Mixing Ratio (Fraction of effluent to receiving water) =		0.00
Water Depth (m) =		
pH =		Optional
Far Field Flag, 0 - Use MeHg/T-Hg, 1 - Use pH, DOC, 2 Use MeHg/R-Hg		0
Outputs	Near Field	Far Field
T-Hg concentration (ng/L) =		
MeHg concentration (ng/L) =		
MeHg % =		
DOC concentation (mg/L) =		
pH =		
Dissolved Inorganic Hg of Receiving Water (ng/L) =		
Dissolved Reactive Hg of Receiving Water (ng/L) =		
Mixed Reactive Hg (ng/L) =		
Reactive Hg (% of Total) =		
Total Bioavailable Hg (ng/L) =		
Bioavailable Inorganic Hg (% of Bioavailable Hg) =		
Methylmercury (% of Biaovailable Hg) =		
WWTP Bioavailable Hg Mass Flux (% of Total Flux)		

Figure 2-4. Bioavailability Tool Inputs and Outputs for Example 2.

2.3.2.1 Description of Inputs and Outputs

The inputs and outputs for the worksheet in this regard are the same as for the comparison of two effluents. The difference arises in that the characteristics of effluent #2 would be those of the non-point source of interest.

2.3.2.2 Differences between Fresh, Estuarine, and Marine Worksheets

As with the question #1 analysis, DOC is not needed in for the estuarine or marine analysis if reactive mercury is entered as an input parameter (see the discussion in Section 2.3.1.1).

2.3.2.3 Guidance for Input Parameter Estimation

The chief difference in the guidance for estimating parameters for non-point versus point sources is that the non-point source chemical fluxes are generated by hydrologic events and other natural processes. This requires that they be characterized differently from point source data and presents some complications when attempting to compare them to point source data.

The first three non-point sources covered here are all rainfall-runoff driven, although it all three cases there may be a "wet-weather" and "dry-weather" component to the chemical flux. Since the bulk of the solids and, in the case of mercury or other chemicals strongly sorbed to solids, moves during wet-weather storm events, wet-weather concentrations and dry-weather concentrations are often measured and reported separately. Obviously, wastewater fluxes will be a smaller percentage of the overall watershed flux when compared to wet- versus dry weather flows. Needless to say, significant misinterpretations could occur by using too little data to characterize highly variable non-point source inputs.

Another difference to be aware of when sampling for mercury species is that the flow condition can affect speciation. During high flows when solids are abundant, much of the mercury will be sorbed to particles, and by our definition, unreactive. During base flow conditions, however, when solids concentrations are low, dissolved and reactive concentrations will increase as a percentage of total mercury in the water column. Thus, it may be prudent to either utilize existing data or collect data that recognize these differences. So, for instance, when comparing WWTP effluent mercury data to mining, urban stormwater, or non-urban stormwater runoff, it would be reasonable to compare both wet-weather and dry-weather conditions. Or, if enough data is available, they could be compared to long-term average concentrations and fluxes from non-point sources, where such averages take into account both wet- and dry-weather conditions.

If data is to be collected, samples of the various mercury species should be collected at the same time to the extent possible.

Similar to sampling for non-point runoff driven sources, much has been written on the sampling of rainfall for mercury characterization and quite a lot of data has been collected for T-Hg, less for MeHg, and far less for R-Hg. It is beyond the scope of this document to provide comprehensive guidance on sampling rainfall for mercury. Suffice it to say it is somewhat logistically simpler than stormwater sampling. Rainfall collection is normally accomplished with stationary collectors and samplers are taken periodically of the accumulated contents.

Sampling of contaminated sediments in an aquatic environment borders on research at this stage. Currently the most defensible way to make such measurements is to sample sediment pore water and some research of this type has been done. Collected pore water is normally filtered prior to being analyzed for T-Hg or MeHg, so that exactly the same calculations may not be possible for pore water as they are for surface water.[8] In addition, it has been shown that redox conditions affect the value of the R-Hg measurement in sediments and any such measurements would need to take this into account. For these reasons, it is less likely that the Bioavailability Tool in its present form would provide a robust comparison of the bioavailability of WWTP effluent to pore water in contaminated sediments and the researchers do not recommend its use for such an application.

Total mercury (T-Hg) - Total mercury will be much more variable during wet-weather flows than in dry weather due to fluctuations in solids concentrations. The number of samples one might wish to collect will be more dependent on the number of wet-weather events that occur during the rainy season or throughout out the year rather than the anticipated variability in the data, which would be the main consideration in sampling a WWTP. Although it is beyond the scope of this document to go into detail concerning the optimal ways to sample stormwater, it is important to stress that, in order to accurately depict T-Hg concentrations in wet-weather flows, representative depth-duration storms should be sampled and that inter-storm, as well as storm-to-storm, differences are taken in account. The "first flush" effect that occurs in rainfall-runoff events is well-documented. Thus several samples during a storm would be recommended in order to accurately characterize "event mean" T-Hg concentrations. Seasonal effects are not as likely to affect T-Hg as some other mercury species.

In dry-weather flows, T-Hg concentrations are apt to be less variable. Due to the typically lower TSS concentrations and steady flows, the variability and sampling regime under these conditions should not be unlike that for WWTP effluent. Therefore, the researchers recommend that a CV of 50% be utilized to determine the number of samples to characterize T-Hg during these periods.

Representative values of T-Hg in these sources can be found in Section 4.2 of the Phase I report.

Methylmercury (MeHg) - The same caveats apply to sampling methylmercury as to total mercury, in terms of the need to sample representative storms and representatively within storms. Seasonal effects are much more likely to manifest themselves with methylmercury however. It has been noted that seasonal phenomena such as autumn leaf fall can affect MeHg levels in water bodies (Balogh et al., 2002). It has been observed that MeHg can be formed under warm late summer low flow conditions in slow moving streams, wetlands and the hypolimnia of lakes when dissolved oxygen levels are low and the activity of sulfate-reducing bacteria is stimulated. These elevated levels of MeHg may be "flushed" from these systems during subsequent high flows. Therefore, it is important that representative samples for MeHg be collected with care. While it is probably more important to consider potential seasonal effects when sampling for MeHg, it is equally important to collect samples for T-Hg and MeHg simultaneously. This argues that a year-round sampling program for characterization of non-point sources would be best.

Representative values of MeHg in these sources can be found in Section 4.2 of the Phase I report.

Reactive Mercury (R-Hg) - Limited information on R-Hg in these sources can be found in Section 4.2 of the Phase I report. In general, there is too little data available to provide

[8] The methylation/demethylation constant ratios used in this study to assess the methylation potential of surface waters uses the ratio of MeHg (unfiltered sample) to R-Hg (filtered sample).

statistical guidance on the measurement of reactive mercury from non-point sources and there is little data in the literature. Given this situation, the best guidance is to take samples for R-Hg on the same frequency as T-Hg and MeHg samples. The little information available for receiving waters suggests that R-Hg may exhibit about the same variability as filtered samples for T-Hg (about 50%) (see Section 2.3.4.3).

Dissolved Organic Carbon (DOC) - Representative values of DOC in these sources can be found in Section 4.2 of the Phase I report. However, there are few DOC data for mining, urban, or nonurban sources that would support a statistical analysis for the determination of sample numbers. Available information for receiving waters (see Section 2.3.4.3) suggests that DOC has a lower sampling variability than T-Hg and therefore could be sampled less frequently. However, if R-Hg is not being measured, then a DOC measurement should be made with each measurement of T-Hg for which reactive mercury will be calculated.

Total Suspended Solids (TSS) - Similar to the guidance for total mercury, TSS should be measured from representative storms over an annual cycle and during dry weather flow periods as well. The CV for TSS for dry-weather flows is recommended to be 50%. If sampling is not practical, representative values of TSS in these sources can be found in Section 4.2 of the Phase I report.

Distribution Coefficient (K_D) - If filtered and unfiltered measurements of T-Hg and TSS are made for any of these sources, a partition coefficient can be calculated per Equation 2.3.1.3-3 or 2.3.1.3-4. K_Ds for T-Hg are not well established for mining sites. They have been reported for T-Hg downstream of the New Idria mercury mine (San Carlos Creek, CA) ranging from 4.3 to 4.8 (Ganguli et al., 2000), somewhat lower than the values for WWTPs and for most surface waters. By contrast, K_Ds calculated on samples taken upstream of the mine ranged from 6.2 to 6.5. Distribution coefficients for the Guadalupe River, CA which drains former mercury mining areas were reported to range between 5.5 and 5.7 (Thomas et al., 2002). The lack of available values prohibit an assessment of variability. Similarly, for urban runoff, few values are available. Mason and Sullivan (1998) calculated partition coefficients for T-Hg in the Anacostia River, which drains a primarily urban watershed. K_Ds for the NE and NW Branches ranged from 3.2 to 5.3 with an average of 4.32 and a standard deviation of 0.67 (CV=15.6, n=24). No K_Ds were found for non-urban runoff *per se*, but values for waters receiving primarily non-urban runoff are provided in Section 2.3.4.3.

Flow or Volume - Flow from non-point sources can be difficult to measure. By definition, there is no single location that it can be easily measured, in contrast to a WWTP discharge outfall. Normally on small sites or drainages, it is measured in small conveyances immediately downstream of a particular type of source. This may be done by means of a calibrated flume or by measuring depth in small channels and calculating flow by means of Manning's equation or similar formulae. On larger creeks, there may be a USGS or equivalent stream gage and data may be available from the operating agency. In special studies, velocities may be measured using a current meter and depths determined though the use of staff gages. Channel cross-sections at the point of measurement must be known in order to calculate discharge. Often, flow is not measured when water quality measurements are made as it is time-consuming. However, if fluxes are to be calculated from chemical data, flow must be known or measured at the time the water quality samples are collected.

2.3.2.4 Examples

Example 3, Estimation of Bioavailability in an Effluent and Comparison to a Non-point Source Lacking R-Hg Data. This example makes use of the data from Facility E effluent and compares the results to those from a non-point source, specifically, urban runoff. The data for the WWTP are entered into the Bioavailability Tool, as in Example 1. The urban runoff data are entered as Effluent #2, as in Example 2.

The data for Facility E are as follows: T-Hg = 4.9 ng/L, MeHg = 0.842 ng/L, and R-Hg = 0.844 ng/L. The urban runoff data are lacking R-Hg information, which must then be estimated from other inputs. The urban runoff inputs (Effluent #2 in the spreadsheet) are: T-Hg = 10.4 ng/L, MeHg = 0.363 ng/L, DOC = 5.5 mg/L, TSS = 16.3 mg/L, and log K_D = 5.

The results (Figure 2-5), in the middle of the left hand side of the spreadsheet, show that the %MeHg is substantially higher in the effluent of Facility E than in typical urban runoff from the same geographical area, 17.2% versus 3.5%. Note that this is probably not a representative result as the MeHg in the effluent of Facility E is higher than a typical value, which would normally be in the 1 to 3% range for WWTP effluents. Given the TSS and log K_D values, dissolved inorganic Hg is estimated to be 3.95 ng/L. Given the DOC concentration of 5.5 mg/L, the R-Hg as a percent of ID-Hg is calculated to be around 2% and R-Hg as a percent of total Hg is estimated to be 0.77% versus 17.2% in the WWTP effluent. One of the reasons for this disparity is the low concentration of solids in the WWTP effluent. Still over 34% of the WWTP effluent is estimated to be bioavailable versus 4.3% in the urban runoff. The concentration of bioavailable Hg is also estimated to be higher in the WWTP effluent, 1.69 ng/L versus 0.44 ng/L in urban runoff. This latter result could change dramatically when the effluent were to be mixed with urban runoff with its higher TSS content. Contact with higher suspended solids could drive dissolved inorganic and therefore reactive Hg levels down. However, also note that reactive Hg in the effluent may be methylated when mixed with the runoff water, given that the percent MeHg in urban runoff (82% of bioavailable Hg) is higher than MeHg in the WWTP effluent (50% of bioavailable Hg), indicating that there may be a stronger methylating environment in the urban runoff.

2.3.3 Question #3 User Guidance and Examples

This question addresses the issue of how the concentrations and fluxes of bioavailable mercury in one effluent compare to those of another source.

2.3.3.1 Description of Inputs and Outputs

In terms of answering this question, the inputs to the analysis are the same as they would be for questions 1 or 2 (see Section 2.3.3.1). The difference is that in order to calculate and compare effluent fluxes, a flow rate or volume must be entered.

The principal outputs that address this question are the Bioavailable Hg Concentration (ng/L) and the Bioavailable Hg Flux (μg or μg/day). As such, they are contained within the heavy border in the output section. The output units are chosen to provide a number that is convenient in terms of its magnitude, even though other units may be desirable in certain instances.

FRESHWATER BIOAVAILABILITY COMPUTATIONS

Evaluate Bioavailability of a Single Effluent	Effluent #1	Effluent #2
Inputs Effluent #1		
T-Hg (ng/L) =	4.9	
MeHg (ng/L) =	0.842	
R-Hg (ng/L) =	0.844	
DOC (mg/L) =		
TSS (mg/L) =		Optional
log K_D =		Optional
pH		Optional
Flow rate or Volume (million gallons or MGD)		Optional
Outputs		
%MeHg =	17.18	3.49
Dissolved Inorganic Hg (ng/L) =		3.95
Dissolved Reactive Hg (%) =		2.02
Reactive Hg (% of Total) =	17.224	0.769
Bioavailable Inorganic Hg (% of Bioavailable Hg) =	50.06	18.05
Methylmercury (% of Bioavailable Hg) =	49.94	81.95
Bioavailable Hg (%) =	34.41	4.26
Bioavailable Hg Concentration (ng/L) =	1.69	0.44
Bioavailable Hg Flux (ug or ug/day)		
Compare Bioavailability of Two Effluents		
Inputs Effluent #2		
T-Hg (ng/L) =		10.4
MeHg (ng/L) =		0.363
R-Hg (ng/L) =		
DOC (mg/L) =		5.5
TSS (mg/L) =	Optional	16.3
log KD =	Optional	5
Flow rate or Volume (million gallons or MGD)	Optional	

Mix Effluent #1 With Receiving Water		
Inputs Receiving Water		
T-Hg (ng/L) =		
MeHg (ng/L) =		
R-Hg (ng/L) =		
DOC (mg/L) =		
TSS (mg/L) =		Optional
log K_D =		Optional
Flow rate or Volume (Effluent #1)		
Flow Rate or Volume (Receiving Water)		
Mixing Ratio (Fraction of effluent to receiving water) =		0.00
Water Depth (m) =		
pH =		Optional
Far Field Flag, 0 - Use MeHg/T-Hg, 1 - Use pH, DOC, 2 Use MeHg/R-Hg		0
Outputs	Near Field	Far Field
T-Hg concentration (ng/L) =		
MeHg concentration (ng/L) =		
MeHg % =		
DOC concentation (mg/L) =		
pH =		
Dissolved Inorganic Hg of Receiving Water (ng/L) =		
Dissolved Reactive Hg of Receiving Water (ng/L) =		
Mixed Reactive Hg (ng/L) =		
Reactive Hg (% of Total) =		
Total Bioavailable Hg (ng/L) =		
Bioavailable Inorganic Hg (% of Bioavailable Hg) =		
Methylmercury (% of Biaovailable Hg) =		
WWTP Bioavailable Hg Mass Flux (% of Total Flux)		

Figure 2-5. Bioavailability Tool Inputs and Outputs for Example 3.

2.3.3.2 Differences between Fresh, Estuarine, and Marine Worksheets

DOC is not needed in for the estuarine or marine analysis if reactive mercury is entered as an input parameter (see the discussion in 2.3.1.1).

2.3.3.3 Guidance for Input Parameter Estimation

See sections 2.3.1.3 and 2.3.2.3 above.

2.3.3.4 Examples

Example 4, Comparison of Bioavailable Mercury Concentrations and Fluxes from Two Effluents. This example makes use of data from Facilities A and B in Appendix A. The inputs are as follows:

Facility A (Effluent #1) – T-Hg = 8.75 ng/L, MeHg = 0.134 ng/L, R-Hg = 1.34 ng/L, and
Facility B (Effluent #2) – T-Hg = 2.40 ng/L, MeHg = 0.036 ng/L, R-Hg = 0.252 ng/L

In addition, Q_A = 36.0 MGD and Q_B = 109.6 MGD.

The mercury data is input as in Example 1; however, in addition, the respective flow rates of the two WWTPs are also input. This is input as the last entry in each of the Effluent #1 and Effluent #2 groups of data. The outputs are, again, in the middle of the left hand side of the spreadsheet (Figure 2-6). This shows that the bioavailable mercury as a percentage of total mercury from these plants is not that different, 16.8% for Facility A versus 12% for Facility B. However, Facility A has a greater concentration of bioavailable Hg in its effluent, an estimated 1.47 ng/L versus Facility B, which has an estimated 0.29 ng/L, or about five times less. The fluxes

FRESHWATER BIOAVAILABILITY COMPUTATIONS

Evaluate Bioavailability of a Single Effluent	Effluent #1	Effluent #2
Inputs Effluent #1		
T-Hg (ng/L) =	8.75	
MeHg (ng/L) =	0.134	
R-Hg (ng/L) =	1.34	
DOC (mg/L) =		
TSS (mg/L) =		*Optional*
log K_D =		*Optional*
pH		*Optional*
Flow rate or Volume (million gallons or MGD)	36	*Optional*
Outputs		
%MeHg =	1.53	1.50
Dissolved Inorganic Hg (ng/L) =		
Dissolved Reactive Hg (%) =		
Reactive Hg (% of Total) =	15.314	10.500
Bioavailable Inorganic Hg (% of Bioavailable Hg) =	90.91	87.50
Methylmercury (% of Bioavailable Hg) =	9.09	12.50
Bioavailable Hg (%) =	16.85	12.00
Bioavailable Hg Concentration (ng/L) =	1.47	0.29
Bioavailable Hg Flux (ug or ug/day)	200847.24	119472.768
Compare Bioavailability of Two Effluents		
Inputs Effluent #2		
T-Hg (ng/L) =		2.4
MeHg (ng/L) =		0.036
R-Hg (ng/L) =		0.252
DOC (mg/L) =		
TSS (mg/L) =	*Optional*	
log KD =	*Optional*	
Flow rate or Volume (million gallons or MGD)	*Optional*	109.6

Mix Effluent #1 With Receiving Water		
Inputs Receiving Water		
T-Hg (ng/L) =	10	
MeHg (ng/L) =	0.6	
R-Hg (ng/L) =	0.05	
DOC (mg/L) =	20	
TSS (mg/L) =		*Optional*
log K_D =		*Optional*
Flow rate or Volume (Effluent #1)	1	
Flow Rate or Volume (Receiving Water)	10	
Mixing Ratio (Fraction of effluent to receiving water) =		0.10
Water Depth (m) =	3	
pH =	6	*Optional*
Far Field Flag, 0 - Use MeHg/T-Hg, 1 - Use pH, DOC, 2 Use MeHg/R-Hg		0
Outputs	**Near Field**	**Far Field**
T-Hg concentration (ng/L) =	9.89	
MeHg concentration (ng/L) =	0.56	0.66
MeHg % =	5.64	6.66
DOC concentation (mg/L) =	18.18	
pH =		
Dissolved Inorganic Hg of Receiving Water (ng/L) =		
Mixed Dissolved Inorganic Hg (ng/L) =		
Mixed Dissolved Reactive Hg (%) =		
Reactive Hg (% of Total) =	1.85	0.83
Total Bioavailable Hg (ng/L) =	0.74	
Bioavailable Inorganic Hg (% of Bioavailable Hg) =	24.67	11.11
Methylmercury (% of Biaovailable Hg) =	75.33	88.89
WWTP Bioavailable Hg Mass Flux (% of Total Flux)	18.1	

Figure 2-6. Bioavailability Tool Inputs and Outputs for Example 4.

of bioavailable Hg, shown as the last entry in the output section, are closer, since Facility B has the greater discharge. The flux of bioavailable Hg for Facility A is estimated to be 2.0×10^5 µg/day or 0.2 kg/day, while the flux of bioavailable Hg from Facility B is 1.2×10^5 µg/day or 0.12 kg/day, about ½ of that from Facility A.

2.3.4 Question #4 User Guidance and Examples

Question #4 asks "To what degree is my effluent affecting the bioavailability of mercury in the receiving waters?" or alternatively "To what degree is the bioavailability of mercury in my effluent (or another) affected when it is mixed with my receiving water?"

These two questions are in reality one and the same since mercury from any effluent and mercury native to the receiving water are indistinguishable once they are mixed. The distinction is really one of degree. To the extent that the effluent flow rate or volume dominates the mixing of the two effluents, the characteristics of the effluent will tend to dictate the answer and vice versa.

The inputs and outputs for this analysis are located on the right hand side of the FW_I_O Sheet, EW_I_O Sheet, and MW_I_O Sheet in Area "C", Figure 2-1 and Area "C", Figure 2-2, respectively.

2.3.4.1 Description of Inputs and Outputs

The inputs for the mixing analysis are primarily the same as they are for the effluent analysis. The first six inputs, T-Hg, MeHg, R-Hg, DOC, TSS, and logK_D are identical to those of Effluent #1 except that these inputs are used to describe the receiving water. Note that the same caveats apply when reactive Hg is input. Also note that if R-Hg is input for the effluent, it must also be input for the receiving water; if not an error message (ERROR 1) will appear to the right of the parameter field.

The seventh input is the flow rate or volume of the Effluent #1. It is entered here again if the user desires to input a different value than was used in the effluent analysis or if no value was used in the effluent analysis. The eighth parameter is the flow rate or volume of the receiving water.

The ninth input in the list is the mixing ratio. This ratio (defined in Section 2.2.4.1) is the ratio of the effluent flow rate or volume to the receiving water flow rate or volume. These rates or volumes may be in any units provided they are the same. The mixing ratio may be input in lieu of the flow rates described above or in addition to the flow rates or volumes. However, the user should note an implication of the latter. If the actual flow rates or volumes are input and their ratio is different than the mixing ratio input value, the ratio of the flow rates or volumes will be calculated and used in preference to the input mixing ratio and the value actually used will be shown to the right of the mixing ratio input field.

The tenth input is the receiving water depth and the eleventh is the receiving water pH. Finally, the user is asked to decide how to calculate the methylation potential of the receiving water by choice of the "far-field flag" value.

Outputs for the mixing analysis are found beneath the red "Outputs" banner on the right-hand side (Area "C", Figure 2-2). Notice that the outputs are separated into "near-field" and "far-field" columns. The near field column shows the result of initial mixing and the far-field column shows the effect of net methylation reactions once the system has been allowed to come to steady state.[9]

The first five outputs (T-Hg, MeHg, %MeHg, DOC, and pH) under the "near-field" column are simply the results of initial mixing as described by Equation 2.2.4.1-2. The next two outputs, Dissolved Inorganic Hg of the Receiving Water, and Dissolved Reactive Hg of Receiving Water, are only calculated if R-Hg is *not* entered as an input. If it is, these calculations are unnecessary and the output fields will be blank. The next two outputs, Mixed Reactive Hg, and Reactive Hg, as a % of T-Hg, is calculated in either case. Total Bioavailable Hg (R-Hg + MeHg) is output next. The percentages of Total Bioavailable Hg consisting of R-Hg and MeHg are output next and are centered to highlight them. Finally, in the heavy-bordered box at the bottom, the percentage of bioavailable Hg attributable to the WWTP discharge is shown as a percentage of the total flux from both the effluent and the receiving water.

Under the "far-field" column, concentrations of various species are estimated according to the value of the far-field flag the user has chosen, as described above. If the values are not different in the near-field or the far-field analysis (for instance, in the case of T-Hg or DOC), they are not shown again in the "far-field" column. In any case, the DOC concentration and depth of the mixed waters is used to calculate a demethylation constant as described in Section 2.4.2.1. The far-field concentration of MeHg is then estimated from the demethylation constant, the selected ratio of methylation/demethylation constant, and the mixed total bioavailable Hg

---

[9] Note that the Tool does not "model" the time or distance from the outfall at which steady-state would occur. In the case of a riverine discharge, this would be some distance downstream of the outfall. However, in the case of an estuary, the reactions might approach steady-state either up estuary or down estuary from the discharge, due to the movement of the effluent in both directions induced by tidal action. In a marine environment, steady state with respect to these reactions may or may not occur within the zone of initial dilution (ZID). In this case, the Tool estimates the effects of these reactions were they to reach steady state within the ZID. The modeling of the time and distance from the outfall at which steady-state conditions might be reached are beyond the scope of this study and exceeds the capability of the Bioavailability Tool.

concentration. The far-field %MeHg is then calculated. A far-field R-Hg (percentage of total mercury) is then calculated using the difference between the total bioavailable Hg and the far-field MeHg value and normalizing to total mercury. Percentages of total bioavailable Hg that are constituted of inorganic and methylmercury are shown next.

The key outputs for answering question #4 are the values of %MeHg, %R-Hg, and the relative proportions of inorganic and methylmercury that make up bioavailable mercury in the near-field and far-field. These outputs show how the mixing of the effluent and receiving water affects the distribution of inorganic and methylmercury in the mixture. Furthermore, comparison of these values with the same percentages in effluent #1 on the left hand side of the page indicate the changes that occur between the effluent and initial mixing.

2.3.4.2 Differences between Fresh, Estuarine, and Marine Worksheets

There are two primary differences in the inputs for the three worksheets. In the case of the freshwater analysis, pH is an input parameter used to index the net methylation tables if water quality parameters are used for this purpose. In the case of the estuarine analysis, salinity is used to do this and is substituted for pH as an input in the worksheet. In the marine waters worksheet, this input field is left blank.

The second difference is in the use of the far-field analysis flag. The flag tells the worksheet how the user wishes to assess the methylation potential of the effluent/receiving water mixture. In the case of freshwaters, there are three possible settings:

- 0 – for using %MeHg,
- 1 – for using pH and DOC, and
- 2 – for using the MeHg/R-Hg (methyl to reactive mercury) ratio.

In the case of estuaries, the far-field flag has two options:

- 0 – to use salinity, and
- 1 – to use the MeHg/R-Hg (methyl to reactive mercury) ratio.

In the case of marine waters, the far-field flag also has two options:

- 0 – to use the default ratio value of 0.02, and
- 1 – to use the MeHg/R-Hg (methyl to reactive mercury) ratio.

2.3.4.3 Guidance for Input Parameter Estimation

The guidance for parameterizing receiving waters is much the same as for parameterizing WWTP effluent. In a practical sense, the variability in levels of most of the water quality constituents we are interested in will be higher and therefore a greater number of samples will be needed to characterize them for similar levels of uncertainty.

Much of the information which will be helpful for receiving water characterization can be found in the Section 2.4 of this chapter and there will be frequent reference to that portion of the report.

Total Mercury (T-Hg) - Total mercury levels in fresh water rivers is reasonably well characterized by information contained in Table 2-3 and Appendix B. If the user does not have T-Hg levels for the WWTP receiving water, and does not plan to collect any, then a default value could be selected from a river in Appendix B that is the same physiographic province as the user's. For instance, if the user's receiving water is a blackwater Coastal Plain creek in North Carolina, then a comparable T-Hg value might be selected from the Mobile River, AL (T-Hg = 1.78 ng/L). The same type of guidance would apply to estuaries. Although the data is much more limited, it may be possible for the user to find a similar estuary in Table 2-8. T-Hg values for coastal oceans waters may be approximated by the values in Table 2-13.

If the user decides to sample for T-Hg instead of using literature values, then the number of samples should be estimated using an appropriate CV for the receiving water. CVs for T-Hg in receiving waters tend to be higher than for WWTP effluent because of the influence of natural processes in natural waters. CVs for receiving waters are typically on the order of 50 to 100% (EPRI, 2003), and a reasonable number to use for T-Hg is 75%[10]. Note that Table 2-8 shows CVs for total Hg in estuarine environments ranging from 20% to 130%. Again, the number of samples can be calculated by using Equation 2.3.1.3-1. The analyst should obtain samples from both low and high flow events, and aspire to cover an annual cycle, in order to acquire a representative mean value for T-Hg.

Methylmercury (MeHg) - The same guidance applies to MeHg in receiving waters. Typical values for fresh water systems can be found in Table 2-3 and Appendix B. Typical values for estuaries can be found in Table 2-9. Values for coastal ocean waters may be approximated by the values in Table 2-13.

CVs for MeHg tend to be somewhat higher than for T-Hg. Typical values are in the 60% to 120% range and a reasonable CV to use for calculating sample numbers is 100%. Table 2-9 shows a range of CVs for MeHg in estuaries of 42-150%. As discussed earlier, getting a perspective on the variation of MeHg over an annual cycle is probably more important for MeHg than for T-Hg.

Reactive Mercury (R-Hg) - There are few R-Hg measurements for receiving waters. Some information on concentrations was presented in Phase I of this report (see Section 3.2.5) which suggested that R-Hg measured in receiving waters of the Connecticut River was about 15% of T-Hg. Otherwise, there is information IB-Hg calculated from the data of Balogh et al. (2004) (see Section 2.4.2.1). This data (n=32) suggests that R-Hg is a much lower % of T-Hg, on the order of 1%.

The paucity of R-Hg data suggests that it should be measured if at all possible for the receiving water of interest. The calculations on the data of Balogh et al. (2004) suggest that the overall CV for R-Hg is about 37%; however, the CV for filtered T-Hg is around 75%, and the measured R-Hg data should be at least as variable as that parameter. Therefore, calculating the number of samples to be collected for R-Hg with a CV of 75% is recommended. The data of Balogh et al. (2004) also suggest seasonality in this parameter. Calculated values of R-Hg were highest in summer, lowest in fall, and intermediate in spring. The same pattern was recognized in measured values of both filtered and unfiltered T-Hg.

Dissolved Organic Carbon (DOC) - There is not a good database for DOC; however, typical values for TOC are shown in Table 2-3 and the individual data for TOC are contained in the USGS database shown in Appendix B. These data should be used with caution as a surrogate

---

[10] Note that the CVs shown at the bottom of Table 1-3 indicate the variability across all the rivers in the database, and are not indicative of the CV for any of the parameters *within* an individual system.

for DOC, however. Maybeck (1982) reports that the ratio of DOC:TOC varies widely for world rivers, ranging from 0.1 to 0.9. TOC approximates DOC from in streams draining relatively undisturbed areas, like forests. However, in uses where land disturbance is high, the percentage of DOC is lowest. It tends to be intermediate in urban areas (Hopkinson and Vallino, 1995). Typical DOC values for estuaries are shown in Table 2-5.

Measurement of DOC is suggested. Balogh's data exhibits a DOC CV of 41% overall. Typical values for CVs for DOC in estuaries range from 13-53% (Table 2-5). Based on this figure about the same number of samples would be needed to characterize DOC as to characterize T-Hg. Therefore, it is recommended that DOC be collected with every mercury sample.

Total Suspended Solids (TSS) - Typical values of TSS for various river systems can be found in Figure 4-5 in the Phase I report. Average values vary from <100 mg/L to >500 mg/L and vary by predominant land use. Typical values for TSS in estuaries are shown in Table 2-6, with CVs ranging from 32-190%. If R-Hg is not being measured and the user is relying on calculation to obtain reactive Hg, then TSS should be collected with each mercury sample.

Distribution Coefficient (K_D) - Distribution coefficients for T-Hg in receiving waters have been summarized by Grigal (2002). Values ranged from 4.4-5.7 L/kg with a mean value of 5.1. Values will tend to be lower where sediment values are higher and where the organic content of particulates is higher.

As noted earlier, K_D cannot be measured directly and is a function of several measured variables, all of which exhibit uncertainty. Log K_Ds measured of the Anacostia River by Mason and Sullivan had a CV of 15.6%, which translates to a CV for the K_D itself of 75%. In the San Francisco Bay Estuary, Conaway et al. (2003) calculated logK_Ds for T-Hg on suspended sediments resulting in a mean of 5.6 and a standard deviation of 0.65 (n=126), resulting a CV for K_D of 72%.

Flow, Volume, Mixing Ratio - On receiving water bodies of any size, reliable measurements of flow should be available from the USGS. The analyst should preferentially rely on these measurements as they are difficult to make on large rivers and/or estuaries. On smaller rivers, where effluent may dominate the flow regime, it may be necessary to make one's own flow measurements.

The mixing ratio referred to here is not necessarily the ratio that may be required or available as a result of regulatory requirements (i.e. a dilution ratio). The mixing ratio of relevance here is a number that would result from the theoretical mixing of the effluent of the WWTP with the flow of the river (for a riverine discharge) or the ratio of effluent to dilution water that would result at steady-state from mixing as a result of tidal action in the vicinity of the discharge (for an estuarine discharge). In the first case, the mixing of the plume with the river water approaches this theoretical value as it moves downstream and becomes fully mixed with the receiving water; however, without the influence of additional inflows. An appropriate number to use would be the average annual flow (or volume) of the WWTP divided by the average annual flow (or volume) or the receiving water at the point of discharge.

In an estuary, the mixing ratio could be calculated in the same way, except that the calculation needs to account for further dilution by saline water. Thus, the calculation would utilize average WWTP discharge divided by the sum of the average annual fresh water inflows above the point of discharge, times the fraction of fresh water at the point of discharge (Dyer, 1973), as is illustrated below:

$$C_0 / C_e = Q\,(f_0) / R \qquad \text{Equation 2.3.4.3-1}$$

where:

C_0 is the average cross-sectional concentration of the discharged constituent in the estuary at the point of discharge (mg/L),

C_e is the concentration of the discharged constituent (mg/L),

Q is the annual average WWTP discharge flow or volume,

f_0 is the fraction of freshwater at the point of discharge, and

R is the sum of the upstream fresh water inflows.

The units of volume or flow are not of consequence as long as they are the same dimensionally. The mixing ratio as defined is the ratio C_0/C_e.

Another way to estimate this number would be to measure the concentration of a discharged constituent (e.g. BOD) in the vicinity of the outfall. The value of the discharged constituent would be highest near outfall (assuming the constituent is discharged at a higher concentration than the ambient water) and would decrease in both directions as a result of tidal mixing. The ratio of the receiving water concentration to the effluent concentration should give approximately the same number. If the estuary is not well mixed, the constituent concentration (and the fraction of fresh water) would be measured in the receiving water above the pyncnocline.

For marine discharges, assume that the mixing ratio is the dilution that would occur at the edge of the zone of initial dilution (ZID).

Ancillary Parameters - In freshwater systems, if the far-field analysis depends on water quality parameters to assess the strength of the methylating environment in the receiving water, then pH must be input. In estuaries, the auxiliary parameter is salinity. Both of these are easily measured in the field at the time samples are collected. However, if measurements cannot be made, typical pH values can be found in Appendix B for fresh waters and in Table 2-7 for estuaries. Note that the CV for pH is usually low and thus it is not necessary to measure pH on the same frequency as other more variable parameters. The data of Balogh et al. (2004) show a CV for pH of 1.6%. The data in Table 2-7 for estuaries show typical CVs for pH of 2-7%. The researchers recommend a CV of 10% in order to calculate the number of pH samples required for the level of accuracy desired by the analyst. In effluent dominated streams, the measured pH of the mixed effluent and receiving water would be the most useful, rather than the pH of the receiving water alone. In estuaries, tidally averaged values measured during over the course of an annual cycle would be most useful.

In terms of salinity, for the purpose of determining the strength of the methylating environment for estuaries, it is only important to know whether the value is above or below 15. This number may vary seasonally. High runoff from rainfall or snowmelt will drive it down, while drought and usually hot conditions will drive it up at any particular point in the estuary. As with pH, tidally averaged values over an annual cycle would be beneficial to have.

2.3.4.4 Examples

Example 5, Estimate Bioavailability of an Effluent Mixed with a Fresh Receiving Water Lacking R-Hg Data. This example assumes that both the effluent and receiving water are lacking R-Hg data. Otherwise the effluent data are taken from Facility B in Appendix A. The receiving water characteristics are taken from the data of Balogh (2004) for Trott Brook.

The characteristics of the effluent, entered as effluent #1 in the upper left hand portion of the spreadsheet are:

T-Hg = 2.4 ng/L
MeHg = 0.036 ng/L
DOC = 10.7 mg/L
TSS = 4 mg/L
log K_D = 5.32, and
pH = 7.

The characteristics of the receiving water, entered in the upper right hand portion of the spreadsheet are:

T-Hg = 1.57 ng/L
MeHg = 0.328
DOC = 14 mg/L
TSS = 7.3 mg/L,
logK_D = 5.11, and
pH = 7.6.

The mixing ratio is arbitrarily assumed to be 1 (i.e. 1:1 effluent to receiving water flow) and the far-field flag is set to 0. Note that if the far-field flag is changed to 1, the results remains the same as the predicted ratio of the reaction constants is 16 using either method. A far-field flag of 2 in this case cannot be used since reactive mercury data for the receiving water has not been entered.

The outputs on which this example focuses are in the lower right hand side of the spreadsheet (Figure 2-7). Since the mixing ratio is 1:1, the near-field T-Hg, MeHg, and DOC concentrations are simple averages of the data from the two inputs, in this case 1.99 ng/L, 0.18 ng/L, and 12.35 mg/L, respectively. The resulting %MeHg is 9.17% and the mixed pH is 7.2.

The estimated dissolved inorganic mercury for the receiving water is 0.82 ng/L and the R-Hg is 0.02 ng/L. The near-field mixed R-Hg of the receiving water and effluent is 0.16 ng/L (7.9% of T-Hg). In this case, the WWTP is contributing relatively little MeHg but is contributing relatively more R-Hg; thus the mixed MeHg concentration is diluted by the effluent, whereas the R-Hg concentration is increased by the effluent. The near-field concentrations of MeHg and R-Hg are similar (MeHg = 0.18 ng/L, R-Hg = 0.16 ng/L) as well as the percentages of these two quantities to T-Hg (54% MeHg, 46% R-Hg).

This receiving water represents a strong methylating environment, as evidenced by its high %MeHg. As such, reactive mercury in the effluent/receiving water mixture will tend to

FRESHWATER BIOAVAILABILITY COMPUTATIONS

Evaluate Bioavailability of a Single Effluent	Effluent #1	Effluent #2
Inputs Effluent #1		
T-Hg (ng/L) =	2.4	
MeHg (ng/L) =	0.036	
R-Hg (ng/L) =		
DOC (mg/L) =	10.7	
TSS (mg/L) =	4	*Optional*
log K_D =	5.32	*Optional*
pH	7	*Optional*
Flow rate or Volume (million gallons or MGD)		*Optional*
Outputs		
%MeHg =	1.50	
Dissolved Inorganic Hg (ng/L) =	1.307	
Dissolved Reactive Hg (%) =	22.71	
Reactive Hg (% of Total) =	12.372	
Bioavailable Inorganic Hg (% of Bioavailable Hg) =	89.19	
Methylmercury (% of Bioavailable Hg) =	10.81	
Bioavailable Hg (%) =	13.87	
Bioavailable Hg Concentration (ng/L) =	0.33	
Bioavailable Hg Flux (ug or ug/day)		
Compare Bioavailability of Two Effluents		
Inputs Effluent #2		
T-Hg (ng/L) =		
MeHg (ng/L) =		
R-Hg (ng/L) =		
DOC (mg/L) =		
TSS (mg/L) =	*Optional*	
log KD =	*Optional*	
Flow rate or Volume (million gallons or MGD)	*Optional*	

Mix Effluent #1 With Receiving Water		
Inputs Receiving Water		
T-Hg (ng/L) =	1.57	
MeHg (ng/L) =	0.328	
R-Hg (ng/L) =		
DOC (mg/L) =	14	
TSS (mg/L) =	7.3	*Optional*
log K_D =	5.1	*Optional*
Flow rate or Volume (Effluent #1)		
Flow Rate or Volume (Receiving Water)		
Mixing Ratio (Fraction of effluent to receiving water) =	1	1.00
Water Depth (m) =	2	
pH =	7.6	*Optional*
Far Field Flag, 0 - Use MeHg/T-Hg, 1 - Use pH, DOC, 2 Use MeHg/R-Hg		1
Outputs	**Near Field**	**Far Field**
T-Hg concentration (ng/L) =	1.99	
MeHg concentration (ng/L) =	0.18	0.32
MeHg % =	9.17	16.06
DOC concentation (mg/L) =	12.35	
pH =	7.20	
Dissolved Inorganic Hg of Receiving Water (ng/L) =	0.82	
Dissolved Reactive Hg of Receiving Water (ng/L) =	0.02	
Mixed Reactive Hg (ng/L) =	0.16	0.02
Reactive Hg (% of Total) =	7.89	1.00
Total Bioavailable Hg (ng/L) =	0.34	
Bioavailable Inorganic Hg (% of Bioavailable Hg) =	46.26	5.88
Methylmercury (% of Biaovailable Hg) =	53.74	94.12
WWTP Bioavailable Hg Mass Flux (% of Total Flux)	49.2	

Figure 2-7. Bioavailability Tool Inputs and Outputs for Example 5.

become methylated over time. Since the ratio of the reaction constants is high (16) about 94% of the reactive mercury is estimated to become methylated, leading to a steady-state MeHg concentration of 0.32 ng/L (16% of T-Hg) and an R-Hg concentration of 0.02 ng/L (1% of T-Hg). At this point, 94% of the original mixed bioavailable Hg is estimated to be in the methyl form, and 6% is estimated to be in R-Hg forms.

Note that even though the actual flows are not input, the Tool is still able to compute the *proportion* of the mass flux of bioavailable Hg of the combined flows contributed by the WWTP based on the mixing ratio, in this case 49%.

Example 6, Estimate Bioavailability of an Effluent Mixed with a Fresh Receiving Water Using R-Hg Data. This example uses R-Hg data for the effluent and receiving water to demonstrate this use of the Bioavailability Tool. For the effluent, the inputs for T-Hg and MeHg are the same. This time, however, a value for R-Hg (0.297 ng/L) is input. Likewise, for the receiving water, the T-Hg and MeHg inputs remain the same and a value is input for R-Hg (0.016 ng/L). Note that neither DOC, TSS, logK_D or pH are required in this case. However, if DOC and pH are not input, only the %MeHg method (Far-field Flag = 0) can be used for the far-field analysis.

In this example, with the exception of some roundoff differences, the outputs are nearly identical to the previous example (Figure 2-8).

2.3.5 Question #5 User Guidance and Examples

The relevant question here is "What happens to the concentration of bioavailable mercury when it is mixed with my receiving water?" Given these changes it is also reasonable to ask "What characteristics of the effluent and receiving waters are contributing or enhancing the bioavailability?"

FRESHWATER BIOAVAILABILITY COMPUTATIONS

Evaluate Bioavailability of a Single Effluent	Effluent #1	Effluent #2
Inputs Effluent #1		
T-Hg (ng/L) =	2.4	
MeHg (ng/L) =	0.036	
R-Hg (ng/L) =	0.297	
DOC (mg/L) =		
TSS (mg/L) =		*Optional*
log K_D =		*Optional*
pH		*Optional*
Flow rate or Volume (million gallons or MGD)		*Optional*
Outputs		
%MeHg =	1.50	
Dissolved Inorganic Hg (ng/L) =		
Dissolved Reactive Hg (%) =		
Reactive Hg (% of Total) =	12.375	
Bioavailable Inorganic Hg (% of Bioavailable Hg) =	89.19	
Methylmercury (% of Bioavailable Hg) =	10.81	
Bioavailable Hg (%) =	13.88	
Bioavailable Hg Concentration (ng/L) =	0.33	
Bioavailable Hg Flux (ug or ug/day)		
Compare Bioavailability of Two Effluents		
Inputs Effluent #2		
T-Hg (ng/L) =		
MeHg (ng/L) =		
R-Hg (ng/L) =		
DOC (mg/L) =		
TSS (mg/L) =	*Optional*	
log KD =	*Optional*	
Flow rate or Volume (million gallons or MGD)	*Optional*	

Mix Effluent #1 With Receiving Water		
Inputs Receiving Water		
T-Hg (ng/L) =	1.57	
MeHg (ng/L) =	0.328	
R-Hg (ng/L) =	0.016	
DOC (mg/L) =		
TSS (mg/L) =		*Optional*
log K_D =		*Optional*
Flow rate or Volume (Effluent #1)		
Flow Rate or Volume (Receiving Water)		
Mixing Ratio (Fraction of effluent to receiving water) =	1	1.00
Water Depth (m) =	2	
pH =		*Optional*
Far Field Flag, 0 - Use MeHg/T-Hg, 1 - Use pH, DOC, 2 Use MeHg/R-Hg		0
Outputs	**Near Field**	**Far Field**
T-Hg concentration (ng/L) =	1.99	
MeHg concentration (ng/L) =	0.18	0.32
MeHg % =	9.17	16.05
DOC concentation (mg/L) =		
pH =		
Dissolved Inorganic Hg of Receiving Water (ng/L) =		
Dissolved Reactive Hg of Receiving Water (ng/L) =		
Mixed Reactive Hg (ng/L) =	0.16	0.02
Reactive Hg (% of Total) =	7.88	1.00
Total Bioavailable Hg (ng/L) =	0.34	
Bioavailable Inorganic Hg (% of Bioavailable Hg) =	46.23	5.88
Methylmercury (% of Biaovailable Hg) =	53.77	94.12
WWTP Bioavailable Hg Mass Flux (% of Total Flux)	49.2	

Figure 2-8. Bioavailability Tool Inputs and Outputs for Example 6.

2.3.5.1 Differences between Fresh, Estuarine, and Marine Worksheets

There are no relevant differences between the inputs and outputs of the three spreadsheets in terms of these questions.

2.3.5.2 Description of Inputs and Outputs

The information needed to answer these questions has already been input. The focus is on certain outputs that are generated. The question focuses on bioavailable concentrations rather than bioavailability (the concentrations of bioavailable species relative to total mercury). The comparison of MeHg concentration in the effluent and receiving water, initial mixing, and far-field illustrates how MeHg changes upon mixing and thereafter. Additionally, the comparison of total bioavailable Hg between the effluent and the "near-field" values provides additional insight. These changes and the factors that resulted in the changes will be discussed further in the examples that follow.

2.3.5.3 Guidance for Input Parameter Estimation

No additional parameter guidance is required for this question and the reader is referred to preceding sections, in particular Sections 2.3.1 and 2.3.4.

2.3.5.4 Examples

Example 7, Estimate Far-field Bioavailable Hg Concentrations and Fluxes Lacking R-Hg Data for an Estuary. In this example, the estuary calculation sheets are used. The inputs for the WWTP discharge (Effluent #1) (upper left-hand side of the EW_I_O spreadsheet) are provided below and mimic those of Facility F in Appendix A:

T-Hg = 1.78 ng/L
MeHg = 0.036 ng/L
DOC = 9.6 mg/L

TSS = 4 mg/L, and
log K_D = 4.82 L/kg.

The inputs for the receiving water are as follow:

T-Hg = 14 ng/L
MeHg = 0.03 ng/L
DOC = 2.4 mg/L
TSS = 94 mg/L, and
log K_D = 5.6 L/kg.

The flows are entered in a ratio of 1 (effluent) to 50 (receiving water). The water depth is set at 3 meters and the salinity to 30 parts per thousand. The far-field flag is given a value of 0 so that the salinity will be used to determine ratio of reaction constants for the far-field analysis.

The outputs for the effluent (middle, left hand side) show that nearly all the bioavailable mercury is in the reactive form (91%), with only 9% in the methyl form (Figure 2-9). The near-field mixing outputs (lower right-hand side) show that the mixed effluent/receiving water is dominated by the receiving water, which should be the case as only 5% of the mixture is effluent. Bioavailable Hg in the receiving water is 0.18 ng/L, and is not influenced appreciably by the addition of effluent, the mixture having 0.18 ng/L of bioavailable Hg as well. In the far-field, because the ratio of methylation to demethylation constant is only 0.14 at this salinity, net demethylation is occurring and the amount of methylmercury in the far-field decreases slightly (0.03 ng/L to 0.02 ng/L), while the reactive form experiences a commensurate increase (0.15 ng/L to 0.16 ng/L). In this case, the WWTP has little impact on the results due to the high dilution. Because there is net demethylation in the far-field, the greatest impact of the discharge would be in the near-field. Note that the WWTP contributes only 4.3% of the bioavailable mercury mass flux to the system.

Example 8, Estimate Near-Field Bioavailability of an Effluent Using R-Hg Data for a Marine Receiving Water. In this example, the marine calculation sheets are used. The inputs for the WWTP discharge (Effluent #1) (upper left-hand side of the MW_I_O spreadsheet) are provided below and mimic those of Facility G in Appendix A:

T-Hg = 1.8 ng/L
MeHg = 0.035 ng/L
R-Hg = 0.52 ng/L, and
DOC = 7.3 mg/L.

The inputs for the receiving water are as follow:

T-Hg = 0.36 ng/L
MeHg = 0.005 ng/L
R-Hg = 0.2 ng/L, and
DOC = 0.9 mg/L.

ESTUARINE WATER BIOAVAILABILITY COMPUTATIONS

Evaluate Bioavailability of a Single Effluent	Effluent #1	Effluent #2
Inputs Effluent #1		
T-Hg (ng/L) =	1.78	
MeHg (ng/L) =	0.036	
R-Hg (ng/L) =		
DOC (mg/L) =	9.6	
TSS (mg/L) =	4	*Optional*
log K_D =	4.82	*Optional*
Flow rate or Volume (million gallons or MGD)		*Optional*
Outputs		
%MeHg =	2.02	
Dissolved Inorganic Hg (ng/L) =	1.408	
Dissolved Reactive Hg (%) =	25.99	
Reactive Hg (% of Total) =	20.559	
Bioavailable Inorganic Hg (% of Bioavailable Hg) =	91.04	
Methylmercury (% of Bioavailable Hg) =	8.96	
Bioavailable Hg (%) =	22.58	
Bioavailable Hg Concentration (ng/L) =	0.40	
Bioavailable Hg Flux (ug or ug/day)		
Compare Bioavailability of Two Effluents		
Inputs Effluent #2		
T-Hg (ng/L) =		
MeHg (ng/L) =		
R-Hg (ng/L) =		
DOC (mg/L) =		
TSS (mg/L) =	*Optional*	
log KD =	*Optional*	
Flow rate or Volume (million gallons or MGD)	*Optional*	

Mix Effluent #1 With Receiving Water		
Inputs Receiving Water		
T-Hg (ng/L) =	14	
MeHg (ng/L) =	0.03	
R-Hg (ng/L) =		
DOC (mg/L) =	2.4	
TSS (mg/L) =	94	*Optional*
log K_D =	5.6	*Optional*
Flow rate or Volume (Effluent #1)	1	
Flow Rate or Volume (Receiving Water)	50	
Mixing Ratio (Ratio of effluent to receiving water) =		0.02
Water Depth (m) =	3	
Salinity (parts per thousand)	30	
Far Field Flag, 0 - Use Salinity, 1 - Use MeHg/R-Hg		0
Outputs	**Near Field**	**Far Field**
T-Hg concentration (ng/L) =	13.76	
MeHg concentration (ng/L) =	0.03	0.02
MeHg % =	0.22	0.16
DOC concentation (mg/L) =	2.54	
Dissolved Inorganic Hg of Receiving Water (ng/L) =	0.36	
Dissolved Reactive Hg of Receiving Water (ng/L) =	0.15	
Mixed Reactive Hg (ng/L) =	0.15	0.16
Reactive Hg (% of Total) =	1.12	1.17
Total Bioavailable Hg (ng/L) =	0.18	
Bioavailable Inorganic Hg (% of Bioavailable Hg) =	83.63	87.72
Methylmercury (% of Biaovailable Hg) =	16.37	12.28
WWTP Bioavailable Hg Mass Flux (% of Total Flux)	4.3	

Figure 2-9. Bioavailability Tool Inputs and Outputs for Example 7.

Note that if reactive Hg data is used to characterize the receiving water, it must also be used to characterize the effluent.

The flows are entered in a ratio of 1 (effluent) to 100 (receiving water). The water depth is set at 15 meters. The far-field flag is given a value of 1 so that the methyl to reactive Hg ratio will be used to determine ratio of reaction constants for the far-field analysis.

The outputs for the effluent (middle, left-hand side) show that nearly all the bioavailable mercury is in the reactive form (94%), with only 6% in the methyl form (Figure 2-10). The near-field mixing outputs (lower right-hand side) show that the mixed effluent/receiving water is dominated by the receiving water, which should be the case as only 1% of the mixture is effluent. In the far-field, because the ratio of methyl to reactive Hg in the receiving waters is only 0.025, net demethylation is occurring and the amount of methylmercury in the far-field decreases slightly (2.5% to 2.4%), while the reactive form experiences a commensurate increase (97.5% to 97.6%). In this case, the WWTP has little impact on the results due to the high dilution. Because there is net demethylation in the far-field, the greatest impact of the discharge would be in the near-field. The WWTP contributes only 2.6% of the bioavailable mercury mass flux to the system.

MARINE WATER BIOAVAILABILITY COMPUTATIONS

Evaluate Bioavailability of a Single Effluent	Effluent #1	Effluent #2
Inputs Effluent #1		
T-Hg (ng/L) =	1.8	
MeHg (ng/L) =	0.035	
R-Hg (ng/L) =	0.52	
DOC (mg/L) =	7.3	
TSS (mg/L) =		*Optional*
log K_D =		*Optional*
Flow rate or Volume (million gallons or MGD)		*Optional*
Outputs		
%MeHg =	1.94	
Dissolved Inorganic Hg (ng/L) =		
Dissolved Reactive Hg (%) =		
Reactive Hg (% of Total) =	28.889	
Bioavailable Inorganic Hg (% of Bioavailable Hg) =	93.69	
Methylmercury (% of Bioavailable Hg) =	6.31	
Bioavailable Hg (%) =	30.83	
Bioavailable Hg Concentration (ng/L) =	0.56	
Bioavailable Hg Flux (ug or ug/day)		
Compare Bioavailability of Two Effluents		
Inputs Effluent #2		
T-Hg (ng/L) =		
MeHg (ng/L) =		
R-Hg (ng/L) =		
DOC (mg/L) =	*Optional*	
TSS (mg/L) =	*Optional*	
log KD =	*Optional*	
Flow rate or Volume (million gallons or MGD)	*Optional*	

Mix Effluent #1 With Receiving Water		
Inputs Receiving Water		
T-Hg (ng/L) =	0.36	
MeHg (ng/L) =	0.005	
R-Hg (ng/L) =	0.2	
DOC (mg/L) =	0.9	
TSS (mg/L) =		*Optional*
log K_D =		*Optional*
Flow rate or Volume (Effluent #1)	1	
Flow Rate or Volume (Receiving Water)	100	
Mixing Ratio (Fraction of effluent to receiving water) =	0.01	0.01
Water Depth (m) =	15	
Far Field Flag, 0 - Use 0.02 Default, 1 - Use MeHg/R-Hg		1
Outputs	**Near Field**	**Far Field**
T-Hg Concentration (ng/L) =	0.37	
MeHg Concentration (ng/L) =	0.01	0.01
MeHg % =	1.42	1.36
DOC Concentration (mg/L) =	0.96	
Dissolved Inorganic Hg of Receiving Water (ng/L) =		
Dissolved Reactive Hg of Receiving Water (ng/L) =		
Mixed Reactive Hg (ng/L) =	0.20	0.20
Reactive Hg (% of Total) =	54.29	54.34
Total Bioavailable Hg (ng/L) =	0.21	
Bioavailable Inorganic Hg (% of Bioavailable Hg) =	97.46	97.56
Methylmercury (% of Biaovailable Hg) =	2.54	2.44
WWTP Bioavailable Hg Mass Flux (% of Total Flux)	2.6	

Figure 2-10. Bioavailability Tool Inputs and Outputs for Example 8.

2.4 Assessment of Water Quality Effects on Net Methylation

2.4.1 Methylation Potential and Water Quality

This section discusses the development of the procedure to estimate the net methylation characteristics of an effluent/receiving water for the purposes of performing a "far-field" screening analysis for mercury bioavailability. In order to provide this analysis, it was important to identify databases that contained both mercury and relevant conventional water quality parameters. For freshwater environments, the researchers relied on a national database that had been pulled together by the USGS that, for the most part, contained all the necessary information. For estuaries, the data was much more limited and was not already summarized in one database. Data bases are available for conventional water quality parameters but mercury data is typically only available through special studies that have been conducted, for instance in San Francisco and Chesapeake Bays, or in research papers. Data for coastal waters was available only for one site in the mid-Atlantic Ocean. These data are discussed in the following sections.

2.4.1.1 Relevant Water Quality Characteristics of Freshwater Environments

The United States Geological Survey (USGS) conducted a national scale pilot study examining the relationship of total mercury and methylmercury in water, sediment and fish. Sampling took place at 106 sampling sites in 21 basins (three to eight sites at each location) across the country including Alaska and Hawaii from June 1998 through October 1998. The sample locations represented a variety of land use conditions including urban, agriculture, mining, background, and forest/agriculture mix. According to the study authors, this effort "represents a first step toward a national assessment of mercury contamination of aquatic ecosystems in the United States." The study details are included in Brumbaugh et al. (2001).

The research team believes the USGS data is a good representation of the freshwater water quality characteristics found nationally in waters exhibiting a wide variety of responses to

Table 2-3. Freshwater Water Quality Characteristics including TOC, pH, Sulfate, Methylmercury, Total Mercury, and Percent Methylmercury.

	TOC (mg/L)	pH	SO_4 (mg/L)	MeHg (ng/L)	Total Hg (ng/L)	%MeHg (%)
Mean	4.2	7.6	54.7	0.20	16.8	4.6
Min	0.4	3.3	0.2	0.01	0.27	0.11
Max	24	10	484	5.12	1106	47.2
Median	3.4	7.8	28.0	0.06	2.3	2.7
Number of Samples	104	103	86	103	102	102
Standard Deviation	3.6	0.8	82.9	0.55	111	6.7
Coefficient of Variation	86%	10%	152%	275%	661%	146%

mercury bioaccumulation. The information presented in Table 2-3 (above) is a summary of this work relevant to this study.

Due to the variety of sampling sites (e.g. mines, agriculture, etc.), there is a large degree of variation in water column mercury. Fourteen of the 106 sites represent mine sites. Although not all mine sites had higher levels of mercury, the highest concentration of mercury found (i.e. 1106 ng/L) was found at a mine site location. In general, the percentage of methylmercury of the total mercury concentration is more consistent than actual concentrations of either methylmercury or total mercury. Note that the mean water column %MeHg is 4.6 and the median value is 2.7%.

2.4.1.2 Conventional Water Quality Characteristics of Estuarine Environments

Estuarine environments are characterized by their brackish waters. Unlike freshwaters which have a salinity content of less than 0.5 parts per thousand (ppt) or marine waters which have a salinity of greater than 33 ppt, estuaries are a blend of freshwater and seawater and salinity typically depends on where the measurement is taken in the estuary. In estuaries that are not well-mixed vertically, bottom waters may also have higher salinity that surface waters due to differences in densities of fresh and salt waters.

Each year the U.S. EPA monitors 28 of the nation's estuaries as part of the National Coastal Assessment program and reports on the health of the estuaries in the National Estuary Program Coastal Conditions Report (NEP, CCR).

The U.S. EPA's coastal assessment data is organized according to national region: Northeast, Southeast including the Gulf Coast, and the West Coast. Databases for both the Southeast and Northeast datasets were made available to us by request to the U.S. EPA. The U.S. EPA has data for West Coast sites in a more limited fashion utilizing their Coastal Data Search Engine. Data for Puget Sound and the San Francisco Bay were available through this online resource. To supplement data provided by the U.S. EPA, additional data requests to representatives at individual estuaries programs was made and a literature review was conducted. The following identifies the sources of data used to characterize estuarine water quality characteristics for this study:

Southeast EPA National Coastal Assessment Program. These data were provided by the U.S. EPA upon request. There are approximately 8,900 samples, varying depending on the parameter, from 2000 through 2004. The following parameters were evaluated for this study: dissolved oxygen (DO), salinity, total suspended solids (TSS), and pH. Locations included estuarine environments all over the coastal Southeast and Gulf Coast including in Alabama, Florida, Georgia, Louisiana, Maryland, Mississippi, North Carolina, South Carolina, Virginia, and Texas.

Northeast EPA National Coastal Assessment Program. These data were provided by EPA upon request. There were just under 1000 samples of both surface and bottom water quality taken between July 2000 through October 2002, depending on parameter. The following parameters were evaluated for this study: dissolved oxygen, salinity, and pH. Locations included estuarine environments throughout the coastal regions of Massachusetts, Rhodes Island, New Hampshire, New Your, New Jersey, Delaware, Connecticut, and Maine. According to the U.S. EPA, the data have received a quality assurance review designed to catch obvious transcription and formatting errors, and to some extent, to highlight questionable data.

Puget Sound, Washington. These data were downloaded from the U.S. EPA Coastal Data Search Engine (http://oaspub.epa.gov/coastal/coast.search). Forty dissolved oxygen and salinity samples and 60 total suspended solids samples taken at varying depths along numerous stations in Puget Sound are available from 6/15/00 through 8/17/00. These data are readily available to the public online, and therefore is assumed to have gone through full Quality Assurance/Quality Control (QA/QC) review.

San Francisco Bay, California. These data were downloaded from the EPA Coastal Data Search Engine (http://oaspub.epa.gov/coastal/coast.search). Forty dissolved oxygen and salinity samples and sixty total suspended solids samples taken at varying depths along numerous stations in San Francisco Bay from 6/15/00 through 8/17/00 are available. Again, these data are readily available to the public online, and therefore is assumed to have gone through full QA/QC review.

In addition, a published mercury study by Conaway and others that sampled 26 surface water locations throughout San Francisco Bay during February, April and July 1999 and February and July 2000 was utilized. This data set includes measurements of total and dissolved mercury and methylmercury.

Indian River Lagoon, Florida. These data were provided by St. Johns River Water Management District upon request. There were approximately 3,500 to 4,300 samples taken, depending on parameter, between January 13, 2003 and February 11, 2008. This database supplements the Southeast EPA National Coast Assessment data in that it includes not only dissolved oxygen, salinity, total suspended solids, and pH but also dissolved organic carbon (DOC) and total organic carbon (TOC).

Mobile Bay, Alabama. These data were provided by the Alabama Department of Environmental Management upon request. The dataset includes 1200 samples of dissolved oxygen, salinity, and pH taken from eight stations within Mobile Bay between 3/10/05 and 2/22/06. In this same time frame, 96 total suspended solids samples were taken from the same locations. Additionally, 21 total and dissolved mercury samples were collected.

Long Island Sound, New York. These data are contained in several references (Vandal et al., 2001; Rolfhus and Fitzgerald, 2001; Rolfhus et al., 2003; Balcom et al., 2004). This dataset includes 61 to 79 samples, depending on parameter, taken at 18 locations throughout Long Island Sound during short duration sampling events between 1995 through 2002. Dissolved organic carbon, salinity and total suspended solids were evaluated for this study. Additionally, total, particulate, and reactive mercury and methylmercury samples were collected at the same time.

Hudson River, New York. These data were obtained from two studies (Heyes et al., 2004; Balcom et al. 2008). Eight samples were collected on two separate sampling excursions at both surface and bottom elevations. Dissolved organic carbon, salinity, total suspended solids, and pH were evaluated for this study. Additionally, total, filtered, and particulate mercury and methylmercury samples were available.

Chesapeake Bay. As part of the Chesapeake Bay Water Quality Monitoring Program, TSS and DOC averages at five locations in the Chesapeake Bay were documented in a technical synthesis published by the U.S. EPA. This information provides additional DOC data for the Chesapeake Bay.

Ochlockonee Estuary, Florida. A study was conducted to evaluate the fractionation of mercury and DOC in estuaries and open oceans by Guentzel et al. (1996). This study provides DOC, salinity, and total and dissolved mercury data. Samples were collected between January 1993 and 1994 along the salinity gradient in the estuary.

Galveston Bay, Texas. A study was conducted by Han et al. (2006) to investigate the complexation of mercury by dissolved organic matter in the surface waters of Galveston Bay, Texas. Samples include pH, salinity, and total mercury and methylmercury. Samples were taken in October, 2003 at numerous locations along the salinity gradient.

The following tables (Tables 2-4 through 2-7) present statistics for dissolved oxygen, dissolved organic carbon, salinity, total suspended solids, and pH from each of the data sets. If a dataset is not listed in the table, data was not available for that parameter. Although it turned out that not all the conventional data collected for this study was utilized in the development of the fair-field net methylation assessment, the data are presented as they may be useful at some future time. Because there was relatively little estuarine mercury data available, the effect of some of these water quality variables could not be thoroughly evaluated.

2.4.1.3 Mercury Data in Estuarine Environments

Total mercury and methylmercury were collected for several of these estuaries. The following tables (Table 2-8 through Table 2-10) present statistical summaries of the available data. Based on these data, average %MeHg in estuarine environments ranges from 1.3% to 2.3% (Table 2-10). An exception to this is the Long Island Sound data which averages 29% with values ranging from zero to greater than 100% methylmercury. Note that the typical %MeHg of the estuary data is less than the average for freshwaters of about 4.6%, providing further confirmation that the methylation potential is lower in estuarine waters than in freshwaters.

Table 2-4. Dissolved Oxygen Values in Estuarine Environments (mg/L).

Location	Mean	Min	Max	Median	Std. Dev.	# of Samples	Coef. of Variation
Southeast EPA	5.9	0	21.5	5.8	5.9	8752	31%
Northeast EPA - surface	7.45	1.12	16.4	7.5	7.45	944	23%
Northeast EPA - bottom	6.82	-0.02	15.57	7.0	6.82	935	27%
Puget Sound	6.50	3.6	11.01	6.04	6.50	40	27%
San Francisco Bay	7.47	7.18	7.93	7.43	7.47	8	4%
Indian River Lagoon	6.3	0	14.3	6.4	6.3	4279	29%
Mobile Bay	6.98	0.27	12.36	7.06	6.98	1198	32%

Table 2-5. Dissolved Organic Carbon Data for Estuarine Environments (mg/L).

Location	Mean	Min	Max	Median	Std. Dev.	# of Samples	Coef. of Variation
Indian River Lagoon	11.4	0	55.4	11.4	5.4	3477	48%
Long Island Sound	2.0	1.1	3.6	1.9	0.6	77	30%
SF Bay - Conaway et al.	2.40	0.13	10.68	1.92	1.56	105	13%
Ochlockonee Estuary	9.1	2.8	17.5	10.0	4.8	7.0	53%
Chesapeake Bay	2.68-2.95	0.82-2.13	4.07-8.97	2.61-2.78	0.427-0.744	165-175	153 to 252%

Table 2-6. Total Suspended Solids Data for Estuarine Environments (mg/L).

Location	Mean	Min	Max	Median	Std. Dev.	# of Samples	Coef. of Variation
Southeast EPA	35	0	1573	20	67.5	3490	190%
Puget Sound	4.31	2	9	4.00	1.36	60	32%
San Francisco Bay	94	4	299	14	128	9	135%
Indian River Lagoon	13.0	-1	265	9.6	12.0	4300	93%
Mobile Bay	15	5	50	13	8	96	56%
Long Island Sound	5.8	0.53	20.1	5.8	3.5	61	61%
Hudson River - surface	25.8	7.9	47.8	24.2	11.8	8	50%
Hudson River - bottom	41.3	7.1	106	44.4	28.9	9	70%
Chesapeake Bay	4.59 to 16.8	1.5 to 2.5	10.3 to 108	4.4 to 14.7	1.33 to 12.8	174 to 181	30 to 120%

Table 2-7. pH Data for Estuarine Environments.

pH	Mean	Min	Max	Median	Std. Dev.	# of Samples	Coef. of Variation
Southeast EPA	7.81	1.9	15.9	7.90	0.58	8140	7%
Northeast EPA - surface	7.70	4.03	9.36	8	0.50	673	7%
Northeast EPA - bottom	7.66	4.04	9.34	8	0.49	670	6%
San Francisco Bay	7.8	7.4	8	7.8	0.2	8	2%
Indian River Lagoon	8.0	6.1	9.56	8.1	0.5	4269	6%
Mobile Bay	7.66	5.81	8.35	7.78	0.43	1198	6%
Hudson River - surface	7.9	7.4	8.5	7.9	0.4	8	-
Hudson River - bottom	7.8	7.5	8.2	7.8	0.3	9	-
SF Bay - Conaway et al.	7.8	6.8	8.7	7.9	0.26	105	3%

Table 2-8.Total Mercury Data for Estuarine Environments (ng/L).

Location	Mean	Min	Max	Median	Std. Dev.	# of Samples	Coef. of Variation
Mobile Bay	0.51	0.30	0.92	0.50	0.104	21	20%
Long Island Sound	1.5	0.12	8.8	1.2	1.2	95	82%
Hudson River - surface	12.8	2.9	44.1	5.2	14.3	8	110%
Hudson River - bottom	19.6	1.9	56.7	15.1	17.8	9	91%
SF Bay - Conaway et al.	14.0	0.14	88.3	7.2	17.7	105	130%
Ochlockonee Estuary	2.3	0.62	3.6	2.5	1.1	7	48%
Galveston Bay	0.39	0.24	1.1	0.31	0.27	12	68%

Table 2-9. Methylmercury Data for Estuarine Environments (ng/L).

Location	Mean	Min	Max	Median	Std. Dev.	# of Samples	Coef. of Variation
Long Island Sound	0.279	0.000	0.906	0.226	0.287	16	100%
Hudson River - surface	0.082	0.041	0.142	0.068	0.037	8	45%
Hudson River - bottom	0.109	0.056	0.179	0.088	0.046	9	42%
SF Bay - Conaway et al.	0.087	0.020	0.461	0.040	0.098	79	110%
Galveston Bay	0.008	0.001	0.045	0.004	0.013	10	150%

Table 2-10. Percent Methylmercury in Estuarine Environments (%).

Location	% MeHg
Long Island Sound	29%
Hudson River - surface	2.1%
Hudson River - bottom	2.3%
SF Bay - Conaway et al.	1.7%
Galveston Bay	1.3%

2.4.1.4 Effect of Estuarine Salinity Gradient

As freshwaters merge with marine waters in an estuary, salinity increases seaward. Corresponding changes occur with other water quality parameters. Of relevance to this study is that DOC generally decreases as the salinity increases. This may be due to several factors. First, the DOC content of ocean waters is typically low, on the order of 1 mg/L, and decreases in DOC along the salinity gradient may be due in large part to the dilution of fresh waters with higher DOC by ocean water having lower DOC. The removal of DOC due to coagulation and settling may also occur as discussed in the Phase I report. In addition, it is also well-established that DOC is destroyed by photo-oxidation which may be enhanced as light penetration increases due to reduction of DOC and TSS coincidental with salinity increases. Another change that tends to occur along the salinity gradient includes an increase in pH. As both pH and DOC affect mercury speciation and bioavailability, it is not surprising that mercury concentrations also tend to change along the salinity gradient. Data taken for the Ochlockonee Estuary in Florida (Table 2-11) and Galveston Bay in Texas (Table 2-12) both demonstrate this relationship.

The importance of this discussion is that in estuaries, the assumption made in the Bioavailability Tool that total bioavailable Hg is conservative may not hold as well as it would in fresh or marine waters. At a minimum this needs to be recognized when interpreting results of the Tool calculations. Perhaps an adjustment of some nature will be required when making the bioavailability calculations for estuaries. However, any such adjustment would have to be determined in a more detailed study of the particular estuarine system of interest.

Table 2-11. Mercury and DOC Concentrations with Varying Salinities, Ochlockonee Estuary, Florida.

Salinity (ppt)	DOC (mg/L)	Filtered T-Hg (ng/L)
1.7	17.5	3.8
4.5	12.9	3.0
6.7	11.6	2.7
8.7	10.5	2.2
11	7.8	2.2
19	5.7	1.1
28.5	2.7	0.68

Table 2.12. Mercury and Methylmercury Concentrations with Varying Salinities, Galveston Bay, Texas.

Salinity (ppt)	DOC (mg/L)	Filtered T-Hg (ng/L)	Filtered MeHg (ng/L)
0.1	3.2	1.00	0.0595
1.3	3.5	1.11	0.0446
5.9	3	0.515	0.0115
7.3	3.1	0.319	0.0069
11	2.9	0.345	0.0042
13	2.3	0.345	0.0044
15	2.6	0.301	0.0023
17	2.4	0.273	0.0041
20	1.68	0.237	0.0031
23	1.56	0.241	0.0021
27	1.2	0.243	0.0012

2.4.1.5 Marine Waters

As opposed to estuaries, marine waters are relatively constant in their composition. Although the salinity of ocean waters may fluctuate, with lower salinities being found near the poles and higher salinities near the equator, the ionic proportions tend to remain the same. At a typical value of 35 parts-per-thousand (3.5%) the salinity of seawater is dominated by a few major anions, the most dominant being chloride (19,340 mg/L), the second most dominant being sulfate (2700 mg/L), with bicarbonate a distant third (145 mg/L). The pH of seawater is also relatively constant at about 8.2 ± 0.1. The composition of seawater with respect to DOC is in the 60 to 90 μM range (0.7 to 1.1 mg/L) with higher concentrations being found in surface waters (0 to 1000 m depths). TSS concentrations in the coastal oceans vary but a range of 10 to 50 mg/L is typical.

The mercury content of seawater tends to be more consistent as well than in freshwaters and estuaries. Table 2-13 gives values for various portions of the world ocean. Total Hg values are low averaging about 0.2 ng/L, whereas the average for the estuary data presented above is about 7.3 ng/L and the freshwater data from the USGS database averages 16.8 ng/L. It is relevant to note that these values are for open ocean waters and that one might expect near shore values to be higher due to higher TSS and DOC values. While these comparisons suffer from a disparity in the number of data points, they point out the general trend of total Hg values fresh > estuary > ocean. In terms of methylmercury, the levels also appear to be higher in freshwater, intermediate in estuaries, and lowest in marine waters. The average freshwater value from the USGS data base is 0.20 ng/L, the average of the above estuary data is 0.11 ng/L, and the open ocean values range from 0.02 to 0.07 ng/L. This order is to be expected if the strength of the methylating environment is higher in fresh waters than in saline waters. Therefore it is somewhat surprising that the data shown for open oceans shows %MeHg contents that rival

Table 2-13. Range or Mean Mercury Concentrations in Subsurface Water at Different Stations of the World Ocean.

Location	T-Hg (ng/L)	MeHg (ng/L)	%MeHg
South and Equatorial Atlantic	0.16 – 0.48	0.005 – 0.04	5-10
North Atlantic	0.48 ± 0.08	0.006 – 0.03	2-7
North Pacific	0.24 ± 0.18	0.01-0.1	10-20
Equatorial Pacific	0.10 – 0.8	0.007 – 0.13	2-15
Mediterranean	0.10 – 0.8	0.004 – 0.09	15-35

Data sources: Mason and Gill (2005) and references therein, Sunderland and Mason (2007), unpublished data (Cossa, 2009, personal communication), and from Sunderland et al. (accepted).

freshwater values in some zones. This may be in part due to low TSS concentrations, which would tend to produce lower T-Hg in the water column.

2.4.2 Linkage of Methylation Potential to Water Quality Conditions

This section provides the basis for the methodologies presented earlier for assessing methylation potential based on water quality information. It presents the concept of the ratio of the methylation to demethylation rate constants and there equivalency to MeHg/IB-Hg at steady-state. It then shows how the screening approaches to estimating methylation potential using % MeHg, pH and DOC (for freshwaters), and salinity (for estuaries) were derived, and then tests the screening approaches using several sets of data from the literature. These approaches, which were the result of the development and testing shown below, were presented earlier in the chapter.

2.4.2.1 Net Methylation in Fresh Waters

This section presents an analysis linking methylation potential to ambient conditions, either in terms of %MeHg or, in the absence of this information, to other water quality variables. There have been several approaches and attempts to link the extent of methylation within a particular waterbody to water quality. One of the most recent and extensive approaches has been the examination of databases by the USGS and the resultant production of so-called sensitivity maps, which estimate the potential for methylation within a system relative to the amount of wetlands in the watershed, as well as the pH, DOC and sulfate concentrations in the water. The analysis relied on examining correlations between variables and total and MeHg concentration but the final maps were generated using as much available information as possible, even if there was not Hg and MeHg data for a particular watershed. The data from one extensive USGS study (Brumbaugh et al., 2001; Krabbenhoft et al., 1999) is gathered in Appendix B and this information was used in the evaluation discussed here.

An examination of this dataset, which includes information on water quality variables as well as Hg and MeHg data, leads to the following observations. Table 2-14 is a correlation matrix for the data derived from considering the linear regression relationships between all variables and some representative relationships are plotted in Figure 2-11. A few comments on

Table 2-14. Correlation Matrix of %MeHg and Water Quality Variables for the USGS Dataset.

	%Wet	Water MeHg	Water Total Hg	Water TOC	Water pH	Water SO4	Water %MeHg	Sediment MeHgsed	Sediment Total Sed	Sediment %LOI	Sediment %MeHgsed
MeHg	0.172										
Total Hg	-0.044	**0.918**									
TOC	**0.738**	0.181	0.023								
pH	-0.378	-0.037	0.092	-0.087							
SO4	-0.264	-0.012	0.049	-0.125	-0.093						
%MeHg	**0.502**	0.143	-0.079	**0.445**	-0.297	-0.193					
MeHgsed	**0.529**	0.215	0.096	0.205	**-0.380**	-0.210	**0.357**				
Total Sed	0.006	**0.542**	**0.606**	-0.006	-0.064	-0.019	-0.078	**0.424**			
%LOI	**0.835**	0.093	-0.050	**0.687**	-0.279	-0.190	**0.568**	**0.484**	0.010		
%MeHgsed	0.243	0.140	-0.062	0.069	-0.180	-0.076	0.215	**0.303**	-0.142	0.010	

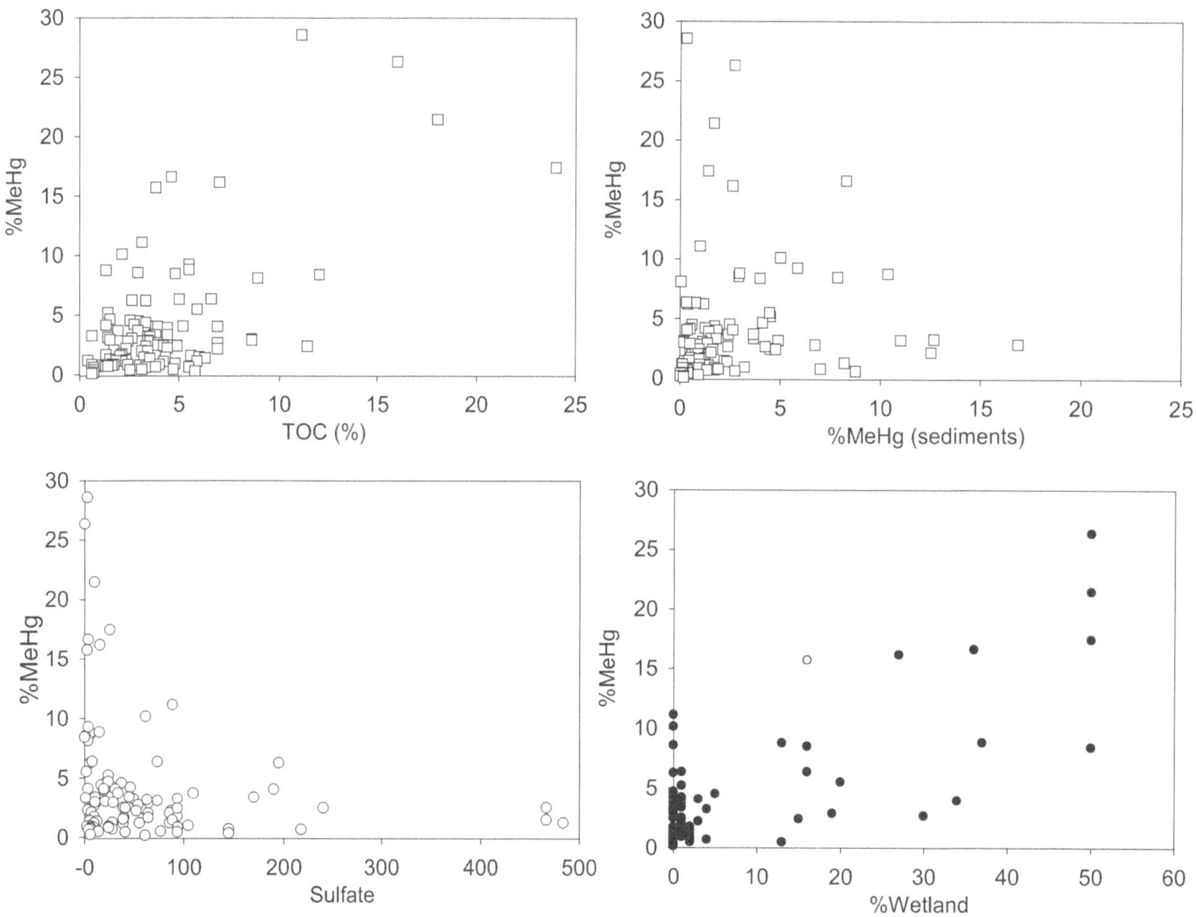

Figure 2-11. Relationships Between Variables Measured During the USGS Sampling of Rivers and Streams Across the USA. Data from Krabbenhoft et al. (1999) and Brumbaugh et al. (2001).

these data and relationships are appropriate. Significant relationships ($p<0.001$) are indicated in blue. MeHg in water was highly correlated with total Hg in water, and with total Hg in sediments. Total Hg in water was correlated with total Hg in sediments. %MeHg in water was correlated with OC in water and sediments, and with sediment MeHg. Sediment %MeHg was also correlated with sediment OC. The %wetland in the watershed was positively correlated with %MeHg in the water column and MeHg in sediments, and OC, both in the water column and in sediments. There were few relationships for pH and sulfate concentration. This may be a function of the overall small variation in these parameters for the rivers. Also, most of the rivers had relatively high concentrations of sulfate. While the concentration of sulfate limiting sulfate reduction, and therefore potentially limiting Hg methylation, varies across ecosystems, the range of values where it is limiting is on the order of < 1 to 3 mg/L sulfate. Therefore, most of the waters considered are above the limiting value and this is likely why there is no strong trend between %MeHg and sulfate in these flowing waters. Thus, the correlation matrix for this dataset does not necessarily indicate that sulfate or pH is of major importance when determining the potential for methylation within a particular waterbody. However, %MeHg was strongly correlated with TOC. While this may reflect the fact that TOC and %wetland were also strongly correlated, there are also likely other reasons for the relationship with TOC. One may be the extent of demethylation, as this is mostly a photochemical process. The extent of the photochemical reaction can be impacted by TOC which influences light penetration, and may also enhance some redox processes.

Overall, based on the evaluation above, it is likely that using %MeHg or an equivalent predictor of methylation potential is a reasonable approach to ascertain the degree to which Hg in an effluent may be methylated or demethylated after mixing of the effluent with receiving waters. As discussed below, the approach to determining the ratio of total bioavailable Hg to MeHg, and its interpretation for natural waters, was derived using the following assumptions and relationships. First, it is assumed that the systems are at steady state and that the interconversions can be represented as first order rate constants. Under these conditions, the following would be true, at steady state:

$$\alpha_{MeHg}[MeHg]/\ \alpha_{Hg}[Hg] = k_{meth}\ /\ k_{demeth} \qquad \text{Equation 2.4.2.1-1}$$

where the rate constants refer to the overall processes that are involved in the interconversions between Hg and MeHg, and the α values represent the relative bioavailability of the inorganic Hg and MeHg within the system being studied. Given the definitions and approaches above, it is assumed that all the MeHg is in a relatively bioavailable form and can be demethylated, but that only the dissolved bioavailable Hg fraction is the fraction that can be methylated. Thus, the equation above can be rewritten as:

$$k_{meth}\ /\ k_{demeth} = [MeHg]/[IB\text{-}Hg] \qquad \text{Equation 2.4.2.1-2}$$

If this relationship is valid, then there should be a relationship between the ratio of bioavailable fractions and the ratio of the rate constants. If this can be demonstrated, then it is reasonable to proceed with using the ratio as a metric for determining the extent to which Hg input into a system will be converted to MeHg. To examine the equations in detail, it was decided that if one of the constants could be determined, the other could then be calculated from the bioavailability ratio. An examination of the factors controlling methylation and demethylation suggests that it is easier to estimate the demethylation constant and this was the approach taken here. It was further assumed that the major factor controlling the rate of demethylation of MeHg in the water column (through abiotic photochemical processes primarily) is related to the penetration of light, and in particular UV radiation, into the water. This approach was taken as the demethylation rate in the water column is the most reasonable to parameterize. The methylation rate constant, as defined here, incorporates the processes of net methylation (primarily in the sediment), diffusion into the water column from the sediment and any additional methylation within the water column. From this the methylation potential can be determined using Equation 2.4.2.1-1. This methylation rate constant is an empirical value for the combination of these processes. The value of the k_{meth} is determined from the estimated value of k_{demeth} and the bioavailability ratio in the water, according to Equation 2.4.2.1-2.

As there is ample evidence of the relationship between water color and organic content, the relationship for the relative rate of demethylation is parameterized using a published relationship between DOC and light penetration depth for lakes (Perez-Fuentetaja et al., 1999). Additionally, the data from published studies measuring the demethylation rate in freshwaters of different DOC content (Sellers et al., 1996; Hammerschmidt and Fitzgerald, 2006), and the fact that the demethylation is restricted to the upper layers due to the extent of radiation penetration (Hammerschmidt and Fitzgerald, 2006; Krabbenhoft et al., 2002), was used to derive relationships for the demethylation function. From these studies, it can be ascertained that the demethylation rate constant under medium DOC concentrations (3-4 mg/L), where most experiments were performed, is around 10^{-6} s^{-1} (~0.1 day^{-1}). Using this value and the information on the relationship between DOC and light penetration, and assuming for the initial calculations that the systems are shallow and well-mixed, the following relationship was established. As the relationship between light penetration and DOC was derived using concentrations above 1 mg/L, it is necessary to define the constant for the low DOC situations. Based on the datasets, a fixed value is chosen (Equation 2.4.2.1-3) which is higher but comparable with the values determined using Equation 2.4.2.1-4:

For DOC < 1 mg/L, $k_{demeth} = 1\ day^{-1}$ Equation 2.4.2.1-3

For DOC > 1 mg/L, $k_{demeth} = (2.4*DOC - 1.6)^{-1}\ day^{-1}$ Equation 2.4.2.1-4

Thus, the demethylation rate decreases as the DOC concentration increases. Above about 7 mg/L, the value is low (< 0.05 day^{-1}). The magnitude of this difference fits with the very limited experimental data. As noted above, this constant takes into account primarily photochemical processes, but there is the assumption of a low background rate of demethylation (a few % of the maximum even for the high DOC waters). It should be noted that the estimations above are relevant for shallow systems where the light penetration depth is of the same order as the water depth and that the system is well-mixed (e.g. most small rivers and streams). For deeper waters, or waters that are stratified, a further parameter needs to be introduced to take the depth dependence of demethylation throughout the water column into account. It is assumed, and the data support this assumption (e.g. Sellers et al., 1996; Whalin et al., 2007), that there is a lower level of demethylation even in the absence of light, and this value is about 1% of the maximum measured value. This fact is therefore also included in the calculation. For waters that are < 5 m deep and are well-mixed the equations above should be used. For all other waters, the equations need to be adjusted as:

For DOC < 1 mg/L, $k_{demeth} = 0.01 + 5/D$ day^{-1} Equation 2.4.2.1-5

For DOC > 1 mg/L, $k_{demeth} = 0.01 + [(2.4*DOC - 1.6)D]^{-1}$ day^{-1} Equation 2.4.2.1-6

where D is the water depth in meters.

This validity of the approach to estimating k_{demeth}, k_{meth}, and for using the bioavailability ratio as a diagnostic variable, was assessed using data from the USGS (Brumbaugh et al., 2001; Krabbenhoft et al., 1999) for over 100 different rivers where a variety of parameters have been measured, including T-Hg and MeHg. This dataset was used as a test of the approach described above, although the dataset does have limitations. One river (Stillwater River near Sterling, MA) which had a very low total Hg (<1 ng/L) and an extremely high %MeHg, was eliminated from the dataset prior to the analysis. Note that the TSS concentrations were not available for this dataset and the measurements of T-Hg and MeHg in the water were for the total (unfiltered) concentration.

The k_{demeth} values were estimated for these rivers using the expression above (Equations 2.4.2.1-3 and 2.4.2.1-4) and then the value of k_{meth} was determined. As the TSS values were not measured, a value of 5 mg/L was used in this illustrative calculation along with a $\log K_D$ value of 5.5 as all the data collected were during baseflow conditions (Krabbenhoft et al., 1999). The relationship with DOC described previously was used to estimate the bioavailable Hg fraction. Relationships of these variables are shown in Figure 2-12. The measured and calculated variables are shown in Appendix B.

The plot in Figure 2-12 shows the relationship between %MeHg and the calculated ratio [MeHg]/[IB-Hg] (or equivalently, k_{meth} / k_{demeth}). There is a strong linear correlation between these two variables ($r^2 = 0.94$; $y = 1.216x-0.907$; $p < 0.0001$) and this partially reflects the fact that both TSS and K_D are fixed, and at high DOC the value of IB-Hg becomes relatively constant. Therefore, the two variables are essentially linearly related at high DOC. Also, given the formulation, a relationship between k_{demeth} and %MeHg is reasonable, but not obvious ($r^2 = 0.55$; $y = 8.96x +0.27$; $p < 0.0001$) (Figure 1-12). It does however support the notion stated in the literature that the difference in %MeHg across ecosystems is more a function of differences in the rate of MeHg production, rather than differences in its rate of destruction.

The estimated values of k_{demeth} ranged from 0.018 day^{-1} to 1 day^{-1}, with an average of 0.26 ± 0.27 day^{-1}. For k_{meth}, the range is 8.7×10^{-3} day^{-1} to 2.31 day^{-1}, with an average of 0.44 ± 0.43 day^{-1}. Overall, there is more variability (larger range in values) for k_{meth}, which is consistent with the literature. As expected given the linear inverse relationship above, the estimated k_{demeth} correlated with the water organic carbon.

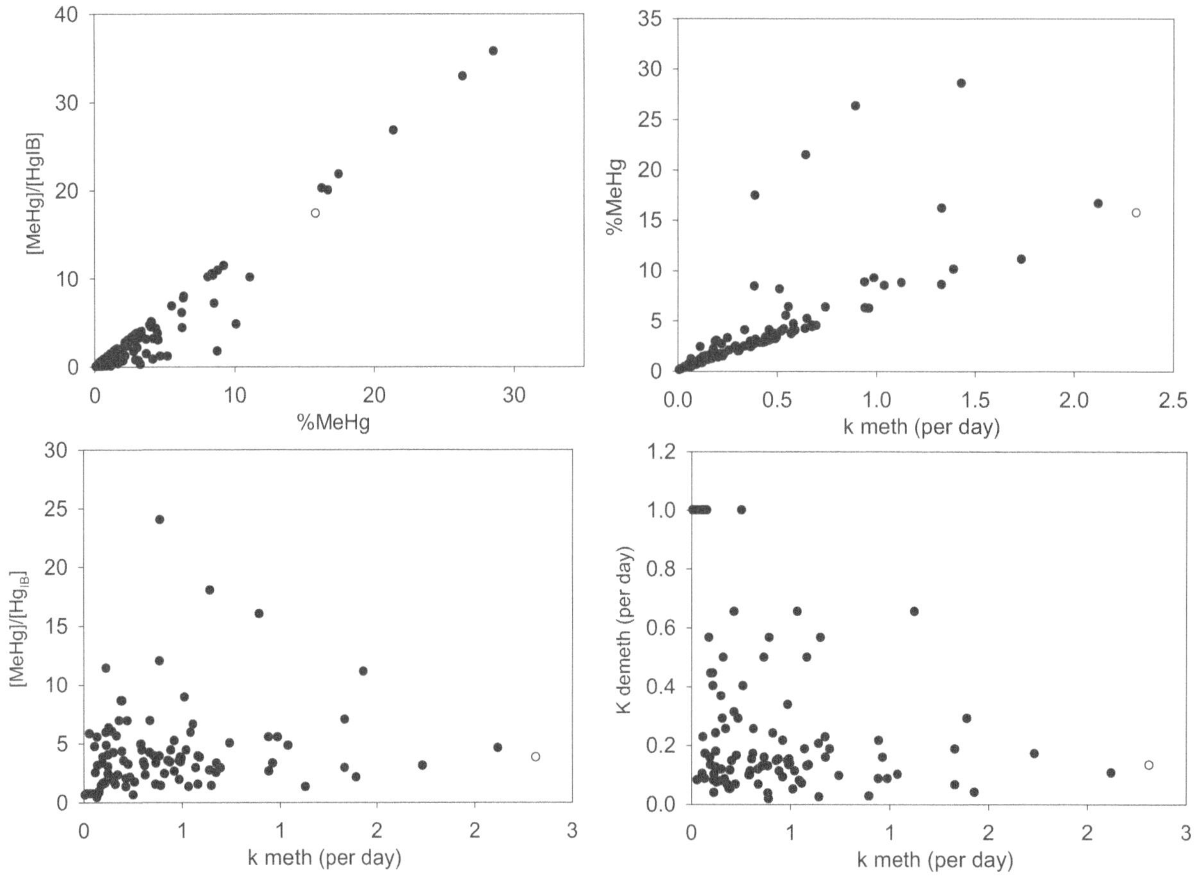

Figure 2-12. Relationships between the Measured and Estimated Methylation/Demethylation Variables for the USGS Data.

In order to provide a linkage to the matrices used in the Bioavailability Tool, some additional analyses were performed. The USGS data were sorted by TOC and then by pH. TOC concentrations and the corresponding MeHg/IB-Hg ratios were sorted in TOC classes of < 2 mg/L, 2 to 7 mg/L, and greater than 7 mg/L. Then, the data in each of these classes were sorted by pH in ranges of < 7.5, 7.5 to 8, and > 8. Once all the data were sorted, the geometric means of the MeHg/IB-Hg ratios for each class were calculated. The results were that for all low TOC classes, regardless of pH, the ratios were uniformly low, with geometric means ranging from 0.1 to 0.4. The intermediate TOC classes had ratio values that were distinctly different from the low DOC classes, ranging from about 5 for low pH waters to 2 for intermediate and higher pH waters. The highest TOC class, lowest pH, had the highest ratio of any class, with a value of 16. The high DOC, intermediate pH class had a mean ratio of about 9. The high TOC, high pH class had a value of 4, which was similar to the intermediate TOC, low pH value of 5. Ultimately, since the ratios in the low TOC class were all quite low, these data were all combined yielding a composite k_{meth} / k_{demeth} ratio value of 0.2 for all fresh waters with TOC < 2 mg/L (see Table 1-2 for the final values).

Since the Bioavailability Tool allows a user to utilize either water quality information or %MeHg to evaluate the strength of the methylating environment, values of %MeHg and corresponding values for rate constant ratios were established, consistent with the above approach utilizing TOC and pH. Each of five classes of methylation potential (very weak, weak, moderate, strong, very strong) were assigned ratio values of 0.2, 2, 4, 8, and 16, respectively. The linear relationship established between %MeHg and the MeHg/IB-Hg ratios was used to calculate the value of %MeHg at the assigned class ratio. When the geometric means of the %MeHg values for each class were calculated, it was found that the actual means were very close to the regression values (i.e. 1.3% vs. 1% for the "very weak" class, 2% vs. 2.5% for the "weak" class, 3 to 4.9% vs. 4 for the "moderate class, 7% vs. 7.5% for the "strong" class, and 13.1% vs. 14% for the "very strong" class. Ultimately, the ranges for %MeHg and ratios shown in Table 1.2 were established for this matrix for use in the Bioavailability Tool.

Overall, there are clearly many different variables that contribute and impact the overall rates of net interconversion of Hg and MeHg, especially in the water column, as illustrated by this example. The overall values for k_{meth} and k_{demeth} are consistent with rates of conversion and fluxes of MeHg between compartments, as measured in various experiments and found in field incubations (e.g. mesocosms). Also, this analysis and the various correlations support the notion that the bioavailability ratio is a reasonable surrogate for assessing the degree of interconversion between inorganic Hg and MeHg, and can be used in this way in assessing the degree of change in speciation upon mixing of an effluent and receiving waters.

Another dataset of Hg and MeHg in Maryland reservoirs (Appendix C) was also evaluated to examine these calculations in a different setting. These waterbodies are all relatively deep and so the calculations used Equations 2.4.2.1-5 and 2.4.2.1-6. Most of these reservoirs stratify in the summer and develop low oxygen waters, and so this fact would tend to enhance relative methylation and the formulations above tend to support such an outcome. In this instance, the dissolved Hg concentration was measured and with the DOC data, all variables could be calculated directly. For these lakes, given their overall high DOC content, and the fact that many are relatively deep, the demethylation constant was relatively low and similar across all the reservoirs. Thus, the k_{meth} value is driven mostly by the ratio of [MeHg]/[IB-Hg]. Again, as most of the reservoirs have a relatively high DOC, there is a strong relationship between the bioavailability ratio and %MeHg, and the calculated methylation constant is again linearly and significantly correlated to the %MeHg (Figure 2-13 and Appendix C)[11].

The existence of the Maryland reservoir data also allowed us to check the validity of the matrices used to establish methylation potential for use in the Bioavailability Tool. The reservoir data was sorted in the same way as the USGS data and the geometric means of the k_{meth} / k_{demeth} ratio were calculated for the available classes. For the intermediate DOC and low pH class, the calculated k_{meth} / k_{demeth} ratio from the reservoir data was 5.05 versus 5.09 for the USGS data, a remarkably consistent value. For the high DOC, low pH class (the only other class available from the reservoir data set), the geometric mean ratio was 7.85 versus 16.45 for the USGS data.

[11] Note that although the USGS database reported TOC and the analysis of the data was done using the TOC data, The researchers used DOC interchangeably with TOC and used DOC data in verification of the method. The researchers feel justified in this for two reasons. First, the TOC values in the USGS database tend to be quite low. Although all the details when or how all the data were collected are not known, the lower the values, the higher the percentage of DOC in the sample should be and the better they should be indicative of DOC. Second, as DOC was used to validate the method to predict methylation potential, its use (in lieu of TOC) results in good agreement with the methylation potential of the waters used for validation purposes. Finally, the use of DOC rather than TOC is preferred because it should have more of an influence on light penetration (affecting demethylation rates) and on dissolved inorganic Hg reactivity than TOC.

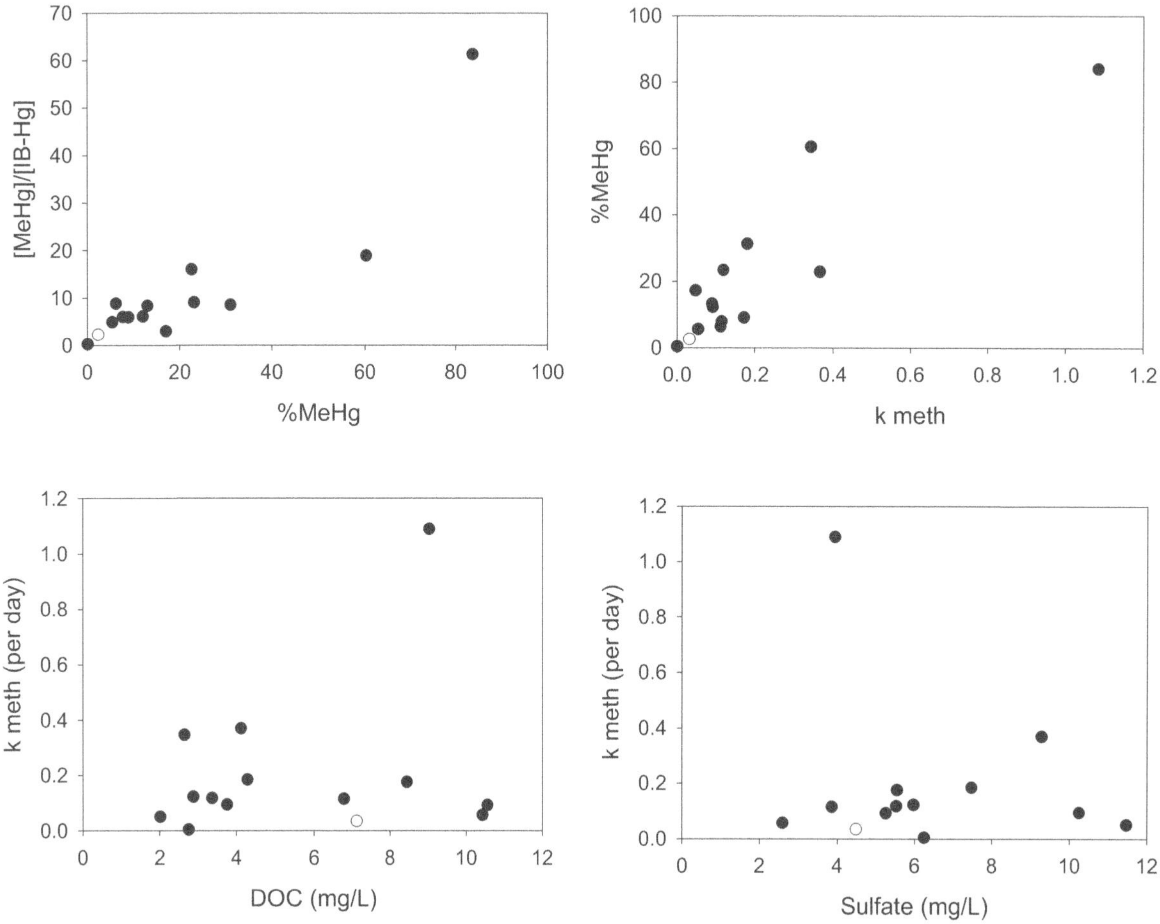

Figure 2-13. Relationships of Methylation/Demethylation Variables for the Maryland Reservoir Data.

However, the range of DOC in the reservoir data was only from 7.12 to 10.57, whereas the range of DOC in this class in the USGS database was from 8.4 to 18. When the ratio data corresponding to DOC values above 11 in the USGS database were eliminated, the calculated geometric mean ratio was 11.09, which is more consistent with the value of 7.85 from in the reservoir data. Based on this comparison of two independent data sets, the researchers feel that the approach used to estimate methylmercury based on methylation potential is reasonably robust for a screening-level analysis. In order to validate the screening approach on additional river systems, calculations were made to compare %MeHg and the k_{meth}/k_{demeth} ratios with those that would be predicted using the results of the USGS database analysis. The validation was performed on three datasets.

Babiarz et al. (1998) provide unfiltered T-Hg and MeHg data from seven Wisconsin rivers. In addition, data is available for suspended particulate matter (SPM) and DOC. Because dissolved Hg concentrations were not provided, a $\log K_D$ value was taken from Hurley et al. (1995) which reported mean $\log K_D$ for mercury in 39 Wisconsin rivers, which included the seven studied by Babiarz et al. Median values of the relevant parameters were used in the calculations. Table 2-15 shows the results. The predicted values are consistently close to those calculated from the data, although they are slightly higher in almost all cases.

In addition, several data points were available for the St. Louis River to which the Western Lake Superior Sanitation District WWTP discharges (Tuominen, 2007 personal communication). Calculations were made for two sampling locations, one upstream and one downstream of the discharge point. For the upstream point, the calculated %MeHg was 6 and the calculated ratio of constants was 4.9. The predicted value is 8. At the downstream location, the calculated %MeHg was 8.3 and the ratio of constants was 6. Again the predicted value was 8. Thus, in this case as well, the screening approach used in the Bioavailability Tool predicts ratios that are in reasonable agreement with calculated values although they are again somewhat higher. Using values of pH and DOC rather than %MeHg, the predicted ratio of constants is 9 rather than 8.

Table 2-15. Calculated and Predicted Net Methylation Constant Ratios for Seven Wisconsin Rivers.

River/Creek	%MeHg	Calculated k_{meth}/k_{demeth}	Predicted k_{meth}/k_{demeth}
Moose	7.8	5.3	8
Wolf	4.7	3.2	4
Tomorrow	4.3	2.8	4
Milwaukee	2.1	2.0	2
Rattlesnake	4.9	2.7	4
Eau Claire	4.1	1.3	4
Lincoln	0.9	0.6	0.2

In the third validation exercise, data from a study by Balogh et al (2004) of Cedar Creek and Trott Brook, MN were utilized. Relevant to the validation exercise, the data set for each stream contained 26 values of unfiltered T-Hg and MeHg, filtered T-Hg and MeHg, DOC, and pH values. These samples were collected weekly from June through October, 2003. For Trott Brook, the geometric mean values of %MeHg, DOC, and pH were 15.9%, 11.9 mg/L and 7.6, respectively. The geometric mean of the ratio of constants was 16.2. Using %MeHg, a value of 16 would be predicted. Similarly for Cedar Creek, the geometric mean values of %MeHg, DOC, and pH were 13.6%, 9.8 mg/L and 7.6, respectively. The geometric mean of the ratio of constants was 16.4. Using %MeHg, a value of 16 would be predicted for both of these streams. Thus, the observed values and those predicted by the screening procedure are quite close.

Using stream pH and DOC data to predict the ratios of methylation/demethylation constants for these two streams gives a slightly different result. The elevated DOC values place one in the 3rd column of the matrix (Table 1-2, DOC > 7 mg/L) but the values of pH place one in the second row of the matrix (Table 1-2, $7.5 < pH < 8$), yielding a ratio of constants of 9. It is to be noted, however, that the pH value is very close to the threshold between the row 1 and row 2 value and the row 1 value is 16. This exercise does point out a limitation of the matrix approach in that the values are discrete. If values of the independent variables (%MeHg, pH or DOC) are close to the interval boundaries, the user of the Bioavailability Tool would do well to evaluate the answer using the two nearest matrix entries[12].

2.4.2.2 Net Methylation in Estuarine and Coastal Waters

Information was also available for estuarine and coastal waters and it was examined. Four data sets were available including two from Long Island Sound and the Hudson River (Balcom, pers. comm.; Balcom et al., 2004; Balcom et al., 2008), one from the Chesapeake Bay and mid-Atlantic shelf (Hollweg et al., 2008) and one from San Francisco Bay (Conaway et al., 2003)[13]. Much of the data from the Hudson River, Long Island Sound and the Chesapeake Bay is either unpublished or is in documents that are currently being prepared for publication, and the data contained in this report is taken from the initial data tables generated from this work. For the estuarine and coastal waters, the results of water column methylation/demethylation studies (Whalin et al., 2007) suggest that the water column demethylation rate is a factor of 10 lower than that for freshwater systems for comparable conditions of DOC. The equations given above have been adjusted to reflect this. Also, as demonstrated by the studies of Lamborg et al. (2004), the relative number of binding sites per gram of DOC is approximately a factor of 10 lower for estuarine/coastal waters compared to that of freshwaters. Since the IB-Hg to DOC relationship used for freshwaters was derived using the data of Miller et al. (2007) (using Everglades DOC), the relationship used for the estimation of IB-Hg for estuaries has also been scaled accordingly.

The results of these calculations are shown in summary form in Table 2-16. Across the different estuarine and coastal ecosystems, the %MeHg is lower than that for freshwaters, except for the data from Long Island Sound. In this case, there are a few very high %MeHg values (>20%) that skew the average data. More investigation is needed to determine the cause of these high values as the actual MeHg concentrations are substantially lower and the high percentage is driven more by the reporting of a low total Hg concentration. However, as the DOC values for the various ecosystems are relatively low and similar (range 1-4 mg/L), the overall calculated values of IB-Hg are similar, and the [MeHg]/[IB-Hg] ratios are similar for the different ecosystems and are lower than those for freshwaters. This ratio provides an indication of the net potential for conversion of bioavailable inorganic Hg into MeHg, and therefore is a measure of the potential for overall MeHg production and transport within ecosystems, which is independent of total Hg concentration. Indeed, the values of the ratio appear reasonable in that they suggest that the lowest relative MeHg is offshore, and for the estuaries, is lowest for the Hudson River,

---

[12] The researchers note that there was also a seasonal variation in the data. Higher values of MeHg in the summer months led to higher ratios in the summer for both streams. In Cedar Creek, the summer versus fall geometric means of the calculated ratios were 18.4 and 13.6. In Trott Brook, they were 17.3 and 14.7. Thus summer values were 20 to 35% higher in these two streams. The "fall" season was indicated by a dramatic drop in DOC values in both streams about the middle of August. Although they were not used in the analysis because stream pH was not measured, ratio values for "spring" samples collected from 4/232003 to 5/27/2003 were lower (8.7 for Cedar Creek and 9.2 for Trott Brook). This may reflect the influence of snowmelt as alkalinities were lowest in the spring in both streams. If these data are combined with the summer and fall data, the overall ratios for the both streams are 14.6.

[13] The dataset of Conaway is part of a larger dataset for the San Francisco Bay maintained by the Regional Monitoring Program (RMP) (SFEI, 2008). Most of the data in the SFEI database (about 70%) are presented in the paper by Conaway et al (2003). However, not all the Conaway data are in the RMP dataset, and some values are different between the published paper and the RMP database. In the course of preparing the final report for this assessment, the researchers evaluated the data from both sources, finding that the results were not vastly different and did not recommend a different approach when all the estuary data where considered.

Table 2-16. Comparison of Net Methylation Characteristics of the Estuaries Evaluated in this Study.

Location	Statistic	DOC (mg/L)	%MeHg	IB-Hg (ng/L)	MeHg (ng/L)	MeHg/IB-Hg	kmeth	kdemeth
Chesapeake Bay - Upper/Mid <15ppt	Average	2.70	2.17	0.65	0.15	0.31	0.03	0.10
	Std dev		2.29	0.64	0.17	0.45	0.05	
Chesapeake Bay - Lower > 15 ppt	Average	2.70	1.08	0.75	0.07	0.23	0.02	0.10
	Std dev		1.07	0.61	0.07	0.37	0.04	
Coastal mid-Atlantic	Average	1.00	2.37	1.76	0.03	0.03	0.01	0.62
	Std dev		2.69	1.13	0.03	0.02	0.01	
San Francisco Bay < 15 ppt	Average	2.89	0.97	0.29	0.17	1.13	0.22	0.21
	Std dev	0.62	1.02	0.26	0.16	1.81	0.33	0.10
San Francisco Bay > 15 ppt	Average	2.73	0.75	0.67	0.06	0.22	0.05	0.21
	Std dev	0.43	0.79	1.27	0.06	0.13	0.03	0.05
Long Island Sound	Average	1.99	12.14	0.46	0.18	0.71	0.25	0.39
(<10 m data only)	Std dev	0.60	11.55	0.56	0.15	0.89	0.24	0.19
Hudson River	Average	2.04	1.20	0.46	0.09	0.16	0.05	0.30
(surface data)	Std dev	0.36	0.70	0.99	0.04	0.03	0.03	1.00
USGS Data Freshwater	Average	4.19	4.64	0.19	0.20	4.40	0.50	0.26
	Std dev	3.57	6.70	0.93	0.55	6.73	0.74	0.27

and is similar for Long Island Sound, the Chesapeake Bay and San Francisco Bay. This is overall consistent, for example, with the differences in %MeHg in sediments in estuarine ecosystems.

A plot of the k_{meth} determined versus the ratio [MeHg]/[IB-Hg] is shown in Figure 2-14. The results show that the relationship falls on the line for that of freshwater ecosystems but that the estuarine and coastal data is at the lower end of the range. This is primarily a function of the formulation - the estimates of IB-Hg are relatively higher resulting in a lower ratio, and overall a lower k_{meth}.

Again, the data for the estuaries and coastal waters were examined to assess an appropriate formulation for use in the Bioavailability Tool. As with the freshwater estimates of the ratio of the methylation/demethylation constants, DOC appears to be an important variable. The coastal water data set had the lowest values of DOC (1 mg/L) and also exhibited the lowest such ratio (geometric mean of 0.02), suggesting much higher demethylation than methylation rates in coastal waters. All of the estuaries had low to intermediate values of DOC, ranging from 1.2 to 3.24 mg/L, except the Ochlockonee, which had higher DOC levels (7.8 to 12.9 mg/L). The only other variable that appeared to have an effect on the ratio of the rate constants was salinity, and since the levels of DOC and salinity are correlated in most estuaries, the apparent salinity effect may in fact be DOC-related. In three instances, the Chesapeake Bay, Galveston Bay, and San Francisco Bay, there were data with significant enough salinity differences to evaluate this effect. In the Chesapeake Bay data, the lower salinity (< 15 ppt) data exhibited a ratio average (0.31) higher than the higher salinity (> 15 ppt) data (0.23) by about 35%. In Galveston Bay, the lower salinity (< 15 ppt) data had a ratio average of 0.21, higher than the higher salinity (> 15 ppt) value of 0.03 by a factor of 7. In the San Francisco Bay data, the lower salinity (< 15 ppt) data exhibited a much higher ratio average (1.13) than the higher salinity (> 15 ppt) data (0.22), a factor of over 5. Thus it appears that the effect of salinity/DOC is relevant to the evaluation of net methylation in estuaries. The Ochlockonee Estuary had two data points, both at salinities less than 10. The ratio geometric mean for these two data points was 0.09, similar to the low salinity value for Galveston Bay. The average ratios of the high salinity data for all the estuaries were fairly consistent, with a range of 0.16 to 0.23, except Long Island Sound, which had an average ratio of 0.71. However, the geometric means for the high salinity data for all the estuaries was more consistent, ranging from 0.1 to 0.26.

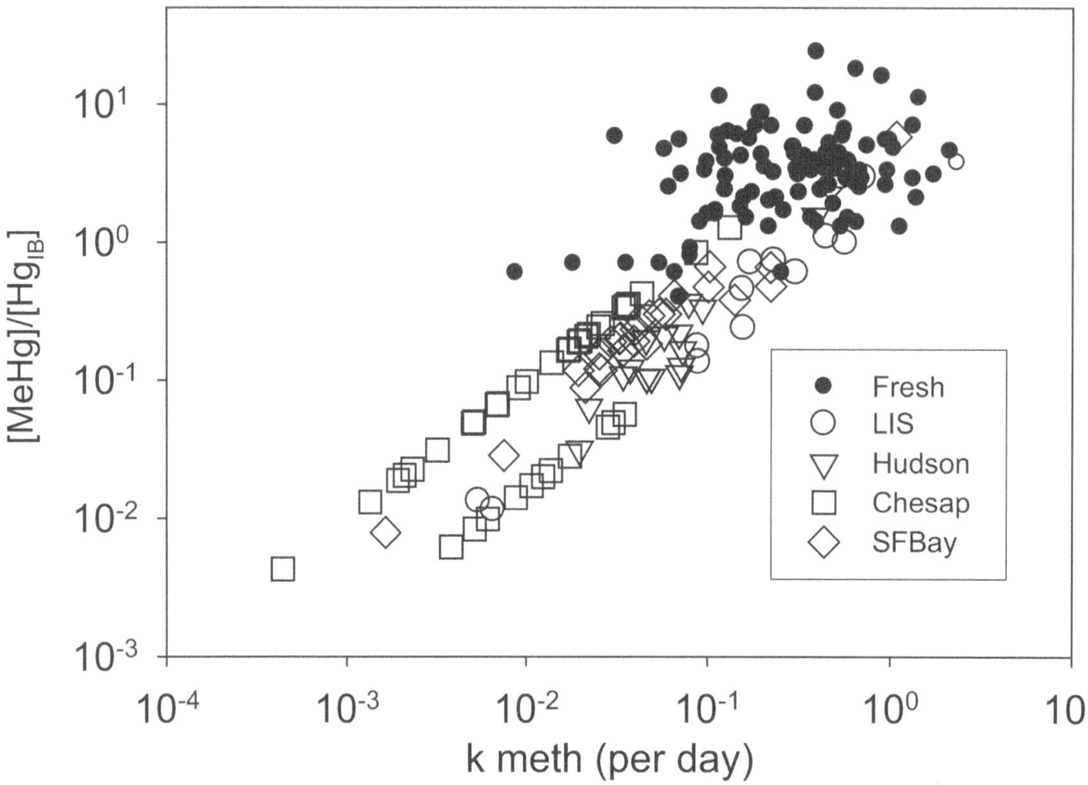

Figure 2-14. Relationship between the Calculated Ratio of Methylmercury to Bioavailable Inorganic Mercury and the Estimated Net Methylation Constant.

In order to be consistent with the freshwater data, the values selected for use in the Bioavailability Tool were geometric means of the methylation/demethylation ratios. Values were selected for low salinity estuary waters (< 15 ppt), high salinity estuary waters (> 15 ppt), and coastal marine water. The three values, which approximate the averages over all estuaries for which data was available, are 0.2 for the low salinity estuary water, 0.14 for the high salinity estuary water, and 0.02 for the coastal marine water. Thus, the value for low salinity estuary water (0.2) is the same as the low DOC freshwater value of 0.2 but significantly lower than the lowest freshwater intermediate DOC ratio of two.

CHAPTER 3.0

DETAILED BIOAVAILABILITY ASSESSMENT

This objective of this chapter is to provide guidance on how to conduct a detailed assessment for estimating (or improving estimates of) the bioavailability of mercury in effluent discharges to receiving waters. There are several steps involved in this process. The first is to gather the available information on the effluent and receiving water of interest and evaluate it in terms of its information content relative to estimating bioavailability. If the available information is adequate, it can be used to estimate bioavailability using the formulae presented in Chapter 2.0 or using the Bioavailability Tool. If information is lacking or is judged to be inadequate in some way, the next step is to formulate a plan for acquiring new data. This new data is then analyzed so that its utility for estimating bioavailability is optimized. At a minimum, the new data would be complete and adequate to perform a screening-level estimate of bioavailability. However, the analyst may wish to go beyond this and make better estimates by reducing some of the uncertainties associated with assumptions in the screening assessment. In general, this chapter follows this format. Section 3.1 discusses preliminary data gathering and Section 3.2 discusses the initial assessment of this data. Section 3.2 spends a significant effort in guiding the analyst in making assessments of the uncertainty in the existing data, as a prelude to making decisions to acquire new data, with which the section concludes. Section 3.3 deals with the acquisition of new water chemistry data in freshwater (rivers), estuaries, and marine waters. Section 3.4 deals with the acquisition of biological data in these same types of systems.

In Chapter 2.0, a procedure was developed for making a screening level assessment of bioavailability of effluents and their mixtures with receiving waters. However, the screening assessment involves a number of assumptions. Chief among these are the following:

- Reactive Hg (R-Hg) is essentially equal to inorganic bioavailable mercury (IB-Hg). If the R-Hg is not known from measurements, it can estimated using alternative procedures,
- Total bioavailable mercury in mixtures (as well as concentrations of other water quality constituents) can be simply derived using the flow ratio of effluent to receiving water (mixing or dilution ratio). No other changes to the aqueous chemistry of these waters occurs on mixing,
- There are well-defined (and robust) relationships between DOC and R-Hg in effluents and receiving waters quantitatively similar to the relationships derived in this report,
- Total bioavailable mercury is constant in the effluent/receiving water mixture. Only the relative amounts of MeHg and IB-Hg change depending upon the methylating potential of the receiving water,
- The ratio of MeHg to IB-Hg in an aquatic system at steady-state is essentially equal to the ratio of the methylation and demethylation constants,
- Estimates of bioavailable (MeHg and R-Hg or IB-Hg) mercury in water are good indicators of bioaccumulation.

Although these assumptions were necessary, made either in the interest of simplicity, or faced with lack of data for developing better, but still simple, approaches, the screening approach has its shortcomings. Nonetheless, it is a useful means of obtaining a preliminary estimate of the bioavailability of effluents and effluent mixtures with a minimum of data. However, if the scientist or environmental manager believes that better estimates should be made, there are definitely ways to improve the estimation procedure.

The first assumption has to do with the measurement of R-Hg, of which few measurements have been made in municipal wastewater. This lack of data is a significant limitation and the researchers encourage the analyst who wants to reduce the uncertainty in the estimates of effluent and receiving water bioavailability to make measurements of R-Hg. In Section 3.3 obtaining representative samples of R-Hg as well as the number of samples required is discussed. Effluent sampling is discussed in Section 3.3.2 and receiving water sampling in Section 3.3.3. In addition, the reader is referred to Appendix A which reports the results of a study of reactive mercury in wastewater samples that was undertaken as a part of this study. The analytical method for measuring reactive mercury is provided in the appendix.

The second assumption has to do with mixing of effluent and receiving waters and the resulting estimates of concentrations, not only of mercury species, but of other enhancer/mitigator water quality constituents as well. The constituents are mixed according to simple calculations based on the flow ratio and the underlying chemical basis for changes which may occur is ignored. For instance, the pH of the mixture is determined by taking the calculating the [H+] concentration in each component, weighting them according to the mixing ratio, and then taking the logarithm of the result. It would be more accurate to mix the waters using alkalinity and total inorganic carbon (TIC), and calculate pH from those results. Obviously this would require yet more data of a different type. Given the appropriate data, models such as MINTEQ or MINEQL can be used to more accurately calculate resulting water quality conditions and inorganic mercury speciation. However, the best way to make these determinations would be through near-field measurements and this approach is encouraged. It is discussed primarily in Section 2.4.3, in which guidance is given for sampling near-field environments in river, estuaries and marine waters.

The third major assumption has to do with the relationship between DOC and reactive mercury. In the screening approach, based on modeling work using literature estimates of the number of binding sites and binding constants for Hg, sulfide and DOC, this relationship was determined to be a simple function of DOC. However, wastewaters (or other receiving waters) may exhibit a different relationship with DOC, with perhaps additional variables besides DOC controlling mercury reactivity. A more advanced investigation of bioavailability would look at this relationship in both the effluent and the receiving waters. This is discussed in Section 3.3.1.3.

The fourth and fifth major assumptions deal with the behavior of mercury in the far-field with regard to speciation and methylation. The first of these assumptions assumes that the sum of MeHg and R-Hg is constant and that the methylation/demethylation reactions simply modify the distribution between these two forms. However, other reactions may affect this outcome. For instance, Hg(II) in surface waters may be reduced to Hg(0) and lost to the atmosphere. The second of these assumptions assumes that in the far-field these reactions come to steady-state and that they follow first-order kinetics. In this, there may also be other mitigating factors. Therefore, if a better far-field assessment is desired, the researchers suggest measurement of bioavailable mercury in the aged mixture, either *in situ* or in the laboratory. This is discussed primarily in Section 3.3.3.2.

The final major assumption is that aqueous measurements, in particular of MeHg and R-Hg, are good indicators of potential bioaccumulation. However, to confirm this in the waterbody of interest, the analyst may want to conduct actual bioaccumulation experiments on effluent, mixed effluent, and mixed aged effluent. This is the subject of Section 3.4 and the remainder of the chapter.

3.1 Preliminary Information Gathering

The first step towards assessing bioavailability is to gather the existing information available about the discharge and the environment of the receiving water. Environmental sample databases, site-specific studies, and research on both receiving waters and sources should be gathered and reviewed. This information may come from internal programs and studies or it may be the result of research conducted by others such as local watershed groups, universities, or other dischargers. By assessing the available information, data gaps and deficiencies can be identified so that a study plan appropriate to the situation can be developed. The use of existing information may also mitigate the need to collect additional samples. The Bioavailability Tool (Chapter 2.0) can assist the user in making an assessment of the existing data and to identify data gaps.

3.1.1 General Information on Receiving Waters

After treatment, point sources discharge their treated effluent back into the environment. Both effluent and receiving water data may already exist as a result of the requirements set forth in National Pollution Discharge Elimination System (NPDES) permits. As part of a NPDES permit, owners of point source discharge facilities (e.g. WWTPs) are required to sample not only their effluents, but often their receiving waters as well. Additional data on the receiving waters may have been collected by other research groups such as the USGS which has conducted water quality studies of water systems across the nation or NOAA which researches ocean waters. Watershed groups, research institutes, and academic research are other potential sources of data on the receiving water. The following subsections provide discuss useful information for various types of riverine, estuarine, and marine receiving water systems.

3.1.1.1 Freshwater (Riverine) Systems

In a freshwater riverine system, the general movement of water is from upstream to downstream, with a measurable horizontal velocity in the direction of flow. These systems may range in size from small creeks to very large rivers. Water quality and flow data may be available for locations either or both upstream or downstream of the discharge location. It is fundamental that the analyst know something of the hydrology of the watershed and waterbody to which the discharge is taking place. At a minimum, effluent and receiving water flow rates should be known. In addition, it will be useful to have basic geometric information about the receiving water. Such information includes cross-section data (so that velocities and travel times can be estimated using flow rates) and the location of significant downstream point sources and tributaries. The water quality of these sources is important to the extent that they might affect the near- or far-field bioavailability assessment. Data or information from the river that is not specific to the discharge should be examined to see if this may be the case.

Upstream water quality data are appropriate for use in the Bioavailability Tool because the Tool calculates the estimated downstream mixture of the upstream receiving water with the effluent. However, data used as representative of the receiving water should be from the environment in which the water will "age"; that is, the environment in which methylation/ demethylation reactions will ultimately occur. Data from downstream waters is more appropriate to use when estimating far-field bioavailability based on actual measurements. This is especially true if downstream waters represent a significantly different environment for mercury speciation.

Depending on geographic location, freshwater flows can vary greatly depending on the season. In some arid locations, the effluent can comprise of the entire flow in the summer. It is important that sufficient data exist to represent conditions that occur year round.

Lakes, or wetlands, *per se* have not been included in the discussion of freshwater systems. The main reason for this is that most inland wastewater discharges do not occur directly to lakes and wetlands, rather to flowing waters in rivers, which have better dilution and mixing characteristics. It is possible however, that a lake or wetland may be the ultimate receiving water environment for an effluent and receiving water mixture, and thus the environment in which methylation and demethylation reactions would ultimately occur[14]. The analyst must take care to recognize that this may be the case and take it into account in a bioavailability study and in using the Bioavailability Tool. For instance, it would be possible to estimate the extent of methylation/ demethylation that would occur once the effluent is mixed with the receiving water at a point downstream of the point of discharge. A subsequent application of the Bioavailability Tool could be used to mix the effluent/receiving water mixture with waters characteristic of the wetland (or other downstream waterbody). Similarly, if a detailed study were being undertaken, measurements could be made at those same representative points in the system.

3.1.1.2 Estuarine Systems

Estuaries represent a mixture of freshwater inflows and tidally influenced saltwater. An important characteristic is their salinity, which changes longitudinally within the estuary. Changes in the chemistry of estuaries occur seasonally due the seasonality of inflows from freshwater sources. Changes also occur diurnally due to the twice daily (semidiurnal) tides that occur in most coastal areas.

The type of estuary to which the discharge occurs (or more specifically the degree of stratification of the estuary) is also of importance to the estimation of bioavailability in estuarine receiving waters. As will be seen later in this chapter, many of the calculations and decisions to be made depend on the estuary type. In terms of their flow/salinity regime, estuaries will be classified as either well-mixed, partially mixed, or stratified, and it is important to know the type of estuary to which the discharge occurs.

Therefore, in addition to the basic geometries useful in riverine systems, some additional data is useful for bioavailability assessment in estuarine systems. Good salinity distribution information is essential. It is especially helpful to have vertical salinity profiles at a number of

---

[14] Wetlands are likely to have very different water quality characteristics from flowing rivers. Whereas flowing rivers would tend to be well-aerated and well-mixed, the same may not be true for wetlands. Levels of DOC and sulfides are likely to be higher in wetlands, and dissolved oxygen is likely to be lower as a result of reducing conditions. All of these factors would have a major influence on the methylation/demethylation potential in the wetland versus a flowing river. The same would be true for lakes, especially if the incoming river water were to mix with hypolimnetic waters below a thermocline, where reducing conditions may develop seasonally.

stations up and down the estuary. It is important to know if a sharp pyncnocline (density gradient) is present and whether it is seasonal or more or less permanent. The presence of a distinct and persistent pyncnocline is indicative of a stratified or salt-wedge estuary. These types of estuaries usually have large fresh water inflows and mixing is poor between the upper (fresh) and the lower (saline) levels. Tide-induced flow reversals normally do not occur in the upper zone.

Good information on tides and tidal range of the estuary is also important. Tidal range, along with good cross-sectional information (both up- and down estuary from the discharge) will be helpful in well- and partially-mixed estuaries to assess the extent of transport of the discharged water, as tidal action tends to move it in both directions. Estuaries of these types typically have more uniform salinity gradients between the surface and the bottom.

Receiving water data available for estuaries may be near the discharge for near-field conditions, or from monitoring stations that represent the far-field background conditions. Data collected near the point of discharge may exist as a result of NPDES permit requirements. For instance, in the San Francisco Bay, a regional monitoring program has been established so that sampling occurs at a limited number of monitoring stations rather than having each discharger conduct their own sampling.

3.1.1.3 Marine Systems

Marine waters are fully saline waterbodies influenced by tides and currents. Relative to estuaries, the water quality characteristics of ocean waters are fairly consistent. Seasonal cycles throughout the marine environment affect the direction of the currents throughout the year. In addition, the tides induce changes diurnally. Existing data may be either in the zone of initial dilution (ZID) or far-field background samples. Available data should be gathered and compiled.

Because of the complexity of the movements of waters discharge into marine environments due to currents, winds, density differences, and other factors, it is nearly essential to have detailed assessment of the extent and migration of the discharged plume. In particular, the extent location and extent of the zone of the ZID is of importance. In addition, it is useful to have knowledge of nearby areas which may be sensitive to mercury loadings.

3.1.1.4 Flow and Dilution

A dilution factor (or mixing ratio) defines the amount of ambient water that is mixed with the effluent in the near-field following discharge. The area in which this mixing occurs is sometimes called the *mixing zone*. Models are typically used to estimate the dilution factor associated with the mixing zone. Factors that are considered in dilution models vary depending on the type of receiving water, yet all receiving water models consider the location, depth, and configuration (e.g. type and number of diffusers on the discharge pipe) of the discharge pipe. Many point source dischargers have modeled the dilution of their effluents because the dilution factor is used in calculating discharge permit limitations.

It is important to note, however, in the bioavailability assessment, that the definition of the near-field mixing zone may be different from that used in modeling studies. Normally, these studies are done for regulatory purposes and with certain objectives in mind. For instance, the regulatory mixing zone usually consists of two areas; an acute mixing zone, and a chronic mixing zone, defined by the extent to which discharged concentrations of toxic chemicals exceed acute and chronic toxicity criteria. For the purposes of a bioavailability study, the near-field zone is defined as that in which the effluent is completely or maximally mixed with the receiving water, subject to constraints having to do with the methylation and demethylation reaction times. Only in the case of marine waters is it considered that the near-field mixing zone corresponds exactly with the regulatory ZID.

Flow in riverine settings may be determined by measuring channel velocities of the cross section at the point of discharge. In estuarine and marine systems, measuring flow is much more complex. In estuaries, tidal flushing, seasonal freshwater outflow, circulation patterns, and relative location of the discharge point, etc. all need to be considered. In oceans, seasonal currents, diurnal tidal influences, distance and depth of the discharge location, etc. need to be considered. For anything more than the simplest analysis, the analyst will need to rely on modeling studies or detailed measurements to develop the necessary understanding of flow in estuarine and marine systems.

3.1.2 Chemical Parameters

The parameters useful to make estimates of bioavailability based on water chemistry vary depending on the complexity and the level of detail of the assessment. Key parameters include MeHg, T-Hg, and R-Hg. Other key parameters for estimating bioavailability may include TSS, DOC, and pH. Data for all of these parameters should be collected for both the source water and receiving waters. More complex assessments may include collecting additional water quality information, biological monitoring and perhaps even the development of a chemical speciation/fate and transport model.

The use of literature values in lieu of site-specific values might greatly reduce the sampling effort, yet the results will not be as robust as they would be if data specific to the particular location of interest are used. Key parameters such as MeHg, T-Hg, R-Hg, TSS, and DOC should be specific to the analyst's location.

3.1.2.1 Conventional Water Quality Parameters

The major parameters useful in making estimations on the bioavailability include TSS, DOC, and pH. In estuarine environments, salinity is also a consideration.

TSS is important in assessing the bioavailability of mercury in source and receiving waters because the inorganic mercury adsorbed to solids is not considered to be bioavailable. Sampling and measurement data for TSS in source waters is common, especially for wastewater discharges where monitoring in effluents is required in NPDES discharge permits. Sampling and testing for TSS is relatively easy and it is a commonly monitored parameter in receiving waters as well.

Higher organic content can potentially decrease bioavailability due to strong complexation with mercury; however, DOC may also keep more mercury in the dissolved state, which may explain the higher bioavailability observed in waters with high DOC. DOC is the indicator of the organic content in water columns used in this assessment. DOC is not typically required to be sampled in effluents or receiving waters by NPDES permits. However, DOC is a useful water quality parameter in characterizing pollutants in water and is often included in sampling studies and research programs. Consequently, existing DOC data may be available through monitoring programs conducted by local watershed groups, in local studies conducted near your discharge site, or part of a special study conducted by the discharge facility.

pH is an important water quality parameter in this assessment as it may influence the degree of methylation in receiving water environment. Recording pH measurement data in effluents is common, especially wastewater discharges as monitoring in effluents is required in NPDES discharge permits. Similarly, receiving water pH data both upstream and downstream of the discharge is also common and is often also required to be monitored by NPDES discharge permits. In marine environments, pH is fairly consistent. In contrast, in ephemeral streams in arid locations, wide swings in pH over the course of a day may be observed, and in streams where algal respiration may induce large pH swings.

The presence of major inorganic ions such as sulfide and chloride can have an influence on mercury complexation. In this assessment, there is no direct dependence on these two water quality parameters. However, some knowledge of their magnitude is advisable to ensure that they are within normal ranges (see Phase I, Chapter 4.0, Table 4-8). If data exists on other major ions, this data should be complied as it may ultimately be useful. Mercury has a very high affinity for sulfides. Thus, even non-detectable levels of sulfide ($< 1 \mu M$ but greater than 0.1nM) will alter Hg speciation. Close proximity of sulfidic sources of water (e.g., wetlands, some wastewater effluents) may contribute these levels of sulfide.

3.1.2.2 Mercury Measurements

The two key mercury species that make up the total bioavailable mercury concentration are methylmercury and reactive mercury. Other important mercury data include total mercury and dissolved inorganic mercury. It is only recently that analysis of mercury species other than total mercury become more common. For instance, in the San Francisco Bay Area, point source dischargers have recently been required to monitor their effluent and receiving waters for methylmercury as part of a special regional study. Although not as frequently analyzed, data for dissolved mercury is sometimes available as well. Ideally, a methylmercury and reactive mercury samples will correspond to total mercury samples (i.e. taken at the same time) in order for the methylmercury and reactive mercury fractions to reflect actual conditions. Data should be recent enough to represent current conditions and so that method detection limits will not limit or affect the dataset. Any data collected prior to the early 1990s are suspect as contamination of samples may have occurred due to the sampling techniques employed.

It is not necessary that data for all mercury species and fractions be available. In the following sections, the user can assess the best approach to estimating bioavailability based on the sufficiency of the existing data and the opportunity to collect missing information. Approaches provided earlier in this report allow mercury values to be calculated using existing or easier to obtain mercury information. For example, dissolved inorganic mercury can be estimated by calculating the difference between the dissolved total mercury and dissolved methylmercury concentrations or estimated by using total mercury, TSS, and the partition coefficient. Once the data gaps are identified, a sampling program can be focused on acquisition of the any additional required information.

3.1.2.3 Distribution Coefficient (K_D)

The partition coefficient (K_D) is a measure of the relationship between the adsorbed and dissolved phase of a material in the water column. K_D is an important factor because it provides an indication of the relative strength of binding of mercury to solid material. K_Ds may be found from the literature although a more accurate assessment would result from calculating them using information on the specific effluent and receiving water of interest. If dissolved mercury concentrations are not available, the K_D and total mercury concentration may be used to calculate the dissolved inorganic mercury concentration.

3.1.2.4 Mercury Methylation and Demethylation Rates

Measured mercury methylation and demethylation rates are not required in order to use the screening methodology. In this assessment, the ratio of k_{meth}/k_{demeth} is assumed to be the same as ratio of the steady-state concentrations of MeHg and IB-Hg (or R-Hg). The ratio of k_{meth}/k_{demeth} (where k_{meth} is a first-order methylation reaction constant and k_{demeth} is a first-order demethylation reaction constant) is used as an indicator of the strength of the methylating environment. Using the approaches developed in this report, this ratio can be estimated using either %MeHg or pH and DOC values. Therefore, either pH and DOC data or MeHg and T-Hg data will need to be available if R-Hg is not available. However, it is also possible that these reaction rates can be measured independently. Measurement of these rate constants and comparison to measured MeHg and R-Hg would allow the analyst to tell whether the system is at steady-state. If it is not, then the rate constants would allow an assessment of how far the system is from steady-state, the time required to achieve it, and the relative concentrations of bioavailable mercury forms when steady-state is achieved.

3.1.3 Biological Parameters

The ultimate purpose of a detailed study is to better understand the availability of mercury in source waters to organisms within the food chain present in receiving waters. Measuring the bioaccumulation of mercury within organisms directly is a useful measure of bioavailability. The U.S. EPA recently recommended changing the basis for expressing the ambient water quality criterion for mercury from a numeric water quality concentration to a methylmercury fish tissue concentration. This change requires that fish-based bioaccumulation factors (BAFs), or equivalent approaches, be available to calculate water quality concentration limits for mercury.

The bioaccumulation of methylmercury in fish is a complex process affected by many site-specific environmental factors. The accumulation process involves the biological sequestering of substances that enter the organism through respiration, ingestion, direct contact with a mercury-containing substrate, and/or other means. The level to which a given substance is bioaccumulated depends on an array of biological and physical factors such as substrate type and composition, speciation of mercury in the water column or substrate, mode of uptake, uptake rate, depuration rate, and metabolic processes occurring within biota. Bioaccumulation is usually not directly correlated with total mercury concentration and the assessment of the potentially bioavailable forms of mercury has been a major part of this study. Even so, the strong expectation that a chemical species is potentially bioavailable falls short of confirming its actual uptake and bioaccumulation. On the other hand, measurement of mercury in biota alone is of limited value when attempting to assess the impact of a particular source on those concentrations. Therefore, assessing biological data in concert with chemical analysis is advantageous.

Studies in close proximity to a particular discharge location may have been conducted to analyze the bioaccumulation of mercury in biological organisms present in the receiving waters. These studies may have been conducted in the lab or directly in the field. The species of primary focus in the study may have been larger fish or smaller organisms such as bacteria or zooplankton. Any studies of this kind will provide much useful information. However, in this chapter, the assumption is made that such studies have not been done (few have been) and that the scientist, engineer or manager is making a first effort at such an undertaking. Even if previous studies have been done, the guidance provided in this chapter may give some insights into their scientific merit.

3.1.3.1 Biological Species of Interest

As discussed in detail later in the report, there are five general groups of aquatic biological indicators, which in decreasing size are: 1) piscivorous (top predator) fish, 2) prey fish, 3) benthic invertebrates, 4) zooplankton, and 5) phytoplankton, other algae (e.g. periphyton) and bacteria (Fisher and Reinfelder, 1995; Wiener et al., 2007; Wiener et al., 2003). Piscivorous fish are a good indicator for human health concerns and are typically used for regulatory purposes. Consequently, there may be available fish tissue data for your receiving water. However, organisms on the lower end of the food chain (e.g. phytoplankton, bacteria, zooplankton, and filter feeders) provide a much more direct link to the local water quality conditions and in turn, may be a better indicator of bioavailability at a specific location. Nonetheless, any available data will be useful and should be reviewed.

3.1.3.2 Mercury Measurements

Mercury concentration in organisms is measured by analyzing their tissues. However, there are a variety of ways that biotic samples can be acquired, handled, and processed. These various methods can affect the outcome of the testing and the reviewer would do well to have a thorough understanding of the way the samples have been handled and processed in order to assess data quality and to compare the results to other samples. For instance, fish tissue may be analyzed on whole fish, using skin-on fillets, or skin-off fillets. Concentrations may be reported on either wet- or dry-weight basis. Many states have standard protocols for these types of analyses but some do not. Furthermore, individual research studies may utilize a wide variety of approaches. On the other hand, bioaccumulation assays are typically done using standard methods and standard test organisms and may be easier to compare from study to study. These types of studies are becoming increasingly more available as regulatory agencies demand them of dischargers.

Testing for total mercury is more common and many environmental laboratories are equipped to provide this analysis. Methylmercury testing in tissues, on the other hand, is more complex and not all labs are equipped to provide this analysis. Consequently, methylmercury tissue results are not as common. One very useful aspect of having studies or tissue analyses for a species of interest is that they will give a very good indication of the levels to which bioaccumulation will occur and point out particular problems with the use of a particular biological species.

3.1.3.3 Bioaccumulation Factors

The bioaccumulation factor (BAF) is the ratio of mercury concentration in an organism to the concentration of mercury in the water the organism resides. BAFs are essentially a measure of an organism's bioaccumulation of mercury. For regulatory purposes, BAFs may used to translate fish-tissue criteria to water quality concentration data.

BAFs vary widely depending on location and species used. In the literature, generic BAFs have been derived for various trophic levels of species and for general water systems such as estuaries, lakes and ponds, and rivers. However, literature values may not be very useful except in the most general sense. Obviously, BAFs depend primarily on two measurements, tissue concentrations and water or sediment mercury concentrations. One of the common pitfalls of past studies which have produced BAFs for various species is the use of very limited data of which to base them. In some TMDLs studies, sample size has been limited to one fish tissue sample and one water quality sample. In research studies, this situation is often improved but not to the point of providing a statistically valid estimate. Bioaccumulation and therefore BAFs are very waterbody, species, and food-chain dependent. Thus, it is risky to place too much confidence in such numbers developed for other water bodies. Later in this chapter, a means of assessing the uncertainty in BAFs will be discussed.

The utility of a species-specific BAF is to provide an idea of the extent of bioaccumulation in a species is more relevant and local tissue data are not available. Obviously, for this to be useful, measurements of mercury in the water column must be available. Care must be taken when using BAFs from other studies in this regard as well. Some BAFs are calculated and reported using the water column concentration of total mercury, some using methylmercury. Still another complication is that the BAF may have been calculated using either unfiltered or filtered water samples. BAFs calculated based on filtered water samples may be orders of magnitude higher for the same species than if computed using an unfiltered water sample.

There are other caveats on the use of BAFs. The BAF is essentially an equilibrium constant which shows, at steady-state, the relative distribution of mercury between the water, or sediment, and biotic tissues. It is certainly possible that in studies where BAFs have been reported, the systems were not at steady-state. This may due to several reasons. First, the water column or sediment concentrations may be changing over time. In most water bodies in the US, due to the recent emphasis on mercury elimination in the environment, concentrations are likely in decline. Second, changes to the food-chain may have the system in disequilibrium; this may be the case when new species are introduced or water quality conditions change so that the species present are naturally evolving. Finally, mercury concentrations in larger fish change slowly over time and some large predators may never come to an equilibrium concentration. For some fish species, this is the reason that BAFs are reported for individuals or composites in a certain age or size range.

Ideally, better information will come from one's own study area, providing BAFs for species that reside in your receiving waters or in an environment similar to your receiving waters. However, as suggested above, the origin of any such numbers should be thoroughly understood and they should always be used with caution.

3.1.3.4 Species-Specific Mercury Accumulation and Depuration Rates

When assessing biological data, mercury accumulation and depuration rates for the test organism are helpful in understanding the bioaccumulation of mercury in the species. Mercury concentrations in one species can vary significantly from another species found in the same receiving waterbody. Therefore understanding these rates for the test species is important.

In addition to differences among biological species, accumulation and depuration rates are also dependent on the chemical species of mercury and the method of intake. In general, inorganic mercury is easily taken up by aquatic life, yet is also depurated relatively quickly. Methylmercury, on the other hand, accumulates quickly but depurates very slowly, and is the primary reason methylmercury more readily biomagnifies up the food chain. In addition, uptake and depuration rates also vary in differing tissues within an organism.

The value of uptake and depuration rates is that they give the analyst an idea of how long a bioaccumulation experiment may take to reach to reach steady-state. If the either uptake or depuration rates are slow, it may take some time for the system to come to this condition. Of course, result may always be reported as of the end of a certain date (i.e. the 15 day or 30 day average tissue concentration). Even so, preliminary estimates are of great value in designing an experiment. Data for both accumulation and depurations rates for varying species may exist in the literature or for specific studies that are pertinent to your study area, but the latter it is not likely. The BAF (which is for instance the ratio of MeHg in tissue to MeHg in the water column) can provide a steady-state estimate of the ratio of the uptake and depuration rate. Thus, if one of the rates is known, the other can be calculated from the ratio.

3.2 Evaluation of Existing Data

This section will discuss the evaluation of data that may be available to the analyst to assess bioavailability with a focus of assessing whether additional data collection is required. These techniques can be used not only initially, but as more data becomes available they can be used to assess the need for further refinements. The drivers behind the decision to collect additional data are, on the one hand, cost, and on the other, the reduction in uncertainty that would result from the acquisition of additional data. Because cost may change over time, the number of samples is used as a surrogate for cost, and the analyst may obtain total cost by multiplying the number of samples by the unit cost per sample[15]. The unit cost per sample includes the cost of planning for data collection, sampling, materials, shipping samples to the laboratory for analysis, analytical cost, and the cost of data reduction and reporting.

In general, the reduction in uncertainty can be measured by the reduction in the standard error of the mean. The standard error of the mean is defined as:

$$SE_x = s_x / SQRT(N) \qquad \text{Equation 3.2-1}$$

where:

SE_x is the standard error of the mean of the variable x,

s_x is the sample standard deviation of the mean of x,

[15] The cost of analytical procedures provided herein are in 2008 dollars.

N is the number of samples, and

SQRT() is the square root function.

Note that SE_x decreases as the square root of the number of samples collected. Therefore, the number of samples required to reduce SE_x by the same percentage grows by 1/SQRT(N). So, for instance, to achieve a 29% reduction in SE_x when N = 5, on would double the number of samples to 10 (collect five additional samples). However, to achieve a reduction of 29% in SE_x when the number of samples is 20 would require that an additional 20 samples be collected.

3.2.1 Initial Data Assessment

The initial data assessment should be performed in a step-wise fashion, with each step aimed at the answering the question important to the analyst. If the question is to compare the bioavailability of one effluent to another, the data requirements are different than if the question is to assess what happens to the bioavailability of an effluent when it is mixed with a receiving water.

3.2.1.1 Sample Size and Uncertainty

A major decision that the analyst, environmental manager, or regulator must make with regard to this process is how accurate the answer needs to be. In general terms, the answer needs to be more accurate the closer the two things being compared are numerically and the higher their uncertainty or variability. Let one suppose, that it is desirable to assess whether the bioavailability of effluent "A" is greater than that of effluent "B". Figure 3-1 demonstrates this schematically. In panel (a), the distributions of two variables are shown where the population means are quite different, and the uncertainty in the two variables is relatively low. In this case, it is easy to intuit that small number of samples would be required to test whether the population means are different. In panel (b), the values of the population means are closer together and the uncertainty in the variables is higher. In this instance, it is easy to see that many more samples would be required to get a sufficiently "accurate" values of the two means, so that the hypothesis that one is greater than the other could be tested with an appropriate degree of certainty.

Typically, then, the analyst will state the need for accuracy in terms of a probability. For instance, the analyst may state a desire to be 95% confident of being within 10% of the mean. Recalling Equation 2.3.1.3-1, the number of samples required to achieve a given tolerance on the mean with a given level of confidence is given by[16]:

$$N = [z_{\alpha/2}\, \sigma / d\,]^2$$

where:

N is the number of samples,

z is the standard normal deviate at probability level σ/2 (for a two-sided test),

σ is the standard deviation of the variable and

d is the error tolerance.

[16] Note that the "standard error of the mean" term σ/SQRT(N) appears in this equation, which is readily apparent if it is rearranged to solve for "d" rather than "N". Thus the standard error of the mean, introduced earlier, is an integral part of the calculation for the required number of samples to achieve a desired error tolerance or uncertainty about the mean.

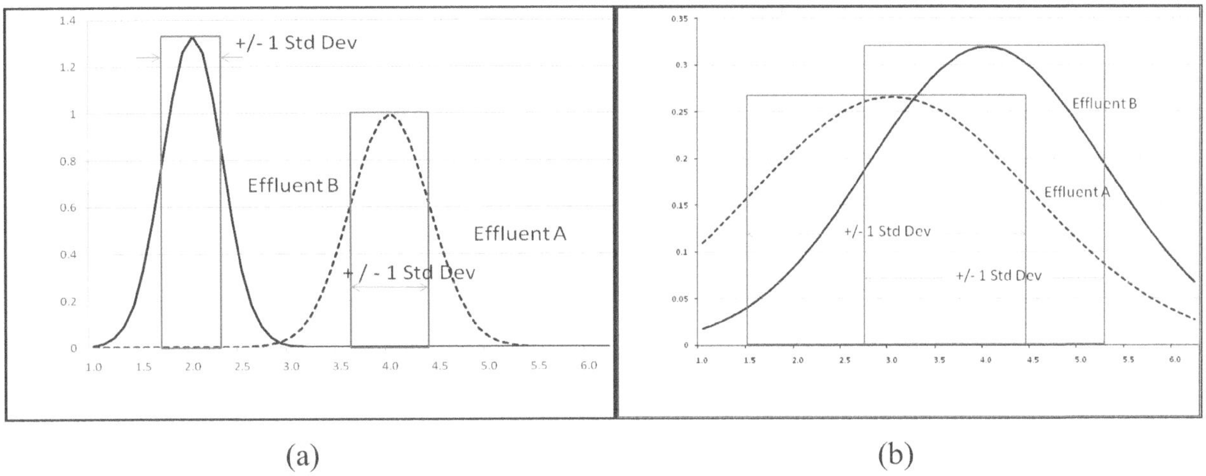

Figure 3-1(a) Example Distributions with Large Differences in Means and Low Variability (b) Example Distributions with Small Differences in Means and High Variability.

Since the error tolerance can be defined as some fraction of the mean, the quantity $\mu[f_{U(x)}]$ can be substituted for d (where μ is the population mean and $f_{U(x)}$ is the allowable fractional uncertainty). Furthermore, since σ/μ = CV, the above equation can be rewritten as:

$$N = [z_{\alpha/2}(CV) / f_{U(x)}]^2 \qquad \text{Equation 3.2.1.1-1}$$

Thus, the number of samples required to achieve a desired fractional tolerance on the mean with a given confidence can be estimated directly using the CV, assuming the measurements of the variable of interest are normally distributed[17]. Conversely, the fractional uncertainty of x, $f_{U(x)}$, can be estimated by rearranging Equation 3.2.1.1-1 if the number of samples is known. Obviously, it is beneficial to have some preliminary idea of the variability in the parameters of interest, either through previous measurement or other experience, such as literature estimates. It is certainly best to have estimates of the CV from direct measurement in the effluent or body of water of interest, if available. Lacking this, however, literature estimates can provide a reasonable starting place. As the data acquisition proceeds, if the analyst sees that too few or to many samples are planned to be collected given the initial estimate of the CV, the collection program can be augmented or reduced accordingly. For most of the water quality parameters necessary to determine bioavailability and bioavailable concentrations of mercury,

[17] If the data are severely positively skewed, as they would be if the data are log-normally distributed, the CV of the data in arithmetic space will be smaller than it should be, due to the presence of a large number of small values in the data set. If this CV were to be used, the calculated number of samples required would be too low. Log-transforming the data results in a higher CV, whose use more accurately reflects the required number of samples. Many environmental variables tend to be log-normally distributed (i.e. exhibit strong positive skewness), especially those whose values are impacted by flow, such as TSS and particle-reactive metals like mercury. Appendix D provides an example of how to test a distribution for normality. Even though the example is for fish tissue data, the procedures apply equally to water chemistry data.

typical uncertainty estimates have been provided in Chapter 2.0 in the form of the coefficient of variation (CV) expressed as a percent.

3.2.1.2 Evaluating the Normality of a Dataset

The CV of the data is required in the equation used to calculate the required number of samples. In the previous section, it was stated that the CV can be affected by the non-normality of the underlying statistical distribution of the variable of interest. Therefore, it is important that the analyst be capable of testing the data to determine if they are in fact normally distributed. Two methods to accomplish this evaluation are presented in Appendix D, one parametric (the Chi-Square test) and one non-parametric (the Kolmogorov-Smirnov test).

If the data are judged to be normally distributed, the analysis may proceed. If not, the data should be transformed (using an appropriate transform, e.g. taking the by logarithms) so that they are, and subsequent analyses should use the transformed data. Once a distribution of the data has been determined to be normal, or has been transformed so that it is normal, the CV for the distribution can be calculated. This is done by calculating the sample standard deviation of the data and dividing by the sample mean. The CV can then be used in Equation 3.2.1.1-1 to determine the appropriate number of samples.

3.2.1.3 Error Propagation and Uncertainty

An important concept in the evaluation of uncertainty in the data and the decision to collect more is that of error propagation in calculations involving the measured variables. According to Taylor (1982) when quantities added or subtracted in a mathematical operation, the uncertainties always add. When quantities are multiplied or divided in a mathematical formula, the <u>fractional</u> uncertainties add. Another convenient fact to have at one's disposal for dealing with error propagation in our bioavailability calculations is that the uncertainty associated with the result of an arbitrary function of one variable is the product of the absolute value of the derivative of the function and the uncertainty in that variable. The utility of these facts will presently be demonstrated.

Another fact concerning error propagation that will assist in process of performing it is to break the expression down into terms and perform error propagation on the terms in the expression that one would first perform in mathematical sequence. Consider the expression

$$y = a + bx^c$$

in which 'x' is a random variable and 'a', 'b', and 'c' are constants. To evaluate this function mathematically (i.e. to obtain a value of 'y' for a given value of 'x'), one would first evaluate the term 'x^c', multiply this result by 'b', and then add this result to the constant 'a'. Error propagation would proceed in exactly the same sequence, eventually arriving at the value of the uncertainty in 'y'.

Fractional uncertainty can be expressed in a variety of ways but a convenient one is the CV (expressed as a fraction). So, for instance, if the mean value of MeHg with a CV of 0.25 is divided by a mean value of T-Hg with a CV of 0.25 to obtain the fraction MeHg, the result would have a CV of 0.5 (0.25 + 0.25 = 0.5). However, if the variables are all independent then the fractional uncertainties may be added in quadrature, i.e.

$$CV_q = SQRT[(CV_x)^2 + (CV_y)^2 \ldots + (CV_z)^2] \qquad \text{Equation 3.2.1.3-1}$$

where 'q' is any result of an operation involving the multiplication or division of random variables x, y, …z. In the case of addition and subtraction, the uncertainties themselves may be summed in quadrature if the variables are judged to be independent.

Thus, if MeHg and T-Hg are independent (as they typically are), the uncertainty in their quotient may be estimated by adding their fractional uncertainties in quadrature. In the example provided above, rather than CV of 0.5 resulting from their division, a value of 0.35 would result ($SQRT[0.25^2 + 0.25^2] = 0.35$).

The relative independence or dependence of two random variables can be determined by calculating a correlation coefficient between the two measurements. If the correlation is weak, then the variables tend to be independent and adding their fractional uncertainties in quadrature would be appropriate. However, if they tend to be strongly correlated, then simple addition of the fractional uncertainties would be more appropriate.

Another measure of fractional uncertainty that will be used in this discussion is the fractional tolerance about the mean which was referred to earlier from Equation 2.3.1.3-1. This fractional uncertainty can be used in the same way as the CV. However, care must taken that when uncertainties or fractional uncertainties are added to propagate error in a series of calculations that the uncertainties are consistently defined.

The importance of error propagation is that uncertainty in the mathematical result of interest is a function of the uncertainties in the variables that go into its calculation. Thus, while the uncertainties in the dependent variables may be acceptable, the resulting calculated quantity may have an uncertainty that is unacceptably high.

Although it is possible to obtain an estimate of the uncertainty in a calculated quantity using these methods, the answer may be inexact for several reasons. First, the method is strictly only applicable to small uncertainties around the mean (see Taylor, 1982; pages 32-34) and in some cases, the uncertainties in environmental variables may be quite large (on the order of 50 to 75%). Second, the actual uncertainty depends on the relationships among the dependent variables in the calculated quantity. Therefore, the actual uncertainty can only be estimated to be between the limits provided by strict addition of the fractional uncertainties (appropriate when the variables may be highly correlated) and their addition in quadrature (appropriate when the variables are independent). The objective, then, may be more typically to identify which variables are the most likely candidates for additional data collection in order to reduce the error in the calculated quantity to the extent possible, and not to reach an absolute goal with regards to the uncertainty in the answer. The best candidates for additional data collection will be those variables with relatively higher uncertainties than the others (and therefore whose reduction in uncertainty will have the greatest impact on lowering the uncertainty in the estimate) and which can be measured at a relatively lower cost.

However, it is possible to make planning level estimates of the number of samples to collect using this approach. For instance, if the goal of the study is to have a mean value of a calculated quantity with a CV of 30% and there are three independent measured variables multiplied to obtain the result, one can allocate this uncertainty to the three measured variables and estimate how many samples might be required to obtain the desired uncertainty in the result. In the following sections, examples of these approaches are provided.

3.2.1.4 Water Chemistry

Armed with these concepts, the assessment of data that may already be in the hands of the analyst can proceed. The best information the analyst might have is direct measurement of bioavailable mercury as MeHg and as R-Hg. Furthermore, if T-Hg measurements are available, an expression of bioavailability can be made, the sum of MeHg and R-Hg as a percent of total mercury.

Assume that 10 measurements (i.e. 10 samples) each are available for these variables in the analyst's effluent. Furthermore, assume calculated CVs for MeHg of 50%, R-Hg of 75%, and 25% for T-Hg (recall that the sample CV is simply the sample standard deviation of the variable divided by the sample mean). The calculated means are 0.5 ng/L, 0.3 ng/L, and 15 ng/L, respectively. Additionally, assume that a determination that the sampling distributions for these variables are normal and that they are uncorrelated has been made. Assuming a desired confidence level of 90%, the tolerance on the mean of these measurements can be estimated using Equation 3.2.1.1-1. For example, in the case of MeHg,

$$f_{U(MeHg)} = (CV)\, z_{\alpha/2} \,/\, SQRT(N) \qquad \text{Equation 3.2.1.4-1}$$

or

$$f_{U(MeHg)} = 0.5(1.645) \,/\, SQRT(10) = 0.26.$$

Therefore, one can conclude that the estimate of the mean available for MeHg is within 26% of the true population mean with 90% confidence. Similar results can be obtained for R-Hg and T-Hg (29% and 13%, respectively). Suppose however, that the analyst wants the estimate of MeHg to be within 10% of the true mean. Then,

$$N = [(CV)\, z_{\alpha/2} \,/\, f_{U(x)}]^2 \qquad \text{Equation 3.2.1.4-2}$$

or

$$N = [0.5(1.645) \,/\, 0.1]^2 = 68$$

Thus an additional 58 samples would have to be collected (i.e. 68 minus the ten already collected).

Now, suppose one wants to determine the overall uncertainty in the estimate of bioavailability. The definition requires that one sums the fraction of MeHg and the fraction R-Hg. The fraction MeHg is calculated from

$$fMeHg = MeHg/\underline{T\text{-}Hg} \qquad \text{Equation 3.2.1.4-3}$$

Because the fractional uncertainties add in this division operation and the variables are considered independent, the results can be added in quadrature so that

$f_{U(MeHg/\underline{T\text{-}Hg})} = SQRT[(f_{U(MeHg)})^2 + (f_{U(\underline{T\text{-}Hg})})^2]$ Equation 3.2.1.4-4

or

$f_{U(MeHg/\underline{T\text{-}Hg})} = SQRT[(0.26)^2 + (0.13)^2] = 0.29$

Therefore, the fractional uncertainty for the fraction MeHg is 29%. Similarly, for R-Hg, the fractional uncertainty for the fraction R-Hg would be 32%. The calculated fractions of total mercury are 0.033 for MeHg and 0.02 for R-Hg. Because, according to Taylor (1982) the uncertainties of sums or differences always add (and for independent variables they add in quadrature) one can also calculate the uncertainty in the sum of fMeHg and fR-Hg, which yields fB-Hg, the fraction of bioavailable mercury. The uncertainty of the fraction of bioavailable Hg, U(fB-Hg) is[18]

$U(f\underline{B\text{-}Hg}) = SQRT[((0.033(0.29))^2 + (0.02(0.32))^2] = 0.012$

Since the sum of fMeHg and fR-Hg is 0.053, the fractional uncertainty of fB-Hg is 0.012/0.053 or 23%. Note that the resulting uncertainty in fB-Hg (0.012) is larger than the uncertainty in either of terms that make it up (0.01 and 0.006). A more conservative estimate of the uncertainty of fB-Hg could be made if it were assumed that the uncertainties of MeHg and R-Hg are not independent. In this case, they would be added arithmetically, the result would be 0.016, and the fractional uncertainty of fB-Hg would be 30%.

The same procedure applies to the assessment of bioavailable mercury if the quantity IB-Hg is calculated rather than being measured as R-Hg. However, the assessment is more tedious since there are more variables involved. The concentration of dissolved inorganic Hg (ID-Hg) must be known as well as the reactive fraction of ID-Hg in order to calculate IB-Hg. Dissolved inorganic Hg can be calculated from the difference of dissolved T-Hg and dissolved MeHg. In this case, the uncertainties in these two variables would be summed. The fractional uncertainty of their difference would then be multiplied by the fractional uncertainty in the reactive Hg fraction (estimated from DOC; see Equation 2.2.1-6) in order to obtain the fractional uncertainty of IB-Hg[19]. The analysis would then proceed as above, assuming that IB-Hg ~ R-Hg. As can be seen, while this procedure can be performed, it is not likely to be useful as it is involved and the answer, as pointed out above, is somewhat uncertain itself.

If dissolved measurements are not available for T-Hg and MeHg, then dissolved ID-Hg must be calculated from T-Hg. The calculation involves the measured (or estimated) quantities of TSS, K_D and T-Hg (see Equation 2.2.1-5). Since the calculation of ID-Hg involves products and quotients exclusively, the fractional uncertainty of ID-Hg would be estimated by adding the

---

[18] Note that since in order to calculate the uncertainty of this sum, the uncertainties of the component variables themselves must be used, rather than the fractional uncertainties, the fractional uncertainty of each component variable is multiplied by its mean to obtain the uncertainty in each of the component variables.

[19] According to Taylor (1982) the uncertainty of any arbitrary function of one variable (such as the relationship of DOC to the fraction of reactive mercury), is the derivative of the function times the uncertainty in the independent variable (in this case DOC). In this case the derivative of the reactive Hg function is $-1.48(73.5)e^{-1.48(DOC)}$, or $108.8\ e^{-1.48(DOC)}$, which would be evaluated at the DOC sample mean value. Recall that the absolute value of the derivative is used to multiply the uncertainty in DOC. To obtain the fractional uncertainty in IB-Hg, the uncertainty obtained from the product of the uncertainty in DOC and the derivative of the fraction reactive Hg function would need to be divided by the value of IB-Hg at the same DOC level.

fractional uncertainties of TSS, K_D, and T-Hg. Since these quantities should be relatively independent of one another, the fractional uncertainties may be added in quadrature.

Due to the complexity involved in the calculation of uncertainty in a result given these functions, it is more likely that the usefulness of error propagation analysis will be confined to the examination of the relative magnitudes of the uncertainties of the key variables and the relative cost of reducing the uncertainty in one variable versus another. However, for simpler calculations, the application error propagation theory may be useful.

3.2.1.5 Aquatic Biota

There are several relevant issues to discuss with regard to the initial assessment of tissue concentrations. The first is the number of samples required to be collected to achieve the desired level of accuracy in the result. The second is error propagation as it applies to calculations of biological parameters like bioaccumulation factors (BAFs).

As with water chemistry data, preliminary and relevant tissue data can be evaluated as to whether they are adequate for the purposes of assessing bioavailability. Of course, the main drawback of trying to use fish tissue or other ambient biological tissue data to draw conclusions about the bioavailability of a particular source or effluent, is that it may be difficult to trace the bioaccumulated mercury to that particular source, as other point or non-point sources in a watershed will tend to confound the analysis. Even so, biological assessment is an important tool to confirm the potential of an effluent or source to be bioaccumulated and, combined with water chemistry data, is a powerful one.

The details of planning and executing a sound biological assessment will be covered in Section 3.4. With that in mind, the focus of this section is on the evaluation of existing data to determine whether additional data would be beneficial to collect. As was discussed in the water chemistry section, the number of samples required to achieve a desired level of accuracy can be calculated from knowledge of the variability of the data and the desired confidence level. The same equation is relevant and the same assumptions for the use of that equation apply below.

Number of Samples. Once a distribution of the data has been determined to be normal, or has been transformed so that it is normal, and the CV of the tissue concentration (or other biological parameter) is known, Equation 3.2.1.1-1 can be used to determine the appropriate number of samples.

If the analyst has no access to measured data in the effluent of receiving water of interest, CVs may be estimated from the literature. An overview of variability of mercury in fish tissue samples is provided in EPRI (2003). Typically, CVs for mercury in fish tissue range from 50 to 100%. Higher CVs tend to result when different species are sampled over multiple locations, seasons, and size ranges in a given waterbody. When sampling is more focused, for instance, by species and by location, CVs are typically lower and may be as low as 25%. In nine Maryland lakes, CVs for largemouth bass ranged from 33% to 60%[20], however, the fish sampled cover a large range of sizes. It has also been observed that lower CV tend to be associated with higher means (see EPRI, 2003), probably due to the fact that measurement of mercury in fish tissue close to detection levels tends to be more variable. Thus, CVs in species that tend to bioaccumulate less mercury may tend to be slightly higher. To be realistic, a CV of 30% is not

[20] Data was taken from TMDL documents for nine Maryland lakes identified in Section 3. Document citations may be found in Appendix A.

unreasonable if the sampling is focused on a single location, species, and size class or age group of fish.

Uncertainty in Tissue Data and Bioaccumulation Factors. The treatment of uncertainty in data and error propagation for calculations are the same as for biological data as they are for water data. Bioaccumulation factors are a measure to which the organism bioaccumulates a chemical present in the water column and in relevant trophic levels below it and are calculated by dividing the tissue concentration by the water column concentration. In this way, they have the same units as the distribution coefficient (K_D) of L/kg and function in the same way, being essentially an equilibrium partition coefficient. The expression is:

$$BAF = 10^6 \, (C_t / C_w) \qquad \text{Equation 3.2.1.5-1}$$

where:

C_t is the tissue concentration of the chemical in mg/kg, and

C_w is the water column concentration in ng/L.

Like K_D, the BAF is occasionally expressed in log units due to its magnitude.

In the case of mercury, the bioaccumulated chemical species of concern is normally MeHg. This is especially true in higher trophic levels, where MeHg typically may account for 90% or more of the tissue burden. However, in lower trophic levels (e.g. in phytoplankton, zooplankton and small herbivorous fish), the percentage of MeHg is lower and inorganic mercury may make up a substantial proportion of the total Hg in the tissue. In such cases, BAFs for inorganic mercury may also be of interest. They may be calculated in the same way, substituting either T-Hg, R-Hg or IB-Hg for MeHg in the water column. It is typical for MeHg that unfiltered water column concentrations are used to calculate BAFs. However, for inorganic Hg accumulation, for organisms at the bottom of the food chain, the dissolved concentration may be more appropriate.

The use of error propagation theory is fairly straightforward in this case. Knowing the CV and the number of samples one has for both tissue and water column concentrations, one can calculate the fractional uncertainty of the mean for each variable. The uncertainty in the BAF can then be calculated by adding the fractional uncertainties. It is important to check the correlation between water column concentrations and concentrations in tissue, as in this case, they may be significant (i.e. MeHg in biota is often strongly correlated with MeHg in the water column). If so, the fractional uncertainties would be added directly rather than in quadrature.

If data are lacking for fish tissue, estimates of the CV may made using information contained in the previous section.

3.2.2 Bioavailability Screening Assessment and Relevant Uncertainties

The analyst may also at this time wish to use to Bioavailability Tool to conduct a screening assessment with the available data. This report already discussed the approaches which would enable the analyst to:

- assess bioavailable concentrations and bioavailability of Hg in an effluent,
- assess the relative proportions of MeHg and R-Hg or IB-Hg in the effluent,
- compare the above quantities to those of another effluent,
- assess the uncertainties associated with these calculations, and
- determine the number of samples required to achieve desirable levels of accuracy in measured and calculated quantities.

In addition, the analyst may wish to evaluate the effects of mixing an effluent with a receiving water and perform a near- or far-field assessment of those effects.

In evaluating these effects, it is possible to use a variety of flow regimes or time frames. In this discussion, average annual flows, as they are perhaps the most relevant, will be used, but the discussion would apply equally to any flow regime or time frame the analyst might select. The near-field mixing calculation is governed by Equation 2.2.4.1-2 and involves the multiplication of the effluent and receiving water concentrations of a given water quality parameter by the mixing ratio. Recall that the mixing ratio is defined as the ratio of the effluent flow or volume to the receiving water flow or volume (Q_{eff}/Q_{rw}). If the analyst has a record of sufficient length, the average annual effluent flow and the average annual receiving water flow past the discharge point may be calculated for a number of years along with the corresponding CVs. The mixing average ratio, R_m, can be calculated from their quotient. The uncertainty in the mixing ratio, $f_{U(Rm)}$, can be estimated by adding the fractional uncertainties in the annual effluent and receiving water flows as follows:

$$f_{U(Rm)} = f_{U(Qeff)} + f_{U(Qrw)} \qquad \text{Equation 3.2.2-1}$$

The uncertainty of the mixed concentration, $U(C_m)$, can be estimated by:

$$U(C_m) = U(C_{eff}) + C_{rw}[f_{U(Crw)} + f_{U(Rm)}] \qquad \text{Equation 3.2.2-2}$$

where:

$U(C_{eff})$ is the uncertainty in the effluent flows,

C_{rw} is the average receiving water flow, and

$f_{U(Crw)}$ is the fractional uncertainty of the receiving water flow.

The fractional uncertainty in the mixed concentration is

$$f_{U(Cm)} = U(C_m)/C_m \qquad \text{Equation 3.2.2-3}$$

in which C_m is the average mixed concentration of a given constituent. Using this formulation, the uncertainties of any near-field concentrations can be estimated.

In the far-field, additional uncertainties are brought to bear on the calculations. These depend upon which method the analyst chooses to use to assess far-field effects. For freshwater systems, if the pH_DOC method is used (see Section 2.2.4.2), then the uncertainties in the pH_DOC matrix entries are relevant. If the %MeHg method is used, then the uncertainties in those matrix entries are relevant. Finally, if actual [MeHg/IB-Hg] values are used, then the uncertainties associated with these measurements would be relevant.

Tables 3-1 and 3-2 below show the CVs associated with the factors in the matrices for %MeHg and pH_DOC based methodologies[21]. It is of interest that the CVs for the %MeHg-based matrix are significantly lower than the CVs for the pH_DOC-based matrix. What this indicates is that, in general, it is probably better to use the %MeHg approach, because the far-field MeHg concentration calculated using the matrix would have a lower uncertainty associated with it than a result based on the pH_DOC method. However, one must also consider that there may be error associated with selecting an incorrect matrix entry. Because both pH and DOC are likely to be less uncertain than %MeHg, which is the quotient of two highly uncertain variables, the probability of selecting the wrong matrix entry using the %MeHg approach is likely higher.

Table 3-1. Coefficients of Variation (%) for the Matrix of Methylation/Demethylation Constant Ratios Based on %MeHg.

% MeHg	<2%	2 to 3%	3 to 5%	5 to 10%	>10%
k_{meth}/k_{demeth}	87	35	46	46	44

Table 3.2. Coefficients of Variation (%) for the Matrix of Methylation / Demethylation Constant Ratios (k_{meth}/k_{demeth}) based on DOC and pH.

		DOC	
pH	**< 2 mg/L**	**2 to 7 mg/L**	**> 7 mg/L**
<7.5	115	88	64
7.5 to 8	91	87	81
> 8	105	68	-

There are two additional sources of uncertainty in the calculation of far-field MeHg concentrations that remain to be considered. The first is the selection of the demethylation constant, k_{demeth}. This constant is based on DOC and, in deeper waterbodies, on water depth.

In shallow waters, uncertainty in the equation for k_{demeth} is a function of the uncertainty in DOC concentrations alone. According to Taylor (1982) the uncertainty in an arbitrary function of

[21] CVs were determined by computing the mean and standard deviations of the ratios of MeHg/IB-Hg calculated for the rivers in the USGS database for each matrix entry. For instance, the CV for the group pH < 7.5 and DOC > 7 is calculated from the mean and standard deviation of all the MeHg/IB-Hg values falling in that group. Similarly, the CV for the %MeHg group corresponding to the range of 2% to 3% was calculated from the MeHg/IB-Hg values falling into that group. The CVs simply reflect the uncertainty of the MeHg/IB-Hg values in that group, and not the error with selecting the wrong group by either having erroneous pH, DOC or %MeHg values.

one variable is the derivative of the function with respect to the independent variable times the uncertainty. So for the case where DOC < 1 mg/L, this would be zero since the function is a constant. For DOC > 1 mg/L, the derivative of the function is $-2.4/[(2.4*DOC-1.6)^2]$.

For deeper waters, where DOC < 1 mg/L, the derivative of the function for k_{demeth} is $-5/D^2$ where D is the water depth in meters. When DOC > 1 mg/L, the function for k_{demeth} is in two variables. In this case, the general formula given in Taylor (1982) for a function of two or more variables must be used, which involves the sum (either directly or in quadrature) of the uncertainties of each variable times the product of the partial derivative of the function with respect to that variable. In our case:

$$U(k_{demeth}) \sim \left|\partial k_{demeth}/\partial D\right| + \left|\partial k_{demeth}/\partial(DOC)\right| \quad \text{Equation 3.2.2-4}$$

Therefore, the partial derivatives of the function with respect to D and DOC must be computed in order to calculate the uncertainty in k_{demeth}:

$$\partial k_{demeth}/\partial D = -(2.4*DOC-1.6)/D^2 \quad \text{Equation 3.2.2-5}$$

and

$$\partial k_{demeth}/\partial DOC = -2.4/[D*(2.4*DOC-1.6)^2] \quad \text{Equation 3.2.2-6}$$

Note that the analyst may not wish to consider the uncertainty in the water depth D in the calculations. There may be several reasons for this, among them that there may not be multiple measurements of depth. Furthermore, treating depth as a constant provides a depth-averaged analysis which is adequate for our purposes. In this case, the analyst would simply ignore the first term in Equation 3.2.2-4 and use only the second term in the analysis. This amounts to reducing the k_{demeth} function to a function in one variable and the partial derivative with regard to DOC is equivalent to the derivative with regard to DOC.

To complete this analysis, the analyst would estimate the uncertainties associated with summing MeHg and R-Hg (to obtain total bioavailable Hg) by summing the uncertainties of these two variables. There is one final calculation required to estimate the far-field concentration of MeHg (Equation 2.2.4.2-2). This equation involves the sum of k_{meth} and k_{demeth}. Since the ratios k_{meth}/k_{demeth} and MeHg/IB-Hg are equivalent, it will be assumed that the uncertainty in k_{meth} is the same as the uncertainty in MeHg and the uncertainty in k_{demeth} is the same as that of IB-Hg. The analyst would evaluate the fractional uncertainty of the term $k_{meth}/(k_{meth} + k_{demeth})$ and add it to the fractional uncertainty in B-Hg (bioavailable Hg) to obtain the uncertainty in the final value of far-field MeHg concentration.

Once estimates are made of the uncertainties in the measurements or calculated quantities, decisions can be made as to whether additional data collection is warranted. As stated earlier, however, the calculations outlined above are tedious and the uncertainty in the result may be inexact. Therefore, the next section will examine ways to make the decision to collect additional data without resort to these calculations.

3.2.3 Results Evaluation

Although the results of the previous analyses may be used in a variety of ways, the focus here is to determine whether additional data collection is warranted. At this point, the analyst would have either one or two sets of information. The first might be the uncertainty in a desired result, for instance the bioavailability of Hg (i.e. the sum of MeHg and R-Hg, expressed as a percent of T-Hg) in an effluent. The second set of information would be the uncertainties of the components of that result, in this case, the uncertainties in MeHg, R-Hg, and T-Hg. In the following sections, this latter case to illustrate the data collection decision.

3.2.3.1 Uncertainty Assessment

Returning to the example in Section 3.2.1.4, recall that the following results were obtained:

- MeHg = 0.5 ng/L   fMeHg = 3.3 $f_{U(MeHg)} = 0.26$
- R-Hg = 0.3 ng/L   fR-Hg = 2.0 $f_{U(R\text{-}Hg)} = 0.29$
- T-Hg = 15 ng/L $f_{U(T\text{-}Hg)} = 0.13$
- B-Hg = 0.8 ng/L   fB-Hg = 5.3 $f_{U(f\,B\text{-}Hg)} = 0.22$

From these values one can see that least uncertain of the measurements collected is T-Hg and the most uncertain is R-Hg. It is intuitive that reducing the uncertainty in the most uncertain variable would have the best effect on the result and therefore additional samples of R-Hg might be worth collecting.

Another way of postulating which additional samples to collect might be in reduction of the overall uncertainty. Suppose that it is desirable to reduce the fractional uncertainty in the estimate of B-Hg from 23% to 15%. Assuming for the moment that the costs of the analyses are equal or that we wish to collect paired samples, one can make the assumption that equal numbers of samples would be collected for all the analytes.

First, calculate the CV for fB-Hg, given the CVs for the measured parameters using the error propagation rules. The CV for %MeHg would be the sum (in quadrature) of the CVs for MeHg and T-Hg:

$$CV(fMeHg) = SQRT[CV(MeHg)^2 + CV(\underline{T\text{-}Hg})^2] \qquad \text{Equation 3.2.3.1-1}$$
$$= SQRT[0.5^2 + 0.25^2] = 0.56$$

The CV for f<u>R-Hg</u> would be the sum in quadrature of the CVs for <u>R-Hg</u> and <u>T-Hg</u>:

$$CV(f\underline{R\text{-}Hg}) = SQRT[CV(\underline{R\text{-}Hg})^2 + CV(\underline{T\text{-}Hg})^2] \qquad \text{Equation 3.2.3.1-2}$$
$$= SQRT[0.75^2 + 0.25^2] = 0.79$$

The uncertainty for fB-Hg would be the sum of the uncertainties for fMeHg and fR-Hg:

$$U(f\underline{B\text{-}Hg}) = SQRT[(CV(fMeHg)*fMeHg)^2 + (CV(f\underline{R\text{-}Hg})*f\underline{R\text{-}Hg})^2] \quad \text{Equation 3.2.3.1-3}$$

$$= SQRT[(0.56*0.033)^2 + (0.79*0.02)^2] = 0.024$$

This is equivalent to the estimated standard deviation of fB-Hg and therefore the CV would be this value divided by the observed mean of 0.053 or 0.024/0.053 = 0.46

Now that one has an estimate of the CV of the desired result, the number of samples required to be within 15% of the mean with 90% confidence can be calculated:

$$N = [1.645\ (0.46) / 0.15]^2 = 25$$

Therefore, the number of additional samples to be collected for each analyte would be about 15. Note that if one substitutes a error tolerance of 0.23 into the above equation rather than 0.15, a result of 10 to 11 samples is obtained, which is equivalent to the initial number of samples, and which indicates that the calculations, overall, are correct.

3.2.3.2 Cost of Additional Data Collection

The cost of data acquisition and sample analysis is a major factor in the decision to collect new data and what type of data will be acquired. In this study, new data for nearly all the parameters important to the estimation of bioavailability was acquired as part of a study of reactive mercury in wastewater effluents. The prices[22] for sample preparation and analysis, expressed on a per sample basis, were as follow:

- T-Hg - $72
- MeHg - $155
- R-Hg - $90
- In-laboratory filtration - $35
- TSS - $18
- pH - $12
- DOC - $54

It was also determined by the WWTPs collecting samples that the price per event for sample collection would be about $350, using EPA Method 1669 (Clean Hands). In addition, the cost for shipping samples to the analytical laboratory was estimated to be about $120 per sampling event. These costs are obviously subject to change and should be confirmed prior to deciding to undertake additional sampling. However, the relative costs (between one analyte and another) should not change dramatically and this is the way in which these costs in the *pro forma* analysis in the next section will be used.

3.2.3.3 Decision to Collect Additional Data

The decision to collect additional data, and/or the decision as to how much additional data to collect, can be made on the basis of reduction of uncertainty versus cost. Equation 3.3-1 can be used to get a quick estimate of this. By substituting the existing number of samples and a proposed number of samples for ‘N’ in the equation, two values of the standard error of the mean

[22] All prices quoted in this report are in 2008 dollars.

can be generated. The difference in the standard error may be compared to the additional number of samples collected times the units cost of collecting the sample. Realize, however, that the analyst may not know the uncertainty in a calculated result, such as a bioaccumulations factor, unless error propagation is used to assess it from the individual uncertainties in the collected data.

For instance, assume that additional samples are to be collected for MeHg determination. Utilizing from the previous examples the estimate of 50% as a CV for MeHg, a mean value of 0.5 ng/L, and a sample size of 10, the standard error is

$$SE_x = s_x/SQRT(N) = 0.5(0.5)/[SQRT(10)] = 0.079$$

If the number of samples is doubled by adding an additional 10 samples,

$$SE_x = s_x/SQRT(1) = 0.5(0.5)/[SQRT(20)] = 0.056$$

Therefore, the reduction in SE_x expressed as a percent is 100*(0.079-0.056)/0.079 = 29%. The corresponding analytical cost (neglecting for any other costs) for unfiltered MeHg would be 10($155) or $1550.

It is relatively easy to identify those parameters that would provide the greatest reduction in uncertainty at the least cost. In order to do so, simply normalize the cost of each chemical analysis to the cost of the most expensive analyte. The CV for each of the analytes can then be divided by the normalized cost parameter. Comparison of the resulting values indicated those parameters which would the greatest uncertainty reduction at the least cost. In this case, filtered MeHg would be the most expensive to analyze at $190/sample ($155 + filtration cost of $35). The R-Hg cost would be $90 + filtration cost, or $125/sample, and T-Hg would be $72/sample. Thus the normalized costs would be MeHg = 1.0, R-Hg = $125/$190 = 0.7, and T-Hg = $72/$190 = 0.4. Dividing the CVs for each of these parameters by the normalized cost for each yields:

MeHg - 0.5/1.0 = 0.5
R-Hg - 0.75/0.7 = 1.1
T-Hg - 0.25/0.4 = 0.6

Therefore, the parameter that would provide the greatest reduction in uncertainty for the cost involved would be R-Hg.

3.2.3.4 Checking for Existing or Acquired Data for Outliers

Existing (or acquired) data should be checked for the presence of outliers as they can unduly affect the outcome of any statistical analysis performed on the data. An outlier is a data point that is anomalously high or low compared to others in the dataset. Two ways are presented for checking for the presence of outliers. The first, the Walsh Test, is nonparametric and is mainly useful for large data sets. The second, Grubbs Test, is mainly used for smaller datasets and will likely be the one of greater utility to the analyst. The tests are detailed in Appendix E.

3.3 Chemical Estimation of Bioavailability

In this section, guidance is provided for the collection of additional samples for key variables for the purpose of making a detailed estimation bioavailability of mercury in the water column via chemical means. In general, once the decision to collect additional samples has been made, the details of the sampling and analysis plan need to be developed. Issues include where samples will be collected, for which parameters samples will be collected, when and how often they will be collected, and how they will be collected. In the previous section, the issue of the number of samples as been addressed and will not be covered again here.

In the following discussion assume that the objective of the study is simply to make an assessment of bioavailability, bioavailable concentrations, or fluxes of mercury in an effluent or effluent/receiving water mixture. There could be other objectives, however. For instance, the analyst may wish to assess whether the bioavailability of an effluent is statistically greater than another, or whether bioavailable mercury concentrations downstream from a discharge are stochastically larger than bioavailable mercury concentrations upstream of the discharge. Or the interest may be in assessing whether bioavailability or bioaccumulation is greater in a certain season than in another in the receiving water. Any number of additional objectives might be hypothesized. Different study objectives would require additional different data analysis techniques, such as the use of two-sample statistical tests of the means or distributions of parameters of interest, and perhaps different data collection procedures. It is out of the scope of this document to attempt to cover these many possibilities. In all such cases, however, the analyst should be the development of a study plan with the end analysis in mind and work back from there to determine what data should be collected and in what quantities, which is what has been done here for the stated objective.

3.3.1 General Considerations

In this section general sampling considerations will be discussed. Issues related to specific types of water bodies will be covered in later sections.

3.3.1.1 MeHg and SnCl2-Reactive Hg

MeHg and R-Hg are the principal species of mercury involved in assessing bioavailable mercury in a water sample. As such, it makes sense to measure these parameters directly, even with the uncertainty that such measurements entail. Although reactive mercury can be estimated by calculation of inorganic bioavailable mercury, the calculation, as has been seen earlier, suffers from the uncertainty in the component variables that go into the calculation. Assuming that TSS, K_D and T-Hg all have CVs of 50%, the resulting uncertainty in dissolved inorganic mercury could be as high as 150%, but at least 87%, even if all the variables are independent[23], and to this must be added the uncertainty in estimating IB-Hg from dissolved inorganic Hg, using the relationship between DOC and IB-Hg. Therefore, it is easy to see that the most certain way to estimate bioavailable Hg concentrations would be through direct measurement; not mention that the cost of measuring the variables that go into the calculation of IB-Hg would be greater than the cost of the direct measurement of R-Hg. Thus the use of direct measurement of these two variables to the extent possible is strongly recommended.

---

[23] Given the particle reactive nature of mercury, there is likely to be moderate to strong correlation between TSS and T-Hg, which would tend to increase the uncertainty in IB-Hg.

However, there may be circumstances in which the measurement of R-Hg is not desirable or possible and therefore and alternative to direct measurement must be discussed.

3.3.1.2 The Alternative to $SnCl_2$ Reactive Hg Measurement

If it is not possible or desirable to measure R-Hg, then dissolved inorganic Hg (ID-Hg) must be estimated followed by estimation of the reactive portion. In order to make the best estimate of ID-Hg, dissolved (filtered) concentrations of T-Hg and MeHg should be measured and the difference between the two taken as dissolved inorganic mercury. There is the possibility that other organic compounds might be present, but concentrations of these should be slight compared to MeHg. Once ID-Hg is calculated, the reactive portion can be estimated using the relationship developed in Phase I.

ID-Hg can also be estimated from unfiltered water column T-Hg, using a K_D; however, this is a less accurate method and involves the questionable assumption that organic forms are insignificant. In addition, if the K_D is not measured for the particular effluent or receiving water, the estimate may be further in error. Thus, this approach is not recommended unless the K_D for the effluent MeHg is known.

3.3.1.3 Site-Specific Assessment of the Impact of DOC and Sulfides on Hg Reactivity

Another source of uncertainty in the screening approach to assessing bioavailable mercury, which may be improved upon by study of the specific effluent and receiving water of interest, is the relationship between reactive Hg, sulfides, and DOC. In this assessment, this relationship was developed by modeling of the Hg-S-DOC interactions, the results of which may not be relevant to all effluents or receiving waters. Due to the treatment processes that occur in a wastewater treatment plant, this relationship may be different than the relationship that might be observed in the receiving water.

In concept, the study is not difficult to perform, requiring paired measurements of R-Hg, ID-Hg, and DOC. Although sulfides are an important part of the overall set of reactions governing this relationship they may be difficult or impossible to measure, especially at the low levels (μM-nM) that may occur in receiving waters. This is not a fatal flaw, however, because their effect will be empirically accounted for in the analysis of the DOC and reactive Hg data. For instance, in evaluating the relationship between reactive Hg and DOC for an effluent versus a receiving water, even though sulfides are not specifically measured (or for that matter any other constituent besides DOC that may affect mercury reactivity), their effects would be factored in to the regression constants, which would reflect a higher or lower reactivity. It may be however, that the relationship between measured DOC and R-Hg is not that strong. In this case, some other factor besides DOC may be exerting a greater influence on R-Hg. If this is the case, it may be necessary to search for other causative factors and develop an entirely different relationship based on or including factors other than DOC.

The function in the Bioavailability Tool used to estimate the reactive Hg (as a percent of dissolved inorganic Hg) from DOC is a declining exponential of the form (with an asymptotic value “a”, which has a value of about 2%):

$$y = a + be^{c(x)} \qquad \text{Equation 3.3.1.3-1}$$

Before manipulating the equation, it is useful to cast it into the following form:

$$(y - a) = be^{cx} \qquad \text{Equation 3.3.1.3-2}$$

The natural logarithm of each side is then taken, resulting in the linear form:

$$\ln(y\text{-}a) = \ln(b) + c(x) \qquad \text{Equation 3.3.1.3-3}$$

When the specific variables of interest are substituted:

$$\ln(f\underline{R\text{-}Hg} - a) = \ln(b) + c(DOC) \qquad \text{Equation 3.3.1.3-4}$$

where:

ln(fR-Hg - a) is the natural log of the fraction reactive mercury (IB-Hg/ID-Hg) minus "a", the asymptote to which the value of fR-Hg declines,

ln(b) is the "y-axis" intercept, which is determined by regression, and

"c" is the slope of the function, also determined by regression.

The natural logarithms of the fR-Hg data are computed and plotted versus DOC. The regression of these values of DOC will yield the regression constants "b" and "c". These new relationships can be put in the Bioavailability Tool by changing the constants in the "Net Methylation" worksheets.

3.3.1.4 Sampling Methods for Effluents and Receiving Waters

All measurements for mercury in the water column should be conducted using Method 1669 (clean hands). If experience in the use of this technique is limited, then field blank measurements should be used to establish whether the sampling is producing a reliable measurement. If contamination is present in field blanks, then the source of the contaminant mercury should be identified and eliminated or minimized before additional studies are performed.

There are two preferred methods for procuring samples for low-level mercury analysis; by bottle immersion or by pumping. If there is good access to the effluent or receiving water, then the best way to take the sample is by immersing the bottle in the water being sampled. The bottle should be opened underwater if possible. It is then filled by allowing all or most of the air to escape. The bottle is then brought near the surface where the contents of the bottle can be gently poured out, reducing as much as possible "gulping" air into the bottle. This procedure is repeated three times (to affect a triple rinse). The bottle is then submerged a final time, filled completely, and capped underwater. A variation of this sampling technique is to use a sampling pole or stick, to allow sampling if the water surface is greater than an arm's length from the sampler. The stick is normally constructed of an inert plastic material with a bucket or clamp at the end to hold the bottle. The disadvantage of this method is that the bottle cannot be open and capped underwater.

Another acceptable method is pumping the sample into the bottle using a peristaltic pump. Clean tubing should be used at each new sampling station and for each new sampling event. Normally, the bottle is opened, filled to overflowing with the water being sampled, and then the contents are dumped. This is repeated three times to affect a triple rinse. The bottle is then filled for a final time and capped. One advantage of this method is that the tubing can be weighted so that the sample is collected at mid-depth, which may not be possible with normal submersion by hand at the surface.

An excellent general reference guide for water quality sampling, with special reference to collection of low level mercury samples, is the USGS' National Field Manual for the Collection of Water-Quality Data (USGS, 2002). This manual has guidance far too voluminous to be repeated here. It is recommended that the reader become familiar with this guidance prior to going to the field to collect water quality samples of any kind.

3.3.1.5 Analytical Suite and Methods

Method 1631E is the currently approved method for low level analysis of total mercury in water. Method 1630 for methyl mercury has been proposed, but as of this writing, had not been adopted as an approved 303(a) method. The fundamental difference in these methods is that in method 1631, mercury in solution is oxidized to divalent mercury by the addition of BrCl, then reduced to elemental mercury (Hg^0) by the addition $SnCl_2$, whereupon the volatile elemental mercury is purged and collected on a gold trap, desorbed, and analyzed by cold vapor atomic fluorescence spectroscopy (CVAFS). In Method 1630, methylmercury in the sample is ethylated by the addition of sodium tetraethyl borate, whereupon the methylethyl mercury is separated from the solution using a graphitic carbon trap, thermally desorbed, converted to Hg^0 and then analyzed by CVAFS.

The method employed to analyze for R-Hg in this study is provided in Appendix A.

Other EPA Methods and Standard Methods for ancillary parameters that may be of interest to the analyst are as follow:

- Total suspended solids - EPA Method 160.2 or SM 2540C
- Dissolved (filtered) organic carbon (DOC) - EPA Method 415.3 or SM 5310 B, C, or D
- pH - EPA Method 150.1 or SM 4500
- Total Sulfate - EPA Method 375.4 or SM 300.0
- Total Chloride - SM 4500.

3.3.2 Effluent Characterization

Effluent is relatively more constant both in terms of flow and chemical composition than receiving waters, as the objective, in general, is to produce treated water of consistent quality. However, there may be periods of high flows, especially if the system is combined (sanitary sewage and stormwater) or there is significant I/I (inflow and infiltration), and periods of low flows as well. The sampling plan may well depend upon the specific objectives of the study being undertaken. For the purposes of this discussion, it is assumed that the objective is to assess bioavailable concentrations or bioavailability of the typical effluent. While "typical" may be interpreted in a number of ways, the intent here is to obtain the concentration of bioavailable Hg that would occur in the effluent when the plant is operating at its normal efficiency within the normal range of conditions under which it was designed to operate.

3.3.2.1 Frequency and Timing of Sample Collection

With this objective, sampling is recommended for a period of at least one year, as removal and speciation of mercury may vary seasonally. For instance, during colder weather, solids removal may not be as efficient due to reduced settling. In summer under elevated temperature conditions, MeHg may be elevated. As was proposed earlier in this chapter, a set of 16 to 20 samples taken over the course of one year (four samples per season) should provide a reasonable "first cut" for assessing bioavailable mercury in an effluent. These samples should encompass a range of flow conditions during each season, all of which should be within the design operating conditions of the plant. The time of day during which sampling occurs should not be a major concern. However, sampling should be avoided during or immediately following maintenance, plant upsets, batch chemical additions, or other such events that may bias the results. Field notes should be kept on conditions (ambient and plant) and any unusual conditions should be specifically noted. Supporting water quality measurements should be made prior to or following sampling with a good quality, recently calibrated, water quality meter. Typical measurements would include temperature, pH, ORP (oxidation-reduction potential) conductivity (or salinity), turbidity, and dissolved oxygen.

Flows should be measured or recorded concurrently so that estimates of mass flux can be made as well. Since mercury levels may depend on solids concentrations and solids concentrations on higher flows, CVs may be higher at plants that experience a higher range of flows and sample numbers may need to be increased to obtain the desired level of confidence in the sample means. However, this can be determined after the initial one year sampling period.

3.3.3 Receiving Water Characterization

In the following discussion of receiving water characterization, both near-field and far-field assessments are covered. In addition, such assessments would naturally benefit from an upstream (or background) assessment to compare to downstream results in order to assess the impact of a specific discharge. While there is no discussion of upstream or background assessment *per se*, it follows that most of the guidance for performing sampling in the near- and far-field discussions would apply.

The selection of sampling locations in a receiving water is a function of the exertion of methylation/demethylation reactions. The discussion will be aided by the introduction of a simple model. This model describes the competing reactions of methylation and demethylation in a waterbody by the differential equation:

$$d(MeHg)/dt = k_1(\underline{R\text{-}Hg}) - k_2(MeHg) \qquad \text{Equation 3.3.3-1}$$

in which:

MeHg is the methylmercury concentration (ng/L),

k_1 is the first-order methylation reaction rate constant (1/day, k_{meth} in previous discussions),

R-Hg is the concentration of reactive mercury (ng/L),

k_2 is the first-order demethylation reaction rate constant (1/day, k_{demeth} in previous discussions), and

t = time in days.

The equation cannot be solved as it is because it has two independent variables, MeHg and R-Hg. However, if the assumption is made that the sum of the concentrations of MeHg and R-Hg is constant, B-Hg[24], then the substitution R-Hg = B-Hg - MeHg can be made and an analytical solution results of the form:

$$MeHg = (MeHg)_0 e^{-(k_1+k_2)t} + k_1(\underline{B\text{-}Hg})/(k_1+k_2) * [1-e^{-(k_1+k_2)t}] \qquad \text{Equation 3.3.3-2}$$

in which $(MeHg)_0$ is the initial methyl mercury concentration (ng/L) and all other terms are as previously defined. The solution has the property that at steady state:

$$k_1/k_2 = MeHg/\underline{R\text{-}Hg} \qquad \text{Equation 3.3.3-3}$$

and the rate of approach to steady state is governed by the sum of the reaction constants, $k_1 + k_2$.

Figure 3-2 shows an example of the solution under two different scenarios. In the first:

B-Hg = 0.3 ng/L

$(MeHg)_0$ = 0.1 ng/L

k_1 = 1/day, and

k_2 = 0.1/day.

---

[24] This is a reasonable assumption for several reasons. As the reactions proceed, reactive, inorganic mercury is converted to methylmercury, but due to the backwards reaction, some methylmercury is converted back to inorganic mercury. While it is true that not all this reconverted inorganic mercury may all become reactive mercury again (e.g. some may be reduced to Hg(0) and lost from the system), much of it may. In addition, this assumption makes the calculations conservative in terms of MeHg production, resulting in a higher MeHg concentration at steady state than if some of the reactive mercury were assumed to be lost. Second, if the concentration of reactive mercury is assumed constant (i.e. is constantly replenished once it is converted to MeHg), the resulting methylmercury concentrations are much too high. Thus the model which assumes a variable concentration of R-Hg produces results more in keeping with the observation that MeHg production in most receiving waters is limited to a few percent. Finally, it will be used here not so much to estimate what the final methylmercury concentration would be in any particular system, but to estimate the speed of the reactions. The model may also appear to suffer from the fact that the concentration of the total bioavailable mercury, B-Hg, is unknown. Again, however, the absolute value of MeHg is not as important as the estimates of reaction speed in this application, and enough is known about concentrations of MeHg and R-Hg in natural waters to make reasonable estimates as to their magnitudes.

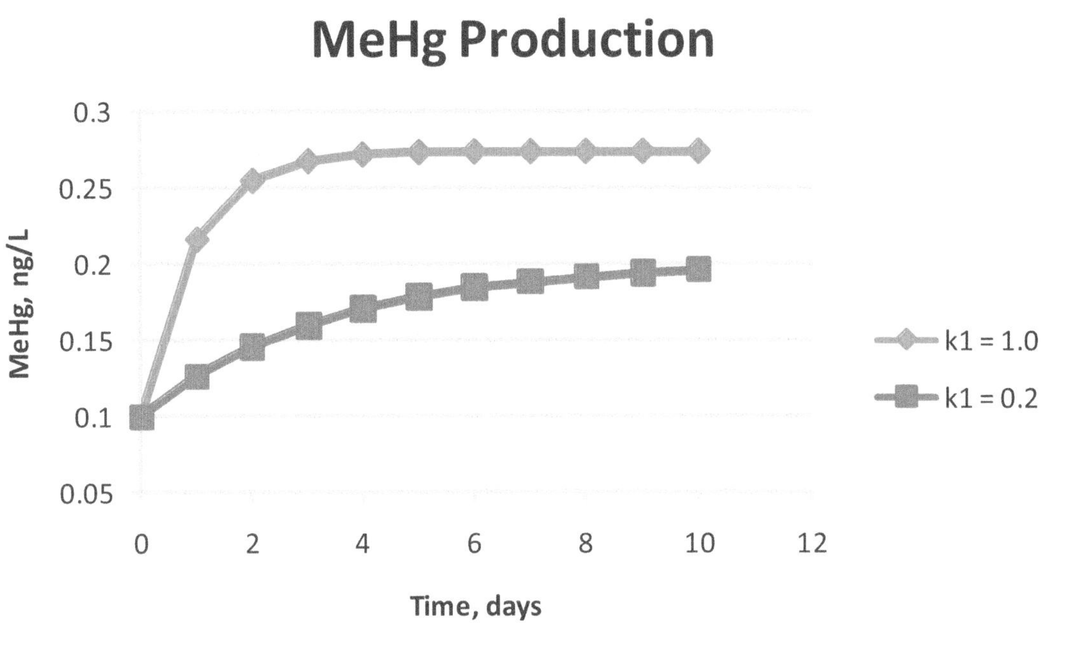

Figure 3-2. Example of Model Results of MeHg Production for Two Different Methylation Rates.

In the second scenario, everything is the same except the methylation reaction constant, which has been reduced to 0.2/day. The first scenario might occur in a fresh water system with a strong methylating environment, where pH tends to be low and DOC higher (> 7 mg/L). In the second scenario, the methylating environment is weaker so the reactions are slower and the amount of MeHg produced is smaller. These conditions might tend to occur in low DOC waters (< 2 mg/L). Note that in the first scenario, the MeHg concentration reaches a near steady-state value in about 2 to 3 days whereas the second scenario requires on the order of 7 to 8 days. In fact, the time required to reach 90% completion of the reactions can be estimated by:

$$t_{0.90} = \ln(0.10) / -(k_1+k_2) \qquad \text{Equation 3.3.3-4}$$

In the first scenario, the calculated $t_{0.90}$ is 2.1 days, while in the second scenario, the $t_{0.90}$ is reached in 7.7 days. In estuaries and marine waters, where the methylating environment is even weaker ($k_1 < k_2$) the reactions would be even slower and the net reaction would favor the loss of MeHg and the production of reactive Hg. However, the speed of the approach to steady state is still governed by the sum of the two reaction constants.

3.3.3.1 Near-Field Assessment

The approach to conducting a near field assessment of bioavailable mercury or bioavailability for a receiving waterbody depends on the receiving water type (i.e. freshwater/riverine, brackish water/estuarine, or marine systems). The term "near-field" means a location where mixing is essentially complete between the discharged effluent and the receiving water and but where time has not elapsed to such a degree that methylation/demethylation reactions in the mixed waters have substantively occurred. As one can easily see, such times and distances might vary among different waterbody types.

Sample Location and Spatial Assessment

Riverine (Freshwater) Environments. In riverine, freshwater environments the challenge is to find a location for sampling that allows one to make measurements after the plume/receiving water is fully mixed. In rivers, the major impediment to achieving thorough mixing is channel width. As the effluent moves downstream from the point of discharge (whether that may be from the side of the channel or via a diffuser in the channel), it becomes mixed to a greater extent across the channel due to lateral dispersion. Thus, the farther downstream from the discharge one looks, the more likely it will be that the effluent and receiving water are well-mixed. However, in areas more distant from the discharge, methylation and demethylation reactions will become more significant as far-field conditions are encroached upon.

A reasonable approach for finding a downstream location to sample to use is to estimate the distance below the discharge where complete mixing across the width of the stream will theoretically have occurred. The expression is (Mills et al., 1985):

$$L_c = 0.4\, W^2\, U / \varepsilon_y \qquad \text{Equation 3.3.3.1-1}$$

in which:

L_c is the distance downstream where complete mixing will have occurred (ft),

W is the width of the river (ft)

U is the velocity (ft/sec), and

ε_y is the lateral dispersion coefficient (ft^2/sec).

According to Fischer (1979), the lateral dispersion coefficient can be calculated by:

$$\varepsilon_y = 0.6\, D\, u^* \qquad \text{Equation 3.3.3.1-2}$$

The friction velocity is calculated from:

$$u^* = \mathrm{SQRT}(g\, D\, S) \qquad \text{Equation 3.3.3.1-3}$$

where g is the acceleration of gravity (ft/sec^2), and

S is the channel slope (ft/ft).

So, for example, in a river with a slope of 0.002 ft/ft with an average depth of 10 ft, the friction velocity would be:

$$u^* = \mathrm{SQRT}(32*10*0.002) = 0.8 \text{ ft/sec}$$

and the lateral dispersion coefficient would be

$$\varepsilon_y = 0.6*10*0.8 = 4.8\ ft^2/sec$$

If it is further assumed that this is a side-entry discharge to a river which has a velocity of 1.5 ft/sec, then the distance downstream of the discharge where complete mixing would occur is

$$L_c = 0.4*(200^2)*1.5/4.8 = 5{,}000\ ft$$

or about 1 mile downstream of the discharge.

If the effluent is discharged to the river through a diffuser, the width used in the calculation above would be the greatest distance to either bank from either the near- or far-end of the diffuser. If, for instance the diffuser is 300 feet long, with the near end 800 feet from the near bank, and the far end 1000 feet from the far bank, the 1000 ft distance would be used in Equation 3.3.3.1-1.

Theoretically, if the sampling location is at or beyond the point at which the effluent is well-mixed with the receiving water, it could take place at any location across the channel and at any depth, since the effluent should be well-mixed both vertically and horizontally. However, the middle of the channel at mid-depth is recommended, as this would minimize any near-shore effects and also any vertical gradients (principally with regard to suspended sediment) in the river.

The selection of sampling location should also consider whether other major point or nonpoint sources (e.g. discharges or tributaries) are present within the river reach from the discharge to the point at which mixing is considered complete. If such intervening sources do occur, and they are likely to affect the near-field assessment, it may be desirable to measure at a location downstream of the discharge but upstream of these interferences. The percentage mixing that would have occurred at an arbitrary distance $L < L_c$ can be calculated by substituting L for L_c in Equation 1 and solving for W. Using the example above, if the percentage mixing that would occur at 2000 ft downstream is desired, simply plug 2000 ft into Equation 5 and obtain a W of 127ft.

Estuarine Environments. In estuaries, the situation may be quite different from that of a river in terms of locating a suitable area for near-field mixing sampling. Typically, estuaries are very wide and therefore long travel times and distances down the estuary to affect complete lateral mixing would be expected. In some estuaries, this may never happen, especially if the discharge is close to the mouth. Even in long, narrow estuaries, where complete lateral mixing may occur, the travel time at that point may be great enough that far-field methylation/ demethylation reactions may be substantially exerted.

Major dischargers to estuaries have likely performed plume migration studies to determine the extent of mixing both up- and down-estuary from their discharge. In this case, a mixing zone may have been defined such that the affected width of the estuary is known at some point up- or down-estuary. If lateral mixing is limited, then measurements would be taken in the mixing zone and several measurements should be taken across the plume width to assure that reliable laterally averaged concentrations can be calculated.

Three mixing scenarios for estuaries will be discussed in the sections that follow:

- the case of a stratified estuary with a discharge above the pyncnocline,
- the case of a well- or partially-mixed estuary, and
- the case of a stratified estuary with a discharge below the pyncnocline.

Stratified Estuary with Discharge above the Pyncnocline - In stratified estuaries, vertical mixing is limited by density differences in the fresher water above the pyncnocline and the denser, more saline water below it. Fresh water inflows are typically high and normally always seaward, while the lower salt water tends to always move landward. In this case, the estuary tends to behave more like a river in terms of chemical transport, with the pyncnocline acting as a "bottom". Since freshwater discharges typically have low conductivity and density compared to salt water, they will tend to disperse in the fresher waters above the pyncnocline if discharged there, much as it would in a river. In this case, the effective depth for sampling purposes should be the depth to the pyncnocline, not to the bottom of the estuary.

If the difference between the time-averaged surface and bottom salinities divided by the mean salinity is greater than 1, the estuary is probably stratified and will have a sharp, well-developed pyncnocline. Samples would be taken across and within the lateral extent of the plume. For a surface discharge, measurements made just above the pyncnocline would demonstrate the greatest degree of mixing.

Well- or Partially-Mixed Estuaries - In a well- or partially mixed estuary, salinity gradients are not as strong as in stratified estuaries and flow reversals occur at all depths with the tide. As pointed out earlier, most major dischargers to estuaries will likely have performed plume migration studies to determine the extent of vertical and lateral mixing. If a plume migration study has not been performed in a well- or partially-mixed estuary, then the extent of the near-field sampling zone around the discharge can be estimated by the following procedure.

In a well- or partially mixed estuary, the velocity induced by tidal action is likely to be greater than that induced by flow. The velocity induced by tidal action is:

$$u_t = V_p/(A_x\, t) \qquad \text{Equation 3.3.3.1-4}$$

where:

u_t is the tide-induced velocity (ft/hr),

V_p is the volume of the tidal prism[25] above the discharge point (ft^3),

A_x is the cross-sectional area of the estuary at the point of discharge (ft^2), and

t is ½ the tidal cycle (6.2 hrs for a semidiurnal tide).

[25] The tidal prism is that volume of water that migrates in and out of the estuary with each tidal cycle. It is typically estimated by segmenting the estuary and adding the increments of volume exchanged in each segment. Segment volumes are estimated by taking the difference between mean high water and mean low water and multiplying by the average water surface areas of the segment at these respective water levels.

The velocity u_t provides an estimate of how far a plug of newly introduced discharge water may migrate up- or down-estuary from the discharge in the time between high and low water or vice versa. The product of this velocity and the 6.2-hour diurnal tidal cycle yields a distance from the discharge point that should contain water recently discharged at its point of maximum mixing with the receiving water. Near-field sampling should occur within this reach.

As an example, it is assumed that an estuary receiving a discharge has a tidal prism above the discharge point of $5x10^8$ ft^3 and a cross-sectional area A_x of $4x10^4$ ft^2. Therefore,

$u_t = 2x10^8/(4x10^4\ 6.2) = 806$ ft/hr.

The distance that the discharged water would have moved by tidally induced flow over ½ the tidal cycle is 806 ft/hr x 6.2 hrs = 5,000 ft or about 0.95 miles. The distance the discharged water moves in a tidal cycle provides the length over which for near-field sampling. Therefore, if sampling is performed within 5,000 feet of the discharge (either up or down the estuary) it should be within a zone where near-field sampling can be accomplished. The nearer the sampling is done to this limit, the greater will be the effects of mixing with the receiving water. If no better estimates of the lateral dispersion of the plume are available, its extent may be estimated in the same way as in the example above for rivers. Sampling should occur across and within the lateral extent of the plume. For a surface discharge, sampling should be done at mid-depth as it will be better mixed with receiving water at mid-depth than at the surface. In addition, a mid-depth sampling location avoids water close to the surface where demethylation might be favored along with lower solids concentrations and also water close to the bottom where methylation and higher solids concentrations may be favored.

Stratified Estuaries with Discharge Below the Pyncnocline - The above procedure applies to a well- or partially-mixed estuary. In a stratified estuary, if the discharge is below the pyncnocline, for instance from a bottom diffuser, the discharge will behave like buoyant plume and a zone of initial dilution (ZID) will develop either within or above it. The preceding analyses are not well-suited to a situation where the plume is trapped or partially trapped beneath the pyncnocline. If the plume emerges into the upper zone above the pyncnocline, the procedure used for a discharge into the upper zone of a stratified estuary can be used to estimate the extent of the near-field mixing zone[26]. Here again, plume migration studies will hopefully have delineated the location and extent of the ZID. Otherwise, a more sophisticated modeling analysis would have to be performed to determine the extent, location, and trapping height[27] of the ZID.

Sampling should occur laterally across the plume and at mid-depth between the surface and the estimated depth of the pyncnocline.

---

[26] In this case, the discharge can be thought to have occurred at the "bottom" of the upper zone, just above the pyncnocline. Vertical mixing will occur from this "bottom" upward and lateral dispersion would be the same as if the diffuser were located there, rather than at the actual bottom. The length of the diffuser, however, would be taken to be the width of the ZID, rather than the actual diffuser length. Lateral dispersion would occur outward from the initial ZID width, and the flow-induced velocity in the upper zone would be used to estimate plume width at the desired distance downstream of the ZID.

[27] As water is discharge from a diffuser at a lower density than the ambient water, it begins to rise to the surface in response to the density difference. This density difference may arise not only from salinity differences but from temperature differences as well. As the plume rise, it entrains saline water into it, and approaches ambient water temperature. At some height, known as the trapping height, the plume becomes neutrally buoyant and ceases to rise. At this point, with its upward momentum lost, it begins to move primarily laterally with the prevailing velocities or currents.

Marine Environments. With marine discharges, it is important that the analyst have access to near-field plume dispersion modeling in order to know the extent, location, and trapping height of the discharged plume. In some cases, the ZID will intersect the surface and sampling can be accomplished there. In other cases, the plume may be trapped beneath the surface and sampling would need to occur within this zone. As in estuaries, measurements should be made laterally across the ZID (parallel to the direction of the diffuser), to obtain a reliable estimates of the concentrations being measured. If the plume has intersected the surface and its thickness is known, measurements can be made at the mid-thickness of the plume; otherwise, they should be made near the surface. If the plume is trapped beneath the surface, measurements would have to be made of the plume at depth to obtain near-field mixed concentrations.

The number of samples required to characterize mercury or other water quality parameters in the ZID can be calculated using the techniques presented in Chapter 2.0.

Sampling Frequency and Timing. Receiving water column measurements can vary widely with season. Total mercury variation is probably more a properly attributed to variations with flow than with season. Total mercury tends to be higher when suspended sediment concentrations are higher and higher suspended sediment concentrations are typically correlated with higher flows. However, in many parts of the country, higher flows are associated with late winter and spring conditions. Thus it may be observed that total mercury will be lower in typically low flow season (late summer, early fall) and will be higher in late winter/early spring.

As reported in EPRI (2003), samples collected during spring high flows in six South Georgia Rivers had higher total mercury concentrations than samples taken at the same locations during summer, low flow conditions. Tetra Tech (2005) observed that total mercury levels were higher in two California lakes in the spring and were associated with higher suspended sediment levels. Rolfhus et al. (2003) found higher total mercury concentrations in Lake Superior in April than in August, which was attributed to tributary inputs during spring snowmelt. Data of the Oregon DEQ for the mainstem of the Willamette River, OR[28] show that total mercury (filtered and unfiltered) where lowest in the fall (0.47, 0.31 ng L^{-1}), increased to their highest levels in winter (3.99, 1.60 ng L^{-1}), and declined gradually back to the fall levels over the spring (1.21, 0.69 ng L^{-1}) and summer (0.69, 0.30 ng L^{-1}) seasons (ODEQ 2004).

Methylmercury varies seasonally as well, probably to a greater extent than total mercury (Domagalski, 1998, 2001; Hurley et al., 1995). Superimposed upon the effects of associations with suspended sediments is the observation that methylmercury increases in the hypolimnion of lakes as summer stratification advances and conditions become more reducing (Tetra Tech, 2005). Similarly, methylmercury concentrations increase in wetlands during the summer, when higher temperatures and lower DO levels favor methylation. Interestingly, as reported in EPRI (2003) however, methylmercury levels can be higher in the spring in rivers. Presumably, this is a consequence of methylmercury being produced in wetlands in the summer and fall during low flows but being flushed into stream channels during higher flows occurring in winter and spring. Similarly, methylmercury concentrations may be lower in the epilimnion of lakes during the summer, but higher after fall turnover when mercury released into the hypolimnion during stratification is redistributed throughout the lake profile. Total (filtered and unfiltered) methyl mercury in the Willamette River exhibited their highest levels in the summer (0.146, 0.038 ng

[28] Geometric means of 7 locations.

L^{-1}) and fall (0.082, 0.071 ng L^{-1}), declining gradually over the course of the winter (0.069, 0.019 ng L^{-1}) and spring (0.040, 0.017 ng L^{-1}) (ODEQ 2004).

Given this seasonality, it is reasonable that total and/or methylmercury measurements should be made over the period of a year with samples taken each quarter to characterize the concentrations in either the waterbody of interest.

3.3.3.2 Far-Field Assessment

By far-field assessment, the researchers refer to the assessment of bioavailable concentrations of mercury measured after a sufficient elapsed time and distance from the discharge such that methylation/demethylation reactions poised to occur as a result of the mixing of the effluent and receiving water would have occurred and in fact have achieved steady-state yielding a new distribution of MeHg and IB-Hg. While this is possible in theory, in practice it may be difficult to identify a suitable location for accomplishing these measurements. The difficulty lies in the probability that other sources may enter the receiving water downstream of the discharge and may change the character of the mixed effluent and receiving water so that the anticipated steady-state conditions resulting from this mixing may not actually be achieved. In such a case, it may not be possible to confirm predicted results in the field and it would be advisable to do the experimentation in a laboratory setting.

Of keenest interest in the far-field is how MeHg and R-Hg have changed relative to their concentrations in the near-field mixed effluent, the effluent itself, or in the receiving water upstream of the discharge. This may be done by measuring MeHg and R-Hg and comparing to those same levels in the other waters/locations mentioned above. The way in which these quantities may be analyzed to assess bioavailable mercury and bioavailability have been discussed in detail.

One additional data analysis technique is worth mentioning, that is the evaluation of methylation and demethylation constants. These can be used directly in calculations or in modeling assessments to make far-field predictions.

It is far easier to do the sort of analysis required in the laboratory rather than in the field as it involves multiple measurements over time, and may involve the use of other techniques such as stable isotopes of mercury. Fundamentally, it is desirable to use mixed effluent, or mix effluent with an appropriate quantity of receiving water in the laboratory, and observe how MeHg and R-Hg levels evolve over time. If one has an idea how long the experiment will take based on prior estimates of reaction times, the experiment can be planned for an appropriate duration and for five or six sampling events during the course of the experiment. So, if the sum the methylation/demethylation reaction constants is expected to be about 0.1/day and the reaction is allowed to proceed to 90% completion (assuming a first-order reaction), then

$$t_{exp} = \ln(0.1) / -0.1 = 23 \text{ days}$$

Thus, the experiment should last for at least 23 days. One set of measurements would be taken at the outset and perhaps another five over the course of the experiment, every four or five days. The analyst may wish to take them more frequently early on, just in case the reactions proceed more quickly than anticipated. As the experiment progresses and the reactions slow down as they reach steady-state, it may be desirable to take them less frequently.

At each sampling, MeHg and R-Hg would be measured at a minimum. However, it would also be desirable to measure T-Hg, and other ancillary water quality parameters which are suspected to have an influence on methylation and demethylation including pH, temperature, DOC, sulfate, chloride, TSS, dissolved oxygen, redox conditions, and nutrients. It may also be desirable to measure both filtered and unfiltered MeHg and T-Hg. As discussed later in this section, the analyst may wish to initially or incrementally change conditions in the experiment as they might occur in the field, or have separate experiments which may reflect these conditions. As the effluent ages and moves away from the discharge point in the receiving waterbody it may be affected by increasing dilution, increasing or decreasing salinity levels, or other changes that may be expected to occur, and it may be desirable to take these changing conditions into account in the experimental design. A number of studies (e.g. Compeau and Bartha, 1984; DiPasquale and Oremland, 1998; Matilainen and Verta, 1995; Miskimmin et al., 1992) have made these types of measurements and a thorough review of the relevant literature in this area would be advised before undertaking one.

Once the data have been obtained, the gist of the analysis to develop rate constants by plotting the changes in MeHg and R-Hg over time. If the reaction is first-order, the natural logarithm of MeHg will plot as a straight line versus time. The reaction constant is the slope of this line. If net methylation is occurring, there should be a corresponding decrease in R-Hg as MeHg increases. If net demethylation is occurring, MeHg will decrease and, if the reaction is first-order, the natural logarithms of MeHg will plot as a straight line versus time, with the slope being the net demethylation constant. In this case, there should be an increase in R-Hg over time.

The basic experiment described above allows the experimenter to assess the changes that occur in the far-field given a particular effluent and receiving water and would yield net methylation or demethylation constants, depending on the direction of the reactions. Other variations on the experiment that must be run to assess separate methylation and demethylation constants. To suppress demethylation, a popular procedure is to run the experiment in the absence of light. Thus photo-catalyzed demethylation is suppressed and only bacterially-mediated demethylation occurs, which is normally at a low level. The methylation constant which would result is closer to the true methylation rate. To assess photo-catalyzed demethylation rates the water can be sterilized to eliminate both microbial methylation and demethylation. Methylmercury is added and the experiment exposed to light. A dark control is used to account for the effects of any other demethylation reactions that may have occurred. More recently popular approach is to add isotopic MeHg and a different isotope of reactive Hg(II), which allows a more direct measurement and assessment of methylation and demethylation rates (Eckley et al., 2005; Eckley and Hintelmann, 2005; Hintelmann et al., 2000; Marvin-Dipasquale, 2003; Mauro et al., 2002; Monperrous et al., 2007; Stordal and Gill, 1995; Whalin et al., 2007).

It is also worth noting that the initial conditions measured in such an experiment represent the conditions which would be prevalent just after mixing occurs and as such would correspond to a "near-field" assessment.

Sample Location and Spatial Assessment. Far-field sampling should occur in a location far enough removed from the discharge that the samples reflect conditions which might exist location after methylation/demethylation reactions resulting from mixing are advanced, versus locations in which they have not yet begun to be exerted. Because of the relatively slow rate constants for methylation and demethylation relative to flows and water movement, it is likely that an appropriate sampling are will be too far removed from the discharge to be practical, except in freshwaters where the methylation reaction can be more rapid. In addition, such an analysis is limited by the dilution that may occur with the effluent in the receiving water. At some point, the effluent will be diluted to the extent that the "impact" would not be capable of being detected analytically. For practical purposes, this could be assumed to be on the order or 10 or 20 to 1.

Riverine Environments. In riverine environments, at low pH and at high DOC levels or in higher %MeHg environments), the methylation reaction may be much faster that the demethylation reaction. As has already been demonstrated, assuming a demethylation rate of 0.1/day, and a methylation rate 10 times the demethylation rate, the time it would take for 90% of the net reaction to be exerted would be:

$$t_{0.90} = \ln(0.1) / -1.1 = 2.1 \text{ days} \qquad \text{Equation 3.3.3.2-1}$$

At a river velocity of 0.5 ft/sec, the distance the water would move in 2.1 days would be about 17 miles. Since it may be impractical to sample this far from the discharge, a closer location by sampling might have to chosen. Practically speaking, the furthest location downstream before the next major source or inflow that would dilute the effluent beyond 10 or 20:1 would the most appropriate sampling point. Given the travel time to this point, the percent completion of the methylation reaction could be estimated. For instance, given the above scenario if a point only five miles (0.61 days) downstream of the discharge was the most practical to sample, the percent completion of the net methylation reaction would be:

$$\% \text{ Completion} = 1- e^{-kt} = 1- e^{-(1.1*0.61)} = 0.49 \text{ or } 49\% \qquad \text{Equation 3.3.3.2-2}$$

In a case such as this, it may be more practical to bring a quantity of mixed effluent and receiving water into a laboratory and allowing it to age for an appropriate length of time. Sampling should occur at intervals throughout the experiment. The water would need to be kept at ambient temperature, spend roughly half the time in a light/dark environment to mimic day/night variations that would occur outside, and be gently stirred to emulate the reaeration that would occur in the natural environment. Such a study might also occur in conjunction with or as part of a biological assessment of the mixed effluent.

Estuarine Environments. In estuaries, methylation reactions will be many times slower than in fresh waters, 0.1 to 0.2 times the demethylation rate, and travel times will have to greater for any exerted reactions to be observed. Practically speaking, if the methylation reaction rate is less than the demethylation rate in the receiving water, then net demethylation will occur and there is little point in performing a far-field analysis, as the highest MeHg concentrations and the greatest impacts would occur near the discharge. As in freshwaters, another very practical consideration in assessing far-field effects is the relative magnitude of the effluent and inflows from other sources, or more precisely, the flux rates of mercury from the discharge and from other sources. Assuming that Hg levels are not much different among the sources, it is relevant to look at the relative magnitude of the flows. If the cumulative flow rates at the discharge point of the other inflows, (R) >> the effluent flow rate (Q_{eff}), then the effects of mixing the effluent with the receiving water will be essentially masked by the conditions existing in the other freshwater sources in the far-field.

Upstream of the outfall, the effluent is distributed according to the salinity profile, and downstream as the fraction of freshwater. If the effluent fraction at the outfall is assumed to be unity, at an arbitrary distance 'x' upstream of the discharge point, the fraction of effluent ($f_{e,x}$) is:

$$f_{e,x} = S_x / S_0 \qquad \text{Equation 3.3.3.2-3}$$

where:

S_x is the average salinity at a distance 'x' above the outfall, and

S_0 is the average salinity at the outfall.

Downstream of the outfall, the fraction of effluent is:

$$f_{e,x} = f_{f,x} / f_{f,0} \qquad \text{Equation 3.3.3.2-4}$$

where:

$f_{f,0}$ is the fraction of freshwater at the outfall, and

$f_{f,x}$ is the fraction of freshwater at a distance 'x' downstream of the discharge.

Assuming that it would be impractical to observe any far-field effects at effluent fractions < 0.1, values of 0.1 can be substituted into Equations 3.3.3.2-3 and 3.3.3.2-4 to estimate the salinities or fractions of freshwater (and hence the distances up-estuary and down-estuary from the discharge) at these points. If the lateral plume dispersion is less than the width of the estuary at this point, a sampling transect reduced to the width of the plume would be appropriate.

The "age" of the discharged water can be no greater (theoretically) than the flushing time of that portion of the estuary above the discharge, as over this time period the fresh water in this portion of the estuary would be entirely replaced[29]. Using a "rule-of-thumb", the "oldest"

[29] The flushing time is the time it takes to entirely replace the freshwater volume within the estuary. However, in this case, the interest is with the flushing time of that portion of the estuary above the discharge point. The calculation is:

$$t_f = V^* f_f / [86400 (R + Qeff)]$$

where:

t_f is the partial flushing time (days),

f_f is the fraction of fresh water in the volume above the discharge point, and

V is the volume water in the estuary above the discharge point.

Obviously, if $Q_{eff} \ll R$, Q_{eff} may be ignored.

effluent would be furthest from the discharge and have an age approximately equal to the flushing time. At the discharge point, the "age" would be zero. Therefore, the average "age" of the discharged water above the discharge should be proportional to the distance from the discharge. At the point where $f_{e,x}$ is approximately 0.1, the age of the water might be expected to be approximately 90% of the flushing time. This age (or elapsed time) can be used in Equation 3.3.3.2-1 with the sum of the reaction rates to estimate the percent completion of the net methylation/demethylation reaction and that time and distance from the discharge.

If it is impractical to measure far-field effects in the field, laboratory measurements may be in order. In this case, mixed effluent may be brought into the laboratory where it would be subjected to the same ambient conditions as exist in the field, to the extent possible. In this case, the analyst would need to consider the change in salinity that would occur as the mixed effluent ages and moves away from the discharge point. Using the techniques described above, the analyst may make an assessment as to how distant the properly aged effluent mixture from the discharge point, either up or down the estuary. At that point in the estuary, the salinity of the receiving water would be assessed. By adding a quantity of fresh or saline water in the laboratory, the experimenter could achieve the conditions that would occur when the effluent reaches $t_{0.90}$. Alternatively, since this effluent, as it ages in the estuary and moves further away from the discharge, sees either increasingly or decreasingly saline water, the salinity of the experiment could be incrementally adjusted over its course to more accurately reflect field conditions.

Marine Environments. Due to the even slower rates of the net methylation reaction and the degree of dilution that will likely occur in the ZID, it is not likely that a far-field assessment will be practical. In addition, since the net reaction will likely be demethylation, the highest MeHg and greatest impact would likely occur in the near-field. Finally, locating an appropriate far-field sampling area in marine waters may be a significant challenge. There are two reasons for this. As alluded to above, the first difficulty is that reaction rates, especially methylation rates, will likely be much lower in marine waters than in freshwater and estuarine systems. Therefore, excursion distances from the ZID to an appropriate far-field sampling area will be even longer. By the time these reactions might be anticipated to be nearly complete, the dilution achieved may be so great that effects of the discharge simply may not be observable. The second difficulty is that the direction the discharged water moves away from the ZID will vary with prevailing currents,

---

Normally, to find the flushing time, the estuary would be segmented. The freshwater volumes in each segment above the discharge would be found by the product of the estuary volume and the fraction of freshwater in each segment. The total volume of freshwater above the discharge would be found by summing the freshwater volumes of those segments. The fraction of freshwater in any segment is determined by:

$$f_{f,i} = (S_s - S_i) / S_s$$

in which:

$f_{f,i}$ is the fraction of freshwater in the ith segment,

S_s is the salinity of the local seawater (parts per thousand), and

S_i is the salinity in the ith segment (parts per thousand).

winds, and other environmental conditions. This is typically not an issue in rivers and estuaries where flow is more constrained; although in some very large estuaries, it could be an issue.

One possibility does recommend itself for study area selection; that is, any environmentally sensitive areas nearby where enhanced methylation due to the presence of wastewater might be an issue - for instance, a shellfishing area. In this case, the discharger is probably already aware of its whereabouts and outfall studies may already have assessed the potential impact of other pollutants (e.g. coliform bacteria) on the area and predicted dilutions of the wastewater in the potentially impacted area may already be available. The recommended means of approaching the study would be to perform it in the laboratory, where the effluent sample would be diluted to the proper ratio with seawater. The study would be conducted until the estimated $t_{0.90}$ time is reached for the net methylation reaction. If the dilutions are such that the effects of the wastewater may be unlikely to be measurable, the study may not be feasible. As a worst case approach, diluted effluent from the ZID might be aged and sampled to assess whether there is any enhanced methylation there versus in a seawater control.

Sampling Frequency and Timing. The same frequency and timing considerations apply to far-field sampling as to near-field sampling for evaluation of bioavailability. While summarized here, the reader is referred to Section 3.3.3.1 for a more detailed discussion. Given the documented seasonality that exists with T-Hg and MeHg levels, and suspected seasonality in R-Hg levels, it is reasonable that total, methyl and/or reactive mercury measurements should be made over the period of a year with samples taken each quarter to characterize the concentrations in either the waterbody of interest.

3.4 Biological Estimation of Bioavailability

The concentration and speciation of mercury in wastewaters, other sources, and receiving waters can be highly variable, impacted by both physical processes (e.g. changes in river flow, particulate loading) and biological processes, primarily at the base of the aquatic food chain. This variability is a function of many factors and is different for different types of water bodies, at different times of the year, and in different locations. Thus, it is often not easy to ascertain the average mercury concentration and exposure for a particular environment, because it is often difficult to obtain the necessary number of samples required to provide a statistically defensible result. For this reason, many approaches have been used that “integrate” the chemical signal over time, and perhaps space, and allow for a limited number of measurements to reflect the average situation. While there are emerging technologies that allow for the *in situ* collection and/or measurement of an integrated chemical sample, these approaches have not developed to a sufficient degree that such approaches can be used in the current situation covered by this project and report.

An alternative approach to the measurement of the bioavailable concentration of a compound in waters is to use a biological indicator (often referred to as bioindicators or biosentinel organisms). Fish are an obvious example and are mostly used as the metric for assessment of impairment under a regulatory framework. In choosing a biosentinel organism, it should be noted that the lower the organism in the food chain, the simpler the link to changes in the water concentration, and, because of organism size, the more rapid the response to changes in water concentration. As noted by Wiener et al. (2007), zooplankton respond to changes in water concentrations over weeks, while juvenile fish can reflect changes in concentration on a seasonal to annual basis. Predatory fish will take years to reflect a change in water concentration, and therefore integrate a signal over a large timeframe.

However, the lower trophic level organisms are removed to a larger degree from the typical human health screening organism, predatory fish. Thus, the choice of organism is clearly driven by the specific aspects of the question being addressed. For example, the U.S. EPA uses a fish concentration as a water quality criterion for Hg and while such an approach simplifies the determination of an impairment and makes this a relatively simple task (i.e. the measurement of the fish Hg concentration in a number of representative species), it is not easy or simple to convert between a fish concentration and a water MeHg concentration. The U.S. EPA has spent considerable time and effort developing the protocols for translating a water concentration from a fish tissue value, and has only provided guidance on this in recent documentation. Thus, there are approaches for interpreting the fish tissue concentration but these often require the collection of additional field chemical data. In the context of this project, for example, it seems possible to determine using a biosentinel organism whether there is more bioavailable Hg or MeHg below an outfall, as this should be recorded in a higher fish concentration. This would give a quick yes or no answer to the question, but not an answer regarding why, and the answer will likely require further confirmation.

For instance, it is possible that this higher fish concentration may be related to differences in organism growth rate or system productivity (i.e. the ecosystem dynamics or the organism's bioenergetics) between sites, which can also have a substantial impact on tissue concentration; and the reverse is also possible - no fish concentration change even if water concentration increases. Furthermore, if an increase is determined, it could be due to 1) increased MeHg from the source; 2) increased bioavailable Hg from the source which has been methylated; or 3) inputs of other constituents (DOC, sulfate) that has enhanced methylation downstream of the source. Designing a mitigation strategy to combat the increase would require a different approach to identify and mitigate against the specific driving factor controlling increased bioaccumulation. Thus, while using sentinel organisms is one useful approach, they cannot always be used in isolation. Indeed, the information gained from biological monitoring differs depending on the type of methodology used and the organisms deployed/analyzed. The various approaches will be briefly reviewed below before detailed guidance is presented on the preferred approaches to biological monitoring of Hg and MeHg bioavailability.

3.4.1 General Study Considerations

There are five general (broad) groups of aquatic biological indicators, which in decreasing size are: 1) piscivorous (top predator) fish, 2) prey fish, 3) benthic invertebrates, 4) zooplankton, and 5) phytoplankton, other algae (e.g. periphyton) and bacteria (Fisher and Reinfelder, 1995; Wiener et al., 2007; Wiener et al., 2003). It is also possible to use aquatic mammals and birds as indicators (Wolfe et al., 2007) and while these will be briefly mentioned, it is not likely that these organisms would be used under the current project scope and questions, as there are many other potential confounding factors that can impact the relationship between bird or mammal tissue concentration and water column MeHg or bioavailable Hg concentration. As piscivorous and other large fish are the species that are mostly consumed, and are a cause for concern in terms of exposure of humans and other wildlife, they provide a useful metric for regulatory control (Wiener et al., 2007). However, because of the many steps in the food chain and other confounding factors, they are indeed not the best choice in terms of determining the bioavailable Hg and MeHg concentration in a particular waterbody, or in the specific context of comparing an effluent's input relative to other sources. As concluded elsewhere (Mason et al., 2005; Wiener et al., 2007) they are not the preferred organism for examining changes in inputs to a particular system over time because of their slow response to changes in concentrations and/or loadings. Using organisms lower in the food chain results in a more direct link between the biota and the water concentrations (Fisher and Reinfelder, 1995; Wiener et al., 2007; Wiener et al., 2003). Thus, the best bioindicators are either microorganisms (phytoplankton, bacteria) or the organisms that feed on them (zooplankton, filter feeders, herbivorous, planktivorous or benthivorous fish). Sampling and collection of phytoplankton is difficult as the suspended particulate phase is a mixture of living and dead microorganisms, inorganic material and other detritus, and therefore *in situ* sampling of this fraction is seldom a useful metric.

In defining the best bioindicator species to answer the questions posed under this project, the researchers suggest it would be an organism that accumulates either inorganic Hg or MeHg or both from its environment (from water or pelagic organisms/particles, preferably) in such a way that there is a significant and relatively simple relationship between concentration of the chemical in the organism and that in the surrounding waters. Controlled studies show that many organisms can fit such criteria (Harris et al., 2007b; Orihel et al., 2006). In addition, the advantage of a bioindicator is that the accumulation is continuous and the average tissue concentration after exposure is related to the average bioavailable concentration in the water rather than the total concentration, or some short-term fluctuation in concentration. Thus, such organisms should be extremely useful in any attempt to ascertain the bioavailable concentration of Hg species in a particular environment. Finally, from a practical point of view, the best organisms should be abundant, hardy, and cosmopolitan in the environment, easy to catch and handle, but be responsive to changes in Hg and MeHg (Harris et al., 2007a; Wiener et al., 2007). The criteria and choice of such organisms has been discussed recently in various publications (Mason et al., 2005; Saltman et al., 2007), and while the application is somewhat different in this case, the overall results and criteria from these studies that have merit will be considered, along with other literature sources, in the development of this guidance document. As will be discussed below, there are two approaches that are typically used: 1) sampling the *in situ* population, and 2) deploying "clean" organisms from elsewhere and monitoring the change in their tissue concentrations over time.

Table 3-3 provides a summary of these discussions and broad conclusions about the relative usefulness of the various approaches to examine Hg bioavailability and bioaccumulation in effluents and receiving waters. These are conclusions for the "average" system and clearly there will be some situations where one of the approaches deemed to be not suitable overall may, in fact, be the best approach. The reader is encouraged to reflect on the various approaches before making a final decision. The details of each approach for recommended organisms are provided in the remainder of this chapter while the pros and cons of each bioindicator or biosentinel species (or group of organisms) for assessing bioavailability are given in Appendix E.

3.4.1.1 Direct Biological Measurements of Bioavailability

Microorganisms. The use of phytoplankton to examine the relative bioavailability of inorganic Hg and MeHg for different waters and the impact of dissolved constituents has been widely used for the examination of metal uptake and accumulation at the base of the food chain. Examples of such studies are provided in Appendix E. While the approach and details of how such studies are done are presented below, as noted in Table 3-3, these types of studies are not recommended, in general, in the context of this project.

Table 3-3. Bioindicator Organism Summary Table.

Organism	Approach	Complexity	Recommend
Phytoplankton, Microorganism	Uptake measurements with water from field. Lab-based study	Medium	Perhaps
Seston	Field collection of in situ organisms/material	Low	No
Bioreporters	Lab examination of bioavailability	High	No
Zooplankton	Collection of in situ organisms	Low	Yes
Filter Feeders	Collection of in situ organisms Transplantation of caged organisms	Low Medium	Yes Yes
Benthic infauna	Collection of in situ organisms	Low	No
Prey Fish	Collection of in situ organisms Transplantation of caged organisms	Low Low/Medium	Yes Yes
Predator Fish	Collection of in situ organisms	Low	No
Mammals, birds	Collection of organisms or blood, feathers or other tissue	Medium	No

In performing such uptake experiments, it is possible, if both Hg and MeHg are monitored in the organism, to determine the relative uptake rates for both forms of Hg. Overall, in terms of uptake into phytoplankton and other unicellular organisms, the process can be parameterized as:

$$\text{Uptake rate, V (mass cell}^{-1}\text{ s}^{-1}) = kC \qquad \text{Equation 3.4.1.1-1}$$

where k is the first-order uptake rate parameter, which incorporates the rate of active or passive transport across the membrane (an uptake velocity, typically in cm sec^{-1}), as well as the cellular surface area (cm^2), and has units of cm^3 sec^{-1}, and C is the concentration of the species of interest in solution (mass cm^{-3}). If there were more than one bioavailable mercury species in solution then the overall uptake would be:

$$V = \sum k_i C_i \qquad \text{Equation 3.4.1.1-2}$$

for each individual complex, i. Thus, if the exposure time and concentration of the bioavailable form(s) is known, then the uptake parameter can be estimated from the results of a laboratory study (Mason et al., 1996). If the cells are growing and dividing at a particular rate, then the steady state cellular concentration is determined by the ratio of the uptake rate to the growth rate, or:

$$\text{Cell concentration (quota), } C_q = V/k_g \qquad \text{Equation 3.4.1.1-3}$$

where k_g is the cellular growth rate ($time^{-1}$). The steady state cellular concentration, C_q, is directly related to the concentration of the bioavailable species in solution, and the values of the various parameters needed to understand the relationship can be either determined in the laboratory using controlled studies, or can be found in the literature (Fisher and Reinfelder, 1995; Mason et al., 1996; Pickhardt and Fisher, 2007). In the absence of such information, it is possible to examine the relative bioavailability of different waters through comparative exposure experiments. As noted above, these exposure studies can be used to examine both inorganic Hg and MeHg bioavailability separately, or just the overall accumulation. In addition, if the experiments are designed to examine only the relative uptake, then *in situ* concentrations, if low, can be amended through the use of stable or radioactive Hg and MeHg spikes (Pickhardt and Fisher, 2007), as long as the concentration used is not substantially elevated above the background concentration, as there is some evidence that uptake may become slower (due to saturation of uptake mechanisms) at high concentrations.

An example of such a study is shown in Table 3-4 (Pickhardt and Fisher, 2007). In this study, the uptake of Hg and MeHg (added as isotope spikes into filtered water collected from different sites) was compared and contrasted. One site (FT) had higher DOC (280 μM, 3.4 mg/L) and higher pH (7.9) than the other (CR) (177 μM or 2.1 mg/L and pH=6.8). It is apparent from the results that MeHg is taken up by all plankton species more readily than inorganic Hg. The ratio between the two sites was relatively similar amongst species, suggesting that uptake rate, when corrected for size (i.e. the concentration factor), was relatively independent of species. Overall, the cellular burden or the concentration factor was more similar for the two sites for inorganic Hg than for MeHg. These results reflect differences in the speciation of inorganic Hg and MeHg in these waters, and differences in their mode of uptake. Such an approach is adapted easily to a bioavailability study and is the basic approach detailed below.

The uptake experiments can be performed using either the *in situ* environmental phytoplankton species, or can be done in the laboratory under more controlled conditions using water collected in the field and laboratory phytoplankton cultures, as done by in the above example (Pickhardt and Fisher, 2007). The latter approach is more controllable and is the approach that generally is used, and is recommended here.

Depending on the environment (fresh, brackish or saline waters), there are a variety of phytoplankton species available for different waters and salinities. The study above examined different species and related their relative uptake rates to the surface area, and their concentrations to the cell volume. Different uptake dynamics and C_q values will be found for different size phytoplankton species (see Table 3-4).

Table 3-4. Mean Concentrations (nmol g^{-1}) and Volume Concentration Factors (VCF) for All Species at Steady State. Adapted with permission from Pickhardt and Fisher, 2007.

Species	Inorg. Hg CR	Inorg. Hg FT	Ratio	MeHg CR	MeHg FT	Ratio
Burden (nmol g^{-1})						
Cyclotella	23.2	9.7	0.4	92.4	197.5	2.1
Chlamydomonas	41.8	38.4	0.9	317.2	602.8	1.9
Cryptomonas	147.1	166.0	1.1	176.3	450.5	2.6
Synechocystis	93.5	114.8	1.2	511.2	347.1	0.7
Concentration Factor	x 10^4					
Cyclotella	0.86	0.43	0.5	14.5	37.8	2.6
Chlamydomonas	2.08	2.39	1.2	31.0	64.4	2.1
Cryptomonas	4.36	4.95	1.1	12.7	31.7	2.5
Synechocystis	1.74	2.19	1.3	144.6	146.1	1.0
Range			*0.4-1.3*			*0.7-2.6*

Notes: *Cyclotella meneghininiana* is a diatom, *Chlamydomonas* reinhardtii is a green alga, *Cryptomonas ozolini* is a cryptophyte and *Synechocystis* is a blue-green alga.

Finally, it is possible to examine not only the steady-state bioaccumulated concentration but also the initial rate of accumulation. Many of the phytoplankton that can be used in such studies have been well studied, exist in laboratory culture and can be purchased, and their growth kinetics and specifics of culturing in the laboratory are well established. It is necessary to determine whether any of the ingredients of the culture medium (e.g. required nutrients, essential trace metals, vitamins), added to the water from a concentrated stock solution can affect the speciation of Hg and MeHg in solution. Of the typical ingredients, only the addition of EDTA, which is added to help control the essential (nutrient) metal's concentration, may impact Hg and MeHg speciation. As noted in the above-mentioned study (Pickhardt and Fisher, 2007), it is possible to culture the phytoplankton in EDTA-free medium, and this is the recommended approach.

Reduced organism growth rates may indicate toxicity but this is unlikely in terms of Hg and MeHg for natural waters and effluents. It is not known if there would be difficulties in growing phytoplankton in wastewater effluents, and this would need to be investigated in each case. Another approach, if this is a problem, would be to do a series of dilution studies where phytoplankton are exposed to different mixtures of effluent and receiving waters. If cellular concentration increased with increased amount of effluent in the mixture, and growth rates did not change, this would indicate bioavailable Hg or MeHg in the effluent stream (Van Veen et al., 2002). Note from the equations above that comparisons need to account for any differences in growth rate as a cell growing at half the rate will have double the cellular concentration at steady state. So, growth rates need to be measured in all cases to ensure that they are the same or to correct for any differences that may occur.

The use of phytoplankton to examine the uptake of MeHg is clearly reasonable but it may be more difficult to translate the uptake of inorganic Hg by phytoplankton into relative bioavailability in terms of Hg methylation. If uptake is dominated by passive processes this should be a reasonable assumption. To recall the premise, the bioavailable inorganic Hg is important as this is the form that is likely to be readily methylated in the receiving waters. Thus, it may be more appropriate to examine bioavailability of inorganic Hg using bacteria, and specifically the bacteria that are important for Hg methylation. This has been done in a number of studies, and in this case the production of MeHg from added inorganic Hg is taken as a surrogate for the relative bioavailability of the different waters (Benoit et al., 2003; Benoit et al., 2001). This approach however is not as directly applicable as most effluents and receiving waters are oxic while the growth of the methylating organisms mostly occurs in anoxic waters and sediments. The use this approach requires substantial manipulation of the waters to be examined and thus is not recommended.

It is possible that a chemical measurement, such as an octanol-water partitioning assay, can be used as a surrogate, as shown by the studies of Benoit and colleagues (Benoit et al., 2001; Benoit et al., 1999b). Again, however, this approach is not recommended here as it requires a complex experimental design.

Biosentinel Organisms. Biosentinels are good indicators of MeHg bioavailability and bioaccumulation as they magnify the signal in the water and integrate it over time. Overall, trophic transfer factors between predator and prey are in the region of 80% for MeHg, and <20% for inorganic Hg (Fisher and Reinfelder, 1995; Mason, 2000). Thus, organisms high in the food chain bioaccumulate most of their Hg as MeHg while the lower trophic organisms have some lower fraction of their Hg as MeHg, and this is especially true for bivalves and other herbivorous/detrital organisms. These lower food chain organisms may better reflect the relative bioavailability of inorganic Hg for uptake into the food chain, but they do not provide any indication or information about the bioavailability of inorganic Hg for methylation. Clearly, however, with the strategic deployment or sampling of sentinel organisms it may be possible to ascertain this, especially if the biological monitoring is coupled with some water column sampling.

The major concern with the use biosentinels in this context is that most of these organisms will reflect bioavailability in terms of MeHg but their response to, and reflection of, bioavailable inorganic Hg is less clear. Recall that by the definition used, all MeHg is considered bioavailable and thus there is a reasonable likelihood that an organism feeding on suspended matter will provide an indication of changes in the overall MeHg concentration. In contrast, only dissolved inorganic Hg is considered bioavailable to and methylated by sulfate reducing bacteria, and thus organisms feeding on suspended matter or algae may not adequately reflect the availability of this fraction to sulfate reducing and other methylating bacteria. The solution to this potential issue probably lies in the sampling strategy. If more than one site is chosen downstream of the source and mixing, then it is possible to obtain an assessment of the bioavailability immediately upon mixing (near-field), and then of the bioavailability after steady-state has been achieved downstream (far-field). Comparison of these two measures would allow an assessment of the degree to which any bioavailable inorganic Hg has been methylated after mixing.

In terms of the overall approach, the simplest method is to sample the *in situ* population of the organisms of interest and assume that these measurements are reflective of the relative differences in concentration between sites. This approach is valid although there could be specific factors related to a particular site that could have influence on the rate of bioaccumulation and tissue concentration. There will be a lag time in the response of the organism to changes in concentration which will reflect the particular response time of the organism to change, which corresponds in most situations to the rate of elimination of the contaminant. The steady state concentration is directly related to the ratio of the uptake and elimination rate constants, and in most cases, and especially for MeHg, elimination is much slower than uptake. Indeed, for MeHg and most organisms, it is not elimination that reduces the burden but growth, the other factor that can result in a decrease in contaminant concentration. Thus, large organisms such as predatory fish, do not respond quickly to changes in water column concentration, having a response time of years (Wiener et al., 2007). In contrast, some zooplankton, such as copepods, can respond to changes within days to weeks given their rapid growth and short lifetime (a few weeks typically).

Another complication with using *in situ* organisms is that the approach relies on examining differences in the steady state concentration, which may be relatively small compared to the actual concentration. To alleviate such problems, it is sometimes advantageous to use organisms taken from a relatively clean site transplanted to the site of interest. In this instance, the rate of accumulation can be followed until steady state is reached and this provides an additional level of information that may be useful in the interpretation of the data. Such transferred organisms are often kept in a defined area (i.e. caged) upon transplantation and this general term will be used mostly to refer to organisms that are transplanted. While transplantation has advantages, there are often problems with the caging for transferred organisms in terms or organism growth and viability. Generally, the organisms most suitable for transplantation are relatively small, short-lived, primary consumers so that their response to the water column concentration is rapid. However, they need to be relatively robust, and for this reason, zooplankton are not recommended and the most used species are either bivalves or other invertebrates or juvenile fish.

As noted above, and this is important in determining the type of organism to use, biotic response to changes in concentration is overall a function of size and growth rate of the organism. Zooplankton will respond in days to weeks to changes in water/food concentration as they grow quickly and have a short lifetime but are not suitable for transplantation (Wiener et al., 2007; Mason et al., 2005). Filters feeding organisms, such as mussels and clams, will take a month to a few months to reach steady state, depending on their size and growth rate. Small fish respond to longer term (seasonal to yearly) changes. Large fish respond on an even slower basis. Birds can show seasonal effects, as blood can be used as a short-term measure. Larger mammals would respond more slowly (Wolfe et al., 2007).

Given this background, some specific comments about each recommended class of organism will be given. These discussions are focused on the organisms deemed most useful for this study.

Zooplankton and Other Small Herbivores. As noted in Appendix F, these organisms are widely distributed, common, and important in pelagic food webs. However, there are only a few zooplankton species that are truly cosmopolitan and can be used for monitoring purposes (Wiener et al., 2007). For freshwaters, daphnia are often used and in marine environments, genera such as *Arcatia* are abundant. The life cycle of zooplankton is short so populations will show a rapid response to changes in concentration in the water column - within hours or days (Herrin et al. 1998, Paterson et al. 1998, Tsui and Wang 2004, Mason and Lawson 1998). Zooplankton can accumulate significant quantities of MeHg from food (typically algae) and/or water (Monson and Brezonik, 1999; Peech Cherewyk, 2002; Tsui and Wang, 2004; Lawrence and Mason, 2001) which is a potential confounding factor.

However, even if zooplankton accumulate both Hg and MeHg from water and food, the dominant uptake route is food, and the overall integrated signal will still be reflective of the water column, except if food sources between sites are mixed. This is a complication if concentrations in food vary widely. Zooplankton are readily sampled but samples should be checked and processed to remove phytoplankton and detritus, and it is recommended for studies of Hg that the organisms are "picked" individually under a light microscope or visually for larger species (Wiener et al., 2007; Mason and Sullivan, 1997; Watras and Bloom, 1992). Zooplankton can be separated from much of other seston using a large mesh filter (>50 μm typically used) and a single bulk sample from a plankton net can be used to characterize the open-water community of zooplankton at a particular time. The processing of samples can require some effort if picking of individual species or genera is done (which would provide the most robust signal).

An alternative to zooplankton in freshwater environments is to use freshwater insect larvae, although these organisms can have a variety of feeding strategies ranging from being herbivores to carnivores (e.g. Mason et al., 2000). However, such approaches have been used in relatively pristine environments to monitor the differences in concentration and bioaccumulation of MeHg between streams in Western Maryland (Mason et al., 2000), in the Sudbury River (Naimo et al., 2000), and in many other studies. Clearly, if the feeding behavior of the organism is well-known, then these organisms which are often widely distributed provide a useful organism for *in situ* studies. There is little information on their use as transplanted organisms.

Another approach worth mentioning is to perform a series of dilution experiments of receiving water with effluent and monitor the increase in accumulation and/or toxicity (Figure 2-2), such as those conducted by Van Veen et al. (2002). Even though these were toxicity rather than bioaccumulation studies, the approach is still valid. These authors were able to demonstrate that the toxicity decreased, as reflected in an increased LC_{50} value, in concert with the increased mixing of effluents with receiving waters, the increase in complexation capacity of the mixture, and the resultant decrease in bioavailable Cu (i.e. more Cu was needed to be added to cause toxicity). In the figure, there is an increase in the LC_{50} (i.e. the concentration required to kill 50% of the organisms) with increasing complexing capacity in the water. In other words, as the complexing capacity increases, more copper is needed to cause toxicity. Such an approach of using a mixing experiment to ascertain the impact of complexation could be useful for examining effluent bioavailability for Hg and MeHg, as the increased complexation would result in less bioavailable Hg.

Filter Feeding Bivalves. As noted above there are two broad approaches to assessing bioavailability using sentinel organisms - using the *in situ* organisms or using transplanted (caged) organisms - and either approach can be recommended, or the approaches can be used in concert. Each approach has advantages and disadvantages, and the results obtained from the two methods may be somewhat different. However, for example, in a recent study of the Sacramento Waste Water Treatment discharge, where both transplanted and resident clams were used in the monitoring, the results were similar for both approaches (LWA, 2007). Another example of the

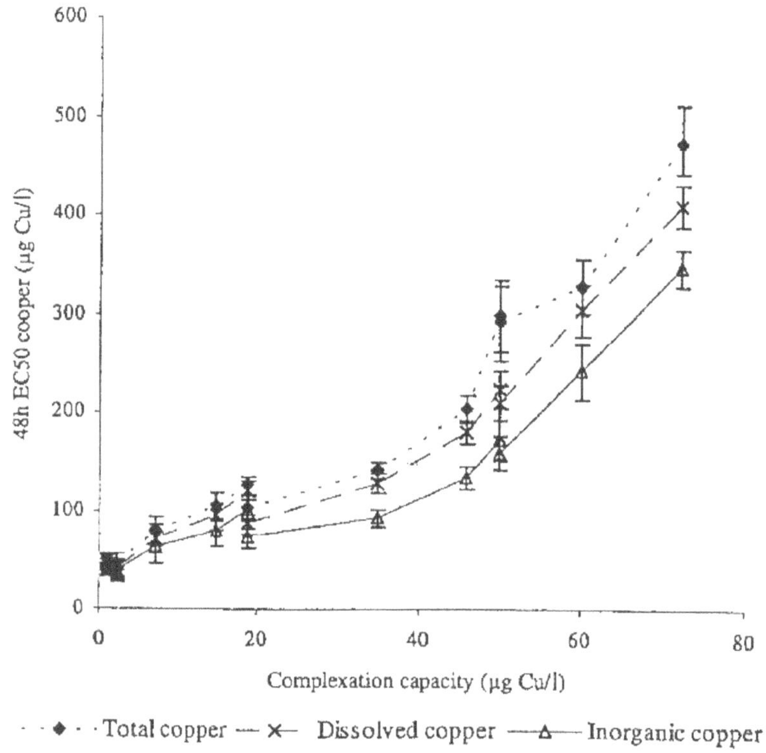

Figure 3-3. The Impact of Copper (Cu) Complexing Ligands on the Toxicity of Cu to Zooplankton. Source: Van Veen et al. (2002). Copyright 2002, Environmental Toxicology & Chemistry, Allen Press Publishing Services. Reprinted with permission.

use of transplanted bivalves is the studies done by Beckvar et al. (2000) using transplanted freshwater mussels (*Elliptio complanata*).

Transplanted and resident clams are likely to be collected by different methods, depending on the deployment strategy for the caged bivalves. It should be possible to have a constant number of organisms for analysis from the caged experiment, unless mortality after deployment is high (which is clearly undesirable). For field collected organisms, while desirable, this may not always be possible. Based on a review of various studies, the recommended number of organisms per sample is 30 for the Hg and MeHg analyses, and triplicate samples should be obtained from each location (i.e. a total of 90 organisms), plus any additional organisms needed to determine growth and other physiological parameters. It may be desirable to add additional organisms to the cages to account for any mortality during deployment. The organism size should be as consistent as possible, and the size range small, to alleviate the impact of size on concentration.

The deployment of caged (transplanted) bivalves can take a variety of approaches and so some basic suggestions are made to guide the approach in a specific situation. For *Corbicula*, clams in the 15-30 mm size range are optimal. For the Mussel Watch program, 5-8 cm is the size range for mussels, and 7-10 cm for oysters, which were collected when mussels were not available. Freshwater mussels are also suitable organisms is present in sufficient abundance (Bekvar et al., 2000). Organisms should also be collected from one or more reference site(s) where there are known or suspected low Hg and MeHg concentrations to contrast the concentrations found at the studied location. Bivalves should be kept cool and damp and be transported and deployed as soon as possible. Depending on the situation, it may be desirable to hold the bivalves in a clean water system (aquarium tank) to allow them to depurate prior to deployment. If this is done, the aquarium should be monitored for ammonia and dissolved oxygen levels to ensure that there is not a toxic buildup or die off of the bivalves. The water should be changed daily, if possible, and aerated. At deployment, bivalves should be thoroughly mixed to ensure that the time-zero samples represent the entire population of deployed bivalves. For deployment, the bivalves should be placed into individual compartments in plastic cages, For *Corbicula*, the methods of Foe et al. are a reasonable guideline (Foe, 2003) as are the approaches of Bekvar et al. (2000).

Depending on water depth and mode of deployment, cages can be deployed using a boat, diver, or by hand. A reasonable configuration is to deploy a suitable weight, use pilings or other permanent structures, or a secure attachment to the sediment surface, with a line running to a surface buoy for site location. The individual cages, containing the required number organisms attached to a line are deployed in the water column near the sediment interface (1 to 2 m off the bottom, again depending on overall water depth). Deploying the organisms above the sediment surface lessens the contribution of sedimentary material to the signal recorded by the bivalves and ensures that they more accurately reflect the water column concentration. Deployment is this fashion has been shown to have little impact on their growth and survival.

At each sampling station, three cages of bivalves are removed to provide replicates of 30 homogenized individuals for Hg and MeHg analyses. All collection, handling, deployment, and retrieval needs to be performed using clean sampling techniques, which involve the use of polyethylene gloves, and pre-cleaned storage and sample containers. The bivalve cages should be removed from the cages onsite and the organisms placed in a plastic mesh bag and stored in a clean cooler, containing river water, for transport to the laboratory, where they should be depurated prior to shucking and storage of the tissue material for analysis. Depuration, which involves placing the bivalves into clean, aerated containers until the water is clear and the organisms are no longer discharging fecal pellets, ensures that the analyses of Hg and MeHg are not biased by the gut contents of the bivalves which could contain elevated Hg and MeHg. It is also probably useful to measure other metrics besides Hg concentration to estimate whether the bivalves grew during the deployment period. A simple dry weight measurement or other more complex measures of growth and body index can be employed to ascertain this. The condition of the organisms can be ascertained by the ratio of the dry tissue weight to shell length. This index integrates changes in clam condition on time scales appropriate to this type of study of weeks to months (Norkko et al., 2005).

A variety of approaches can be used in bivalve collection. *In situ* (resident) bivalves can be collected by dredging sediments, collecting a box core, or hand-collected by divers, or for mussels in the intertidal zone, by hand. If both caged and *in situ* organisms are to be sampled, the caged samples should be taken first as the collection of *in situ* organisms could lead to sediment resuspension. A similar number of *in situ* bivalves should be sampled as outlined above for the caged organisms, and they should be of similar size. Specimens should be removed from the sampling site as cleanly as possible and the organisms should be depurated and processed as discussed above for the caged bivalves.

Removal of tissue should be done in a clean environment and the equipment used for obtaining the samples should be plastic, rinsed with acid and distilled water between samples. Samples should be homogenized in an ultra-clean laboratory space. A blank for the homogenization process can be obtained by running ultra clean (18 megaohm) water through the equipment prior to homogenizing of the actual tissue samples. After being homogenized, samples can be stored frozen in a plastic jar until analyzed.

Small (Prey or Forage) Fish. As noted in Appendix F, the most useful fish species for use as biosentinels are the many small species of fish that forage on zooplankton and other small organisms at the lower end of the food chain (Wiener et al., 2007). These small fish are usually short-lived finfish, such as minnows, perch and the sunfish families (Mason, 2002). In estuarine waters, mummichogs and killifish are typically used. These fish are widely distributed and common. They typically feed on zooplankton, benthic invertebrates, and periphyton (Bodaly et al., 1993). Mercury concentrations in prey fish are a useful indicator of relative MeHg levels in food webs and there is a growing literature on MeHg concentrations in these organisms. Prey fish require low to moderate sampling effort, with samples taken in the early spring or fall, when growth is slow and temporal variation in mercury concentration is small; thereby reducing sample variance. Small fish are recommended for the specific application discussed here and while it is likely better to use *in situ* (resident) species it is also possible to use these fish in caged studies. However, they will respond to changes in concentration more slowly than filter feeding bivalves and therefore will likely need to be deployed for longer periods. This may be a constraint on their usage given that their mortality rate in cages is likely to be higher. Thus, overall, the collection of resident fish is the preferred approach. Many small fish, such as spottail shiners, have strong site fidelity and can be used to examine spatial patterns in accumulation of MeHg in receiving waters at spatial scales less than 1 km (Choy et al., 2008).

As noted above, a sample of 30 individuals gives a level of statistical certainty, such that relatively small differences between sites can be ascertained, and this number of individuals is therefore recommended per site. Fish can be sampled using nets from the shore or from a boat, depending on the particular conditions of the environment. In various sampling programs, a variety of seines and seining techniques have been used and electro-shocking is a possibility. Seining techniques with a variety of different nets are possible: 1) two people wading dragging a wide net; 2) boat-assisted net tows with one person wading and another piloting the boat, with one side of the net attached, traversing deeper waters; and 3) boat-only tows. Boat electro-shocking, customized for small fish collections, can be used in areas and situations where seining is not effective. In small rivers and streams, backpack electro-shocking while wading in the channels is effective. Once obtained, fish should be maintained in aerated water, until field sorted and cleaned. Fish are then sealed into labeled, doubled freezer weight bags with water surrounding and air removed, and if possible, field frozen on dry ice, or at least transported on ice to the laboratory for freezing.

Prior to analysis the fish are thawed, and the total length and fresh weight for each individual determined. Care must be taken to preserve fresh consistency for the initial weighing process, to avoid sample drying or the accumulation of excess moisture. Depending on the analysis regime, the individual fish may be dried to constant weight at 55°C, with this weight recorded for the calculation of the wet/dry ratio. Dried samples can be individually ground to a fine powder. Alternatively, fish can be freeze-dried prior to analysis. Analysis using a solid phase analyzer typically requires a dried sample for analysis while wet digestion followed by chemical analytical methods can be done on an un-dried sample. However, dry weights should still be measured. Details of the analytical techniques for T-Hg and MeHg are given elsewhere in this report.

3.4.2 Study Design Considerations

The research team suggests that most of the biological studies aimed at assessing bioavailability use the fact that Hg and MeHg, and especially MeHg, are readily bioaccumulated into aquatic food chains and organisms are therefore a good and useful measure of bioavailability. This is especially true if the study is designed to examine this in a relative sense i.e. use differences in organism concentration as a metric reflecting bioavailability at the base of the food chain. However, this approach is only valid if there are no other factors besides water concentration that can impact organism concentration sufficiently to negate the linear assumption. As noted, the higher an organism is in the food chain, the less direct the association and the more potential confounding factors. The major impact for larger fish is difference in growth rate. As MeHg is relatively slowly eliminated by organisms, growth rate is the principal mechanism whereby tissue concentration can be impacted, and slow growing organisms will have higher concentrations than organisms of the same size and species, that are growing more rapidly. This effect is termed “growth dilution” and has been documented in the literature in a number of studies. (Mason, 2000; Pickhardt et al., 2002; Pickhardt et al., 2005). Interestingly, while these studies and some modeling studies (e.g. Kim et al., 2008) have suggested the importance of growth dilution, other modeling studies, such as the bioenergetics/ bioaccumulation modeling of Hg in fish in temperate lakes indicate that growth explains a relatively small fraction of the variation of MeHg concentration in fish (e.g. Harris and Bodaly, 1998). In studies on yellow perch and walleye in the upper Midwest, it has been demonstrated that the highest Hg concentrations are found in populations with high individual growth rates (Wiener, personal communication 2008.). Recent work on southern boreal lakes indicates that MeHg concentration at the base of the food web is a key factor affecting Hg in long-lived game fish.

3.4.2.1 Field vs. Laboratory Studies

As noted above, for larger organisms, it is preferable to use field sampling and analysis of both *in situ* organisms and transplanted organisms. Laboratory studies of exposure using large organisms is not recommended given the difficulties in setting up such exposure systems, and the need for large amounts of water to flush the experimental systems. This would be relatively difficult to do for effluent waters. While laboratory studies with controlled exposure conditions and microbes may appear to be the preferred route for the examination of bioavailability, there are logistical difficulties in collecting and preserving the waters until the experiments can be performed, and while there have been a number of studies published using this approach, most of these have not indeed examined the *in situ* concentrations of Hg and MeHg but have enhanced the concentration through the use of radio- or stable isotope spikes (Fisher and Reinfelder, 1995; Mason, 2000). However, if there were a need to examine the potential impact of certain variables, such as for example, the impact of DOC on the bioavailability of Hg and MeHg, then culture experiments with phytoplankton could be a very useful approach. It is not expected that such experiments would be typically done in the context of a bioavailability study and details will not be presented. Details of such an approach can be found in the current literature (Lawson and Mason, 1998a; Mason et al., 1996; Pickhardt and Fisher, 2007).

3.4.2.2 Field Studies - Caged vs. *In Situ* Organisms

For field studies, the main issue in the initial experimental design is whether to use *in situ* organisms or transplanted (caged) organisms. The advantages of using the resident population are relatively obvious, in that the collection of the organisms is straightforward and the measurement of concentration gives a rapid indication of relative changes in concentration along a water course, for example, if a number of stations are occupied. A disadvantage of using *in situ* organisms is that there may not be the same species at all locations to be sampled and it is not advisable to compare different organisms when comparing bioavailability. Also, obtaining the required number of organisms may be problematic in some situations, and may require extensive sampling. In addition, especially for filter feeding organisms, there is the likelihood that the resident organisms will be on or in the sediment and this is a concern in terms of interpreting the signal, as noted above. Bivalves do however grow in the water column if there is suitable substrate for them to attach to and thus bridges, pylons, jetties, and other structures may provide the more desirable water column filtering organisms. As noted, all species may not be found at all locations and such a situation is likely if there are substantial differences in substrate or water flow within the chosen study area. This would also be dependent on the type of organism chosen as the biosentinel. As noted, the organisms chosen should have relatively rapid growth rates and either be short-lived or, for fish, sampled as juveniles.

The advantage of using caged organisms is that there is a larger selection of potential organisms to use, with the only restriction being that they can survive and grow under the *in situ* conditions, although there may be concerns with the deployment of non-resident or invasive organisms. Clearly, there is also more control over the number of organisms deployed at each site, as well as their size. Therefore the major advantage of using caged organisms is the relatively higher level of control over the variables of sample design. However, the organisms used need to be collected from a relatively uncontaminated site, one with a known low and relatively constant concentration, so that they do not have a measurable or variable residual concentration from their previous exposure history at the time of sampling after caging. Collection and depuration of organisms is required prior to, and after, deployment and this requires laboratory equipment and conditions suitable for this – flowing water, and capability to monitor water quality. The largest problem with deployment of caged organisms is the potential "die off" that may occur after initial transplantation if the organisms have been over stressed due to transport and other conditions, and if the conditions in the environment are radically different from those of their previous location. For example, while some estuarine bivalves can live over a wide salinity gradient, rapid changes in salinity are still stressful to the organism and it may be necessary to slowly change the salinity in a laboratory setting prior to their deployment.

As noted above, depuration of Hg and MeHg is much slower than uptake, and organisms with low initial concentration are therefore preferred, as this allows for better resolution of the *in situ* signal. Recall, that additional information can be gathered with transplanted organisms if sampling is done in an intensive way initially as the initial accumulation kinetics can be followed and uptake rate constants can be obtained and compared amongst sites. Techniques for deploying bivalves have been relatively well worked out, and a number of recent studies have used *Corbicula* for monitoring in freshwater environments. However, there can have high mortality upon deployment if conditions are not truly suitable for the organisms. High mortality suggests that the organisms have been stressed in the environment. Therefore the accumulation is likely to be unrepresentative of those of resident organisms and the data should not be used. If it can be ascertained that the mortality is due to conditions prior to deployment (holding time, transport conditions) and that there is not substantial mortality after deployment, this may be sufficient and the data can be used. While there are no hard numbers in the literature, it is suggested that mortalities greater than 20%, i.e. <80% survival, are indicative of compromised results and there should be an investigation into the cause of the high mortality. In the absence of extenuating circumstances, mortalities of greater than 20% are considered to be unacceptable.

Overall, the collection and analysis of *in situ* organisms is the easiest approach but there is a larger potential for incorrect inference based on body burdens given the various factors besides Hg and MeHg water concentration, and water quality (DOC, pH, anions) that may impact tissue levels. The deployment and use of caged organisms has larger logistical difficulties but has the ability to provide a clearer and more robust answer given the additional control of variables that is possible. Finally, it is worth noting that if caged organisms are to be used, there is substantial advantage gained by the collection of *in situ* organisms simultaneously. This will add some small logistical cost, a larger analytical cost, but may provide additional information and comparison between approaches that, while being redundant to a degree, will also provide substantial corroboration of the results from the transplanted organisms.

Indeed, this was the result obtained from the study around a western US treatment plant, where both approaches were used and generally provided the same result (LWA, 2007) (Figure 3-4) that there was relatively little impact of the input on the biosentinel concentration. However, there are small differences with a more marked signal for the resident organisms. These organisms were collected from the sediment while the transplanted clams were deployed off the sediment surface (LWA, 2007). Thus, the differences are most likely a reflection of the impact of sediment concentration, which was higher immediately downstream of the input, on the organism concentration, and not a overall direct reflection of the water column bioavailable concentration, which is more clearly reflected in the signal of the transplanted clams.

This overall result was also confirmed by the results of a small fish biosentinel survey, which covered a larger region and included the receiving waterbody and its watershed. The concentrations in resident fish were slightly elevated downstream of the outfall but overall the differences in concentration found were small compared to other noted differences in concentration throughout the whole study area, and overall the values were not elevated, and were lower than other known impacted sites (LWA, 2007).

3.4.2.3 MeHg and T-Hg Measurement in Organisms - Sampling Methods

A general outline of the sampling methods for each group of organisms was given under the individual discussions above and will not be repeated in detail here. A general overview of the methods will be given as they depend to a large degree on the organisms being sampled and the size and type of waterbody being sampled. Clearly it may be possible to sample small rivers and shallow systems by hand but this is not so for larger waterbodies – large rivers, estuaries and the coastal zone. Zooplankton are readily sampled using plankton nets, either by hand or from being towed behind a boat. As noted above, it is preferable to pick individuals and sort the organisms by taxa at the minimum to minimize errors that can be introduced due to the presence of other organisms that may feed at a different trophic level (Mason et al., 2000), or more likely, to remove the presence of large algae (e.g. filamentous diatoms) that may be present in the samples, and detritus (Mason and Sullivan, 1997). The zooplankton can be separated from much of other seston using a large mesh filter (>50 µm typically used) and a single bulk sample from a

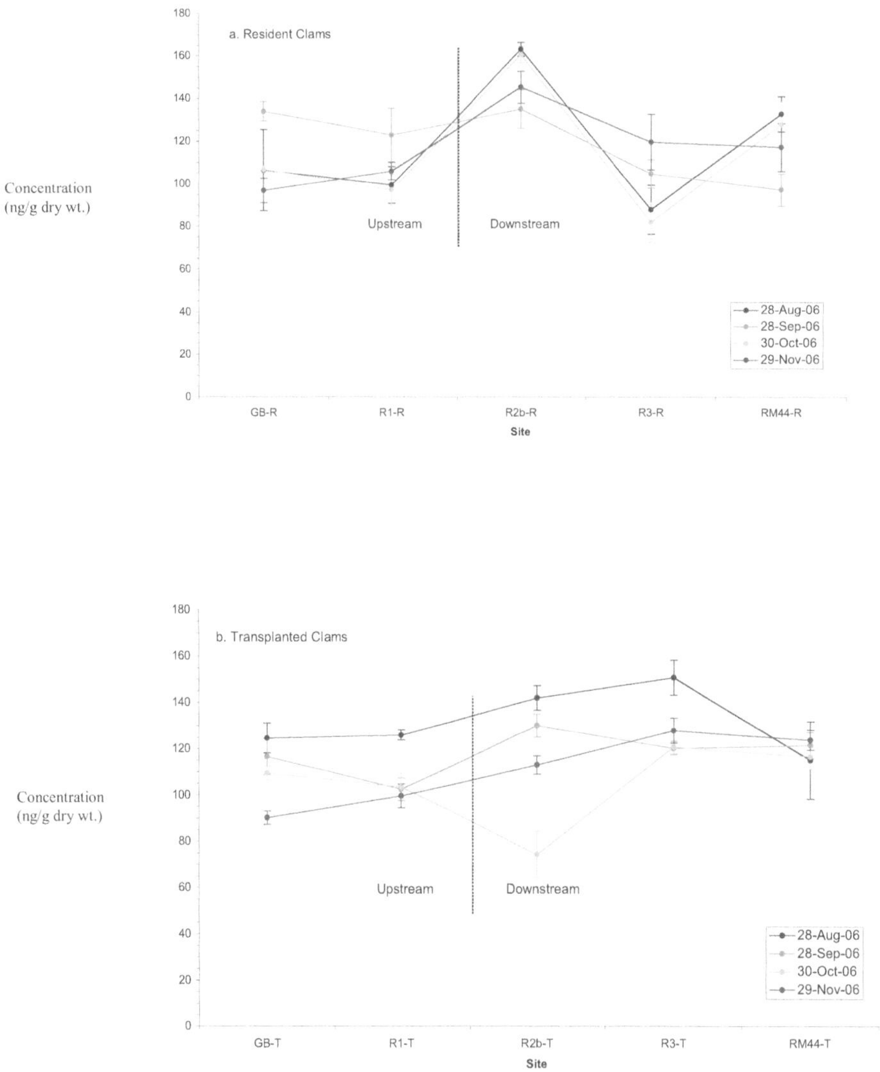

Figure 3-4. Temporal and Spatial Variation of MeHg Concentration (ng/g), on a Dry Weight Basis for Resident (Top Panel) and Transplanted *Corbicula fluminea* (Bottom Panel). Source: LWA (2007)[30]. Sacramento County Regional Sanitation District, 2007. Reprinted with permission.

[30] Data are from five stations on the Sacramento River upstream and downstream of the outfall, collected on four different occasions. The location of the outfall is indicated by the vertical dotted line. Error bars represent the standard deviation of the triplicate samples collected at each site.

plankton net can be used to characterize the open-water community of zooplankton at a particular time. Samples can be stored frozen after collection and dried if necessary depending on the analytical method to be used, as discussed below. Sample weights need to be determined and this is preferable prior to freezing of the samples. While it is possible to determine the concentration of MeHg and total Hg on a relatively small sample (in a recent CalFed study, zooplankton samples of 10 mg dry weight were analyzed for MeHg) it is better to obtain larger samples to reduced variability in the results (Mason and Sullivan, 1997). As typical zooplankton range in weight from <1 mg dry weight, the number of individuals that needs to be collected is dependent on the organism size, but a requirement of at least 100 individuals has been found to yield relatively low RPD values (<30%) in a number of studies (Mason and Sullivan, 1997; Kim et al. 2006, Mason, unpublished data).Invertebrates (bivalves and other mollusks) can be collected using a number of techniques which differ depending on their habitat, and if present in the sediments can be picked on board from sediments obtained using a dredge or other sediment sampling device (Wiener et al., 2007). Samples can be collected by hand or diver if this is possible. Nets can be used for some organisms and nets are one method of catching fish. For larger organisms, electro-shocking is one possible approach (Mason, 2002). As discussed above, the recommendation is that at least 30 individuals should be sampled to obtain representative results with an acceptable RPD (<30%). With larger organisms, there is the potential for the concentration to be a function of the organism size (age) and therefore, to the extent possible, organism size should be kept as constant as possible across the sampling sites to ensure that the comparisons are reflecting changes in concentration rather than changes in organism size. Also as noted above, it is preferable to depurate organisms that feed at or in the sediment as the presence of sediment in the gut can lead to an overestimation of the organism's concentration. In general, and if possible, it is always desirable to depurate biota samples prior to analysis. This may not be possible for fish, but given their relatively large size and the higher concentration, this may be less of a problem.

The researchers recommend that biota samples be homogenized prior to shipment to the analytical laboratory for analysis, although this is not essential. Thawing clams and other organisms after freezing can lead to losses of moisture and therefore an incorrect estimation of wet weight. Consequently, the protocol of field sorting, on site or near on site depuration, and sample preparation and homogenization prior to shipping is the preferred approach. Composites should preferably be homogenates of at least 30 organisms for invertebrates and small fish. If larger fish are sampled, less replication is needed.

3.4.2.4 Analytical Considerations

The methods for the analysis of Hg and MeHg in organisms are based on the U.S. EPA Methods for the analysis of Hg and MeHg in water. Typically the sample is digested in a strong acid solution for the determination of total Hg using cold vapor atomic fluorescence techniques (CVAFS) or by inductively-coupled plasma mass spectrometry (ICP-MS). There are now also various solid phase analyzers on the market that are suitable for the determination of total Hg in biological tissue and sediment and these provide a quick alternative to acid digestion and CVAFS or ICP-MS analysis. The samples for solid phase analysis have to be freeze-dried or oven dried prior to analysis while samples for wet chemistry can be analyzed as a wet sample. It is typical in the scientific literature to report biological tissue concentrations on a wet weigh basis, so if dried samples are analyzed it is necessary to do a wet/dry ratio determination to convert the data to a wet weight basis.

Methylmercury concentration is typically determined by derivitization prior to chromatographic separation and detection of the MeHg, after degradation to elemental Hg, by CVAFS. ICP-MS is also a suitable detector. The samples need to be prepared prior to analysis and given that the concentrations in biological tissue are relatively high it is mostly not necessary to use distillation or other isolation techniques as a small volume of digestate can be analyzed without interference by the ethylation derivitization, the most used approach. Other derivitization methods are possible. Isolation of the MeHg from interfering substances, if necessary for low concentrations that may be found in zooplankton and other lower food chain organisms, can be done by distillation or by liquid-liquid extraction techniques (Horvat, 1996).

Therefore, there are many approaches to the analysis and the details of one particular method are given below. This does not particularly endorse this approach over the others but is detailed to provide the necessary outline to the analytical approach. There are a number of commercial laboratories that can make the relatively difficult MeHg measurements in tissue. In terms of total Hg, there are a larger number of facilities that can perform these analyses.

Total Mercury in Biota. For analysis using wet chemical digestion techniques, approximately a one gram aliquot of tissue is typically weighed to the nearest mg into a 40 mL vial, and digested. This can be done with 10 mL of refluxing concentrated nitric acid for 3 hours. An alternative digestion procedure involves the use of a nitric/sulfuric acid mixture and overnight digestion at 60°C. After cooling, the tissue digests are diluted to 40 mL with a 0.2N solution of BrCl in 1.6N HCl (Bloom and Crecelius, 1983), which is a strong oxidant and ensures that there has been complete digestion and release of the Hg from organic matter, and oxidation of the Hg and its compounds to the Hg(II) oxidation state. After bromination, samples are allowed to sit for at least 1 hour to allow decomposition to occur.

After neutralization of any remaining oxidant by the addition of a mild reducing agent, hydroxylamine hydrochloride solution (which does not reduce divalent Hg to elemental Hg), the samples are quantified using tin chloride reduction, which converts the Hg(II) to elemental Hg (U.S. EPA, 1995). Sparging of the solution in suitable apparatus leads to the release of the elemental Hg, which is concentrated and trapped onto gold-coated sand or glass bead traps. The Hg when amalgamated with the gold is stable at room temperature. To release the Hg, the traps are heated to 500°C, and the released elemental Hg is swept into the CVAFS detector for quantification. The concentration is determined using a calibration curve which can either be a gaseous air standard or a liquid standard. Overall, the approach to analysis is a variant of EPA Method 1631. Batch QC is typically included in the analysis at a level of 10-20% of samples. Typically, three to four method blanks are run, one duplicate, one matrix spike, a matrix spike duplicate, and standard reference material (SRM), such as oyster tissue (NIST-1566) or mussel tissue (NIST-2976) (certified range of 57-65 ng/g dry weight for T-Hg). Field duplicates and field blanks are analyzed and are collected on every field trip.

Methylmercury in Tissues. One gram aliquots of the tissue homogenate are weighed to the nearest milligram into 40 mL borosilicate glass vials. To each sample is added 15 mL of a mixture of 25% KOH in methanol (Bloom, 1992) and the lids are screwed on tightly. Samples are placed into a vented oven, and digested for at least two hours at 70°C. After cooling, the samples are diluted to 40.0 mL with methanol or distilled water, and shaken vigorously. For analysis, a sample aliquot (0.01-0.10 mL) is diluted to 50 mL with ultra clean water, and buffered at pH 5.0 with an acetate buffer. The solution is then reacted with sodium tetraethyl borate, and, after a 10 minute incubation time, the volatile methyl-ethyl derivative formed is separated from the aqueous matrix by purging with argon or nitrogen, and collected onto a solid adsorbent – either Carbotrap or Tenax (Bloom, 1989). The sample is dried with argon or nitrogen prior to heating and desorption of the adsorbed Hg species into an isothermal GC column at 100°C for peak separation. After the organo-mercury peaks elute, they are passed through a pyrolytic column at 750°C that decomposes all compounds down to elemental Hg, which is quantified by CVAFS or ICP-MS. Standard QA/QC typically includes analysis of at least three method blanks, a certified reference material (matrix that is most similar to samples), and a matrix duplicate, matrix spike, matrix spike duplicate for every 20 samples. An SRM that is often used is lyophilized oyster tissue (NIST-1566).

QA/QC. The following very general guidelines are given for the analysis of Hg and MeHg in tissue (U.S. EPA, 1995; U.S. EPA, 1998). The calibration curve should be based on a liquid standard when possible, and especially for MeHg, and should yield a correlation coefficient of >0.995 to ensure accuracy in the analysis. Blank correction of samples is necessary to account for both the analytical blank and for any blank due to field sampling, handling or processing. Field blanks should therefore be incorporated into the sampling protocol. Blanks should not exceed 20% of the sample value but this should not be an issue for tissue analysis. For duplicates, the %RPD should be <15% for total Hg, <20% for MeHg. Recovery of spike additions should fall in the 80-120% range, and similar recoveries should be obtained during the analysis of SRM's. The obtained values for the SRM's should fall within or near the accepted range for the specific SRM (note, for example that the mussel SRM (NIST-2976) has a very narrow certified range and values at the limits of this range are acceptable given the normal variability in MeHg and Hg analysis. Detection limits for total Hg in tissue are on the order of 0.1 ng/g dry wt and for MeHg, 0.25 pg/g dry wt.

3.4.2.5 Number of Samples

As noted above, the desired number of biota samples for organisms such as bivalves and small fish is 30 individuals per sample, and to account for variability, there should be triplicate samples taken at each site, or 90 individuals overall. This is the requirement for total Hg and MeHg, but the amount of tissue analyzed will be a small fraction of the total tissue collected and homogenized and so it is possible to do any ancillary measurements (e.g. dry weight, condition index, etc.) on the same sample. If not, then additional organisms need to be collected for the ancillary measurements.

For large fish and other large organisms, if sampled, it is not often possible or feasible to sample such a large quantity and so the minimum required is often sampled (five fish of approximately equal size). If a size-concentration relationship is to be developed for fish, many more fish over a size range needs to be collected. There are now some biopsy methods available for the collection of small tissue samples from fish prior to their release (Baker at al., 2004; Schmitt and Brumbaugh, 2007). If such approaches are used, then larger samples sizes can be contemplated in many situations, and up to 30 samples should be collected.

For the smaller organisms, the number of individuals should be much greater than 30 per sample if tissue mass is to be relatively large, and to reduce variability between the triplicate samples. While it may be possible under the best analytical conditions to analyze a small number of zooplankton, for example (<10 mg dry wt) using careful analysis and handling of samples, it is recommended that at least 100 individuals are collected for each sample. This ensures that there is sufficient material for both total Hg and MeHg analysis, and also ensures that the RPD of the triplicate analyses are within an acceptable RPD. As noted above, it is suggested that individual zooplankton and other small invertebrates be "picked" and sorted, at least to the genus level, to remove to the extent possible, variability due to differences in the feeding habits of different organisms. Also, it is re-iterated that the use of zooplankton as indicators is limited to the collection of *in situ* organisms and these organisms are not useful for "caged" studies. For phytoplankton/seston, the collection of >10 mg dry wt of material is recommended and thus the amount of water filtered is a function of the TSS content of the specific waterbody. However, the use of seston is not recommended in this application, as noted above.

3.4.3 Effluent Characterization Using Bioindicators

As should be apparent from the foregoing discussions, there are likely limitations in the choices available for the characterization of effluents using bioindicators and other biological approaches. Certainly, it would be feasible to collect water samples and perform phytoplankton uptake experiments as long as the organisms could survive in the effluent medium. The organisms are unlikely to be directly impacted by the Hg concentrations unless they are much higher than the values reported elsewhere in the report. Thus, lack of growth would likely be due to some other inherent property of the effluent.

One approach that may have merit would be the sequential dilution of the effluent water with clean freshwater and the exposure of the organisms to the range of dilutions and comparison of the growth rates of the organisms in each media to assess the impact of the effluent on growth. After the impact of the effluent is known, the exposure experiment could be designed to investigate uptake over a range of different dilutions, which could then allow, assuming a linear extrapolation, an assessment of the potential bioavailability of the T-Hg and MeHg in the whole effluent sample. It is unlikely that effluent outlets and pipes have resident bivalves - these are likely actively removed - and are mostly expected not to have resident small fish. Of course, if resident organisms are found, these can be used. The most likely approach to examining the bioavailability of effluents is to use caged organisms, and, as noted above, bivalves have been used successfully in recent application in examining effluent inputs. However, while *Corbicula* may be a relatively hardy organism and overall more robust than fish, they are likely the preferred organisms to use. It is not true in general that freshwater bivalves are more robust in cages than fish. Thus, the approach outlined above in Section 3.4.1 should be used. It will be briefly reiterated here.

It is recommended that a hardy, resident species in the ecosystem to be studied be used. Candidate species are *Corbicula* for freshwaters and mussels or clams for estuarine and coastal waters. Using organisms that already exist in the environment being studied should ensure that mortality after deployment is low (which is clearly desirable). The recommended number of organisms per cage is 30 (replicates are kept in different cages) for the Hg and MeHg analyses, i.e. for triplicate samples a total of 90 organisms. Additional organisms should be added if required for ancillary measurements and to account for mortality. The organism size should be as consistent as possible, and the size range small, to alleviate the impact of size on concentration. For *Corbicula*, clams in the15-30 mm size range are optimal; for blue mussels, 5-8 cm is the typical size. Bivalves should be kept cool and damp and be transported and deployed as soon as possible. Depending on the situation, it may be desirable to hold the bivalves in a clean water system (aquarium tank) to allow them to depurate prior to deployment. If this is done, the aquarium should be monitored for basic water quality parameters to counteract a die off of the bivalves. The water should be changed daily, if possible, and aerated. At deployment, bivalves should be thoroughly mixed to ensure that the time-zero samples represent the entire population of deployed bivalves. For deployment, the bivalves should be placed into individual compartments in plastic cages.

At each sampling point, three cages of bivalves are collected at each station to provide replicates of 30 homogenized individuals for Hg and MeHg analyses. As the uptake of T-Hg and MeHg into the organisms is relatively rapid, it is suggested that the initial sampling period be one to two weeks, with the second sample being within a month of deployment. After that, monthly or bimonthly sampling is adequate as the organisms have reached steady state and their concentration changes will therefore reflect longer-term variability in the system concentration.

All collection, handling, deployment, and retrieval needs to be performed using clean sampling techniques (wearing polyethylene gloves, and using pre-cleaned storage and sample containers). The bivalve should be removed from the cages onsite and the organisms placed in a plastic mesh bag and stored in a clean cooler, containing filtered effluent, for transport to the laboratory, where they should be depurated prior to shucking and storage of the tissue material for analysis. It is recommended that other metrics besides Hg concentration are measured to estimate whether the bivalves grew during the deployment period. A simple dry weight measurement or growth and body index could be employed to ascertain this.

Removal of tissue should be done in a clean environment and the equipment used for obtaining the samples should be rinsed with acid and distilled water between samples, and should be plastic material. Samples should be homogenized in an ultra-clean laboratory space. A blank for the homogenization process can be obtained by running ultra clean water through the equipment prior to homogenizing of the actual tissue samples. After being homogenized, samples can be stored frozen in a plastic jar until analysis.

3.4.4 Receiving Water Characterization

3.4.4.1 Near-Field vs. Far-Field Assessment

To assess the near-field impact of an effluent stream or other source requires the sampling in the initial impact or mixing zone, and this is likely to be in close proximity to the source (see Section 3.4.3.1). If there are resident organisms in this area, then there is a clear and obvious advantage in using these organisms as an indicator (Choy et al., 2008). Comparison of the concentration of Hg and MeHg in these organisms could then be compared to the concentrations both upstream and downstream of the input region. The downstream samples should be located in such a way that there is at least one site that can be considered to be a far field site i.e. a location where the reactions of methylation and demethylation that are initiated by the mixing of the effluent and receiving waters is complete, and the system is at steady-state (see Section 3.4.3.2).

Of course, this may not be possible if there are other inputs downstream or if there is some factor that is changing downstream that leads to a continual change in the %MeHg in the system. Ideally, there would need to be a minimum of three sampling locations within the receiving waters (upstream, within the mixing zone and downstream), in addition to any sampling that is done in the effluent stream itself. Of course, many systems will not be perfect and ideal, such as locations under tidal influence, or with highly variable flow rates. Thus, as used in the study discussed above (Figure 2-3), use of five or more locations may give a better opportunity to resolve the source characteristics. Clearly, the more sites that are sampled, the better is the resolving power of the data produced. In addition, it is often the tendency to minimize the number of upstream or reference sites but this is statistically ill-advised as the final conclusion will be based on the certainty of the knowledge about the potential increase, which is clearly reliant on the accuracy of the reference value. Thus, two or more reference sites are recommended.

In formulating the study design, it must be remembered the assumption is that bioaccumulation is a good and useful measure of bioavailability, especially if the study is designed to examine this in a relative sense, i.e. differences in organism concentration are used as a metric reflecting bioavailability at the base of the food chain. One needs to recall and assess the validity of the various assumptions that are implicit in this, for each particular ecosystem that is studied. For field studies, the main issue in the initial experimental design is whether to use the *in situ* organisms or transplanted (caged) organisms, and the overall recommendation is that using both may give a more defensible and conclusive answer. The advantages of using the resident population are relatively obvious, the collection of the organisms is relatively straightforward and the measurement of concentration gives a fast relative indication of changes in concentration. The disadvantages are that there may not be the same species at all locations to be sampled, obtaining the required number of organisms may be problematic in some situations, and may require extensive sampling. Interpreting the results, especially for filter feeding organisms, may be confounded by the influence of the sediment in which the animals are residing. All species may not be found at all locations and such a situation is likely if there are substantial differences in substrate or water flow within the chosen study area. This would also be dependent on the type of organism chosen as the biosentinel. As noted, the organisms chosen should have relatively rapid growth rates and either be short-lived or should be sampled as juveniles.

The discussion of the use of caged organisms and their advantages and disadvantages has been outlined in detail above and will not be reiterated here. Suffice to say that the use of caged organisms have some important advantages that suggest their use for studies using bivalves would be advantageous and, as they respond relatively rapidly to changes in water concentration, can be used to obtain a result in a relatively short space of time - three to six months if sampling is done on a biweekly-monthly basis. Again, a number of time points are needed to assess the overall average value and, as noted above, there should be an attempt to sample during the initial rapid accumulation phase as well as during the asymptotic or steady state accumulation phase. To ensure a statistically valid result, at least three time points should fall within this latter phase, and given a time zero sampling and a sampling during the initial growth phase, a minimum of five sampling times is recommended. Thus, with five locations and five time points, and three replicates, the minimum recommended total number of samples is 75, and to reiterate, more is better.

The sampling frequency for studies using bivalves, based on the results of recent studies with *Corbicula*, is recommended to be: at time zero (at or just before deployment); one week later; three weeks later (a month after deployment); two months and four months after deployment. If possible, the timing of sampling should be arranged so that it does not cover a period where there are large changes in physical characteristics, such as temperature, and/or in the flow regime (i.e. during spring runoff, or over the low (baseflow) period). Thus, late summer/fall appears to be an optimal time for sampling.

For marine and estuarine organisms such as mussels, uptake rates are somewhat slower as the organisms tend to be larger than the freshwater species. Steady-state concentrations are reached within one to two months, depending on the organism and size. Thus, the only real difference for the saline environment is that it is probably necessary to space out sampling somewhat and the following regime is suggested for mussels and other bivalves: time zero; two weeks later; one month, two months and four months after deployment. In addition to the use of bivalves, in estuarine environments there are a number of invertebrates that may be suitable alternatives to bivalves, such as crabs and other crustaceans. Studies have also been done using other mollusks. If these organisms are used, it will be necessary to consult the literature to ascertain the most appropriate sampling regime for the organism chosen.

For fish, use of resident species is the preferred approach given the difficulties in keeping caged fish for the time necessary for their accumulation to attain a relatively stable concentration, which could be up to six months for even small fish species. In contrast to bivalves, fish do not effectively depurate the accumulated MeHg and therefore rather than attaining steady state, their concentration continually increases over time. As fish are less robust than bivalves, it is not recommended that caged organisms be used unless the specifics of the situation demand it. Fish sampling should be done on a yearly or seasonal basis, and over a number of years to assess the impact of longer timescale changes. However, while one sampling event may not provide a robust measure of the overall average bioaccumulation, it can provide a relative examination of the concentrations between locations and therefore one sampling event can be used in a quick, screening assessment manner to assess if there is a potential higher concentration or "hotspot" in a particular location (i.e. study area). Obtaining a positive difference could then form the basis of a more detailed and comprehensive study and examination of the factors involved in the enhanced concentration. As noted above, a higher concentration can be caused by a number of factors and these may not all necessarily be related to the concentration of bioavailable Hg or MeHg in the system. As with the use of invertebrates, it is recommended that there are at least five sampling locations for any study using small (prey) fish. Two sites should be located "upstream" of the effluent input or source location, one or even two sites should be in the "mixing zone", and two should be downstream.

3.4.4.2 Sample Location and Spatial Assessment

As noted above, at least five locations are recommended for the receiving waters and an additional site within the source or effluent stream would be highly advantageous. This is the suggested minimum number of sites and more sites will provide better resolution of the information, and potentially a clearer picture, especially if the question which is poised is whether or not an input creates a hotspot for bioavailability and bioaccumulation. Of the five sampling locations, two sites should be located "upstream" of the effluent input or source location, one should be in the "mixing zone" and two should be downstream. Locating these sites is relatively obvious in a non-tidal river, but this is not the case for a tidal system or for the coastal marine environment. Thus, the following sections briefly describe the approach to ascertain the best sampling locations in the three broad instances: freshwater (riverine) systems, estuarine systems and the coastal environment.

Riverine Environments. The theoretical definition for the location of the far-field sampling sites is that where all the reactions have proceeded to steady state. Given the rates of inter-conversion of R-Hg and MeHg, this represents a time period of days rather than hours or weeks, as discussed in the Phase I report. Thus, the location should be defined with this time factor in mind and will therefore depend on the flow rate of the river (see Section 3.3.3.1). As the flow rate is often given in units of volume per unit time, this can be converted into a velocity if the cross-sectional area of the river is known. Then, the distance can be calculated for a specific travel time. Given the reaction rates discussed in the previous sections, it seems appropriate to designate a distance that corresponds to a time period of two to four days. Again, given the experimental design, this should allow for the determination if there is downstream generation of bioavailable Hg, and more specifically, an increase in MeHg in the biosentinel organisms. Thus, an increasing concentration downstream with distance from the mixing site would be indicative of this. Again, referring back to the particular study results shown in Figure 3-4, there is little indication of this occurring in this instance. Indeed, based on the results of the *in situ* clams, there may have even been some net demethylation occurring, although this conclusion is unlikely to be supported by a robust statistical analysis of the data. Examination of the particulars of this study indicate that the locations chosen are less than one day of water travel downstream, and therefore do not necessarily fit the theoretical criteria. The location of the sites was however constrained by the other inputs into the river and other factors. For example, the presence of riparian wetlands is a potential complicating factor given that their longitudinal distribution along river channels can strongly affect the abundance of MeHg and its bioaccumulation in biota (Waldron et al., 2000).

Therefore, while it may be desirable to determine the distances between sites using a theoretical approach, there is likely to be practical limitations in specific situations that may lead to choosing the sites based on other criteria. For example, another downstream source could be present within the region defined by the far field estimate above, and clearly it would be desirable to site the sampling locations within the region impacted only by the particular source of interest. It should still be possible to obtain meaningful data as there should be an indication of an increase in concentration even though this system has not reached the new steady state regime.

Estuarine and Coastal Environments. The information summarized here is taken from the U.S. EPA documentation (U.S. EPA, 1992) and it is reasonable to take the U.S. EPA view in terms of mixing zones in saline waters, which have been defined and used in waste load/effluent allocations. Thus, a particular facility would have already negotiated this definition and can use this to define the region and location for the deployment of the sample within the mixing regime. There clearly is a great diversity of situations and types of environments and a range in the extent of physical mixing and therefore in the initial mixing processes. In terms of the receiving waters, the following need to be considered: 1) the size and flow characteristics of estuaries or coastal water can vary widely; 2) the waterbody may be deep or shallow, stagnant or flowing; and 3) it may exhibit ambient density stratification of various degrees. For the effluent/source, the discharge type and configuration can be: 1) highly variable and may vary greatly relative to the movement of the receiving waters; 2) be denser or lighter than the receiving waters; 3) be located near shore or far offshore; and 4) exhibit various geometric details ranging from single port submerged discharges to multi-port submerged diffusers to surface discharges (see Sections 3.4.3.1 and 3.4.3.2).

Given this diversity of both discharge and ambient environmental conditions, there are a large number of possible flow patterns which will develop as the discharge waste stream mixes in the ambient water, which will determine the configuration, size, and intensity of the mixing process, and the impact of the discharge of Hg and MeHg. Clearly, in any particular situation, a detailed engineering analysis using mixing zone models, would provide the most robust and reliable estimation of the mixing zone, the definition of the near and far field environments. According to the U.S. EPA (1992), the mixing zone is defined as an "allocated impact zone" where numeric water quality criteria can be exceeded as long as acutely toxic conditions are prevented. This is clearly not adequate for the current situation as there is likely to be few exeedances for dissolved Hg and MeHg discharged from a WWTP or other sources. However, a mixing zone can be thought of as the limited area or volume where the initial dilution of a discharge occurs (U.S. EPA, 1992). Special mixing zone definitions have been developed for the discharge of municipal wastewater into the coastal ocean, as regulated under Section 301(h) of the Clean Water Act and the mixing zone is labeled as the "zone of initial dilution" in which rapid mixing of the waste stream (usually the rising buoyant fresh water plume within the ambient saline water) takes place. EPA requires that the "zone of initial dilution" be a regularly shaped area (e.g. circular or rectangular) surrounding the discharge structure (e.g. submerged pipe or diffuser line) that encompasses the regions of high (exceeding standards) pollutant concentrations under design conditions. In practice, limiting boundaries are often defined by dimensions equal to the water depth measured horizontally from any point of the discharge structure. In terms of Hg and MeHg, EPA maintains two water quality criteria for the allowable concentration of toxic substances (U.S. EPA, 1992): a criterion maximum concentration (CMC) to protect against acute or lethal effects; and a criterion continuous concentration (CCC) to protect against chronic effects (U.S. EPA, 1985). The less restrictive criterion, the CCC, must be met at the edge of the same regulatory mixing zone specified for conventional and nonconventional discharges.

The location of the upstream sites may be also problematic in such a situation and therefore these sites should rather be designated as reference stations and should be chosen based on the mixing dynamics of the system and the location of other potential sources of Hg and MeHg. The sites should be located in a region where there is very little likelihood of impact of the effluent on these reference locations. In contrast, the downstream sites should be located outside of the mixing zone but should be in locations that are clearly impacted, based on knowledge of the hydrodynamics and mixing of the system, by the effluent stream.

3.4.5 Data Compilation and Bioavailability Assessment

Given the methods of examining bioavailability using organisms that have been proposed here, it is relatively easy to interpret the results of such investigations. In the case of examining differences in the concentration of Hg and MeHg in organisms collected and/or deployed across a spatial gradient, the decision will rest on whether there is a statistical difference in the concentrations between the different sites at some level of certainly, such as a 99% confidence level, which is the typical level of certainty used in these types of investigations. As there are samples collected at each time point in space, and collections cover multiple dates, it is also necessary to determine if there is some interaction i.e. that the differences are a function of the sampling time rather than the location. Again, statistical tests are available for examining such interactions. Under the proposed sampling approach, it is reasonable to define the null hypotheses as:

- *There are no spatial bioaccumulation gradients in terms of Hg and MeHg in the chosen organisms across the sampling domain, and*
- *There are no temporal bioaccumulation gradients in terms of Hg and MeHg in the chosen organisms across the sampling domain.*

Thus, the weight of evidence must be sufficient to negate this hypothesis, and a negation would indicate a source of bioavailable Hg or MeHg, an input of other chemicals, or a change in the system biogeochemistry, such that there is an increase in the net methylation within the waters after mixing. Thus, as noted above, the study can provide a conclusive examination of the null hypothesis but unless there is additional information collected besides the biota data, it is not possible to conclusively demonstrate the exact cause of this increased bioaccumulation (Wiener et al., 2007). As noted above, the assumption in the estimations and calculations is that the downstream environment can be parameterized using the rate constants and %MeHg in the upstream waters. While this assumption is necessary for the application of the approach, there may be situations where it may not be true. If there is sufficient chemical and biological information available, this could be ascertained, but it would not be possible to determine this based on biological information alone.

A negative outcome, i.e. a demonstration that there is a trend, can be manifest in two forms: 1) if there is a higher concentration at the site within the mixing zone, then the effluent or source must be a source of MeHg, which is readily available and bioaccumulative; and 2) if there is a higher concentration downstream of the mixing zone then the effluent or source is likely a source of reactive inorganic Hg being methylated within the zone downstream of the input zone. A third possibility is that methylation is occurring in the downstream location and not in the upstream location. This again is counter to the basic assumption of the formulation. It could indeed be discerned with adequate information on Hg and MeHg concentrations, and rates of conversion, above and below the source, but could not be ascertained based on biological data alone. It should be possible to statistically determine which, if any, of these outcomes are statistically supported by the data. Clearly, the most definitive outcome is that there is increased bioavailability and bioaccumulation at the site within the mixing zone, compared to the other sites, and the evidence is sufficiently strong (statistically sound) that any reasonable scientist, manager or decision maker would conclude that there is a local source that is contributing to the localized "hotspot" or increase in concentration.

Obviously, while a statistically sound conclusion may be reached, the evidence and the data may however, on some occasions, be weaker and lead to a less determined outcome and a less clear decision concerning the null hypothesis. A strong and conclusive outcome would be possible if the following is found: 1) the null hypothesis is rejected (i.e., spatial and or temporal differences were detected); 2) the spatial differences fall along a definable and scientifically reasonable gradient (e.g. dropping off as a function of distance downstream); and 3) that the differences in organism concentration are associated or correlated with other indicators (MeHg in the effluent or receiving waters, in sediments, or in fish).

Clearly, it is possible for the null hypothesis to be rejected but the trends and distributions are contrary to expectation for a point source input (e.g. the concentrations are higher upstream). Furthermore, the impacts may only occur within the mixing zone and not downstream (as in the example in Figure 3-4 above), or the differences are not associated or correlated with other indicators. Under these conditions, there is some evidence of a potential source but the information is clearly not definitive, and it is likely that the conclusion would not be sufficient to conclude that the source is large and an important contributor to bioavailable Hg and MeHg within the system.

3.4.6 Comparison of Chemical and Biological Measures of Bioavailability

The discussion above has focused on the use of biological data to assess bioavailability without consideration of any chemical assessment or analysis. It has been noted, however, that while biomonitoring can demonstrate an increase or impact, it is not possible to ascertain from the biological concentration measurements alone the actual cause of the increase. Therefore, if it is necessary to ascertain whether the increase is due to an input of MeHg, or an input of bioavailable inorganic Hg, or due to the input of some other chemical that may enhance methylation, or impact the environmental conditions, then it is clear that additional measurements are needed.

Bioindicators cannot provide a conclusive answer in terms of sources and speciation without additional chemical measurements of Hg and MeHg concentration and distribution between dissolved and particulate phases. The use of bioindicators is recommended to answer the following basic questions: 1) is there a local input of bioaccumulative Hg in the study area? 2) does it change in time? 3) can it be linked back to a particulate input or loading to this waterbody? Such questions can be tested via analysis of resident or caged organisms, and clearly is a useful and important screening tool is determining if a particular source is contributing substantial T-Hg or MeHg to the associated receiving water. Use of bioindicators without additional chemical measurements cannot.

Therefore, it is recommended that chemical measurements be made in conjunction with the biological assessment, or the biological assessment can form the first stage of an assessment program. If the biological assessment indicates a substantial increase in concentration, then there would be a necessary secondary examination and chemical analysis that would be aimed at determining the cause of the increase. Again, this could be done in an incremental fashion, with first measurements of T-Hg and MeHg in the dissolved and particulate phases in the effluent/source and in the receiving waters. A simple mass balance mixing calculation could therefore determine if the input was due to a net input of MeHg as a result of the effluent/source, or whether the increase may be due to other factors. If the conclusion is that there is not a net input of MeHg after mixing, then a detailed examination of the chemistry of the receiving waters and the effluent/source would be necessary to ascertain if there is the addition of bioavailable inorganic Hg or if there is some other factor that is leading to an increase in net MeHg formation downstream. The details of such an investigation have been provided earlier in the report.

APPENDIX A

STUDY OF BIOAVAILABLE MERCURY IN WASTEWATER TREATMENT PLANTS

This appendix reports the results of a study of bioavailable mercury in the treated effluents of seven wastewater treatment plants in the Unites States. These plants represent a range of locations and treatment processes. The plants are referred to generically as Plants A though G. A description of each plant is contained in the following paragraphs.

Plant A

This WWTP is a marine discharger that treats portions of 10 communities with a population totaling approximately 360,000 people. It treats an average of 42.5 million gallons of wastewater each day. However, the capacity of the plant is such that 200 million gallons could receive preliminary and primary treatment and 65 million gallons could receive secondary treatment, if necessary. Treatment consists of screenings removal, grit removal, primary treatment, aeration, secondary clarification, hypochlorite disinfection, and bisulfite dechlorination. No hauled-in wastes are accepted for treatment.

Bar racks are employed for screenings removal and grit is removed via aerated grit chambers. The grit slurry is then separated using a screw degritter and disposed of in a landfill along with the debris removed by the bar racks. Approximately 1,150 wet tons of grit is removed annually. The primary and secondary treatments include primary clarification, complete mix, fine-bubble diffused aeration, and secondary clarification. The primary sludge and waste activated sludge are settled in gravity thickeners and then blended with fats, oil, and grease. This mixture is then dewatered by filter presses and biosolids are subcontracted and recycled into a soil conditioner. The POTW produces approximately 8,000 dry tons of biosolids each year. Final effluent treatment includes disinfection with sodium hypochlorite and dechlorination using sodium bisulfite. The following treatment chemicals are utilized:

- Sodium hypochlorite is used for disinfection of the final effluent, disinfection of the wet weather treatment tank influent, odor control in the gravity thickeners, odor control in the press room, and for the control of filamentous bacteria in the return activated sludge,
- Sodium bisulfite is used for dechlorination of the final effluent,
- Liquid cationic polymer is used to promote flocculation of the thickened sludge, and
- Caustic soda is used in unclogging grease lines.

Plant B

This Midwestern facility was designed to fully treat 175 million gallons of wastewater each day (MGD) and is currently treating approximately 120 MGD on average. It also receives for treatment and disposal all the biosolids pumped via force main from a 155 MGD wastewater treatment plant located outside Plant B's wastewater collection system. Wastewater is collected from more than 40 communities covering more than 250 square miles and the plant serves a population of approximately 530,000 people. The preliminary treatment consists of mechanical bar screens and aerated grit tanks for grit removal. The primary treatment consists of settling tanks and skimmings removal. All skimmings removed are incinerated at another facility. Ferric chloride is added at the end of the primary treatment for removal of phosphorus. The next step is a two-stage biological process. The first stage of this activated sludge process consists of aeration tanks and clarifiers, as does the second stage of the process. However, in the second stage, nitrifying bacteria remove ammonia nitrogen. The effluent is then passed through multimedia filters and disinfected by a chlorination/dechlorination process using sodium hypochlorite and sodium bisulfite. The settled biosolids are treated by use of gravity thickeners, gravity-belt thickening, thermal conditioning, and centrifugal dewatering with cationic polymer addition followed by incineration. The resulting ash is transported via the scrubber water to ash lagoons. The water from the ash lagoons then overflows into an interceptor sewer and is returned to the headworks of the plant.

Plant C

Plant C manages the wastewater for many communities in the upper midwest including a medium-sized city, many smaller communities, and two paper mills. The average daily flow is 37 MGD, of which roughly one half is from the paper mills. The wastewater is screened and grit removed and then goes directly to the activated sludge treatment process using high purity oxygen. There are no primary treatment tanks at the facility, but there are primary clarifiers at the paper mills. The biological sludge is clarified in secondary settling tanks, and then passes through polishing clarifiers, and multimedia filters. Sodium hypochlorite is added to disinfect the effluent and sodium bisulfite is used to eliminate any excess chlorine residual.

The activated sludge is thickened, digested in the Temperature-Phased Anaerobic Digester, and dewatered using centrifuges. Biosolids are land applied. Ferric Chloride is added to the digesters to reduce high levels of sulfide in the sludge.

Plant D

This plant's service area consists of only separate sewers and 43% of incoming flow to the plant during the winter is residential, 17% is commercial, and 3% is industrial. The remaining flow comes from stormwater inflow and groundwater infiltration. Effluent grab samples are collected downstream of chlorine addition. The plant treats an average of 81.3 MGD but is designed for an average of 115 MGD and discharges to a saltwater estuary.

Preliminary treatment consists of bar screens followed by aerated grit chambers. After preliminary treatment, the wastewater settles in primary sedimentation tanks. The skimmings and sludge from this process are then sent to the solids handling process, which includes blending and thickening in a dissolved air flotation tank followed by anaerobic digestion. Secondary treatment involves the use of aeration tanks and secondary clarifiers. The secondary effluent is then disinfected using chlorine. Some of the secondary effluent also undergoes advanced treatment (coagulation, filtration, disinfection) and is then reused on-site in plant processes.

Plant E

At this plant, influent wastewater is screened and flows through a primary sedimentation process. This allows most of the solid material to settle to the bottom of the tanks and be removed. Next, oxygen is added to the wastewater to grow naturally occurring microscopic organisms. Secondary treatment is accomplished by means of a pure oxygen activated sludge process. The wastewater then travels to secondary clarifiers where the organisms settle to the bottom and are removed, leaving cleaned water. Next, the water is chlorinated to remove any remaining harmful organisms. The wastewater travels through a two-mile outfall. Before entering the river, sulfur dioxide is added to neutralize the chlorine. The solids train includes dissolved air floatation thickeners (DAFTs), anaerobic digesters, solids storage basins (SSBs), and lined dedicated land disposal units (LDLDs).

Plant F

This is an advanced treatment plant that consists of screening and grit removal, primary sedimentation, secondary treatment (biological nutrient removal or BNR), secondary clarification, filtration, disinfection, and dechlorination before the wastewater is discharged into the environment.

Raw sewage first flows through screens for grit removal and then enters the primary settling tanks. The primary settling tanks have residence times of a little less than two hours, and settle out some of the particulate load carried by the sewage. The supernatant from these tanks, called the primary effluent, is then pumped into activated sludge biological nutrient removal treatment.

There are parallel BNR areas, which are identical in function, although, for historical reasons are called "Secondary" and "Nitrification." Each side is subdivided into banks of four compartments. Wastewater flows from an anaerobic compartment, to an aerobic, anoxic, and a finally aerobic compartment in series. The first anaerobic compartment removes readily degradable chemical oxygen demand (COD) and phosphorus. The second compartment removes biochemical oxygen demand (BOD) and converts ammonia to nitrite and nitrate. Dissolved oxygen concentrations are maintained at 2 to 2.5 mg/l in this compartment. The third compartment is anoxic and converts nitrite and nitrate to nitrogen gas. A fraction of the primary effluent is pumped directly into this compartment to provide a source of BOD for oxidation. The final, aerobic, compartment oxidizes any residual ammonia. Dissolved oxygen in this compartment is maintained at 4 to 4.5 mg/l. The total residence time in the BNR activated sludge process is four to five hours. Water from the process is pumped into clarifiers to settle out the solids.

The sludge stream from the BNR process is sent for thickening via dissolved air flotation, followed by anaerobic digestion (with a residence time of 25-30 days) and delivery to sludge lagoons and solar drying ponds. Water from the clarifiers is then pumped through tertiary filters for additional solids removal. After filtration, the water is chlorinated for disinfection, and then dechlorinated using sulfur dioxide before discharge to the environment.

Plant G

This discharger owns and operates a Publicly-Owned Treatment Works. The treatment system consists of primary sedimentation, activated sludge biological treatment with nitrification and denitrification, secondary sedimentation, inert media filtration, chlorination, and dechlorination. The facility design flow is 25 million gallons per day and about 13.9 MGD is currently treated. About 5 MGD is reused.

A.1 Objectives

The objective of this study was to assess the bioavailability of mercury in the effluents of the seven plants described above. The definition of bioavailable mercury includes not only methylmercury, the form that readily bioaccumulates in the tissues of aquatic organisms, but also inorganic bioavailable mercury; consisting of those species which may become methylated, principally by organisms at the bottom of aquatic foodchains (i.e. sulfate reducing bacteria). Therefore, the principal objective was to measure methylmercury (MeHg) and an operationally defined surrogate for inorganic bioavailable mercury, reactive mercury (R-Hg). Since the definition of bioavailability includes both adsorbed and dissolved MeHg, total MeHg was measured in the samples. By definition, R-Hg is measured in the dissolved fraction. In addition to these two species, both unfiltered and filtered total mercury was measured as well; the former to enable us to calculate %MeHg and %R-Hg as a percentage of the total mercury in the sample. Dissolved mercury (D-Hg), was measured so that partition coefficients for total mercury could be calculated. For one round of sampling, dissolved MeHg was measured so that partition coefficients could be measured for MeHg for each plant. These measurements were required so that a correction could be applied to D-Hg, in order to calculate inorganic dissolved mercury ID-Hg. There is no direct way of measuring ID-Hg. This further enables us to calculate inorganic bioavailable mercury IB-Hg, using procedures developed earlier in the study, and compare it to the measurements made for R-Hg, which should be equivalent. If so, the WWTPs may measure R-Hg as a surrogate for IB-Hg.

While MeHg and total mercury (T-Hg) measurements have been made in a number of WWPTs, this study represents the first to measure bioavailable mercury, in terms of MeHg and R-Hg.

A.2 Sampling Schedule and Procedures

Sampling was conducted weekly for 10 weeks beginning June 26, 2008 and ending August 27, 2008. Samples were collected by the participating WWTPs and submitted for analysis to Frontier Geosciences, Inc. in Seattle, WA. All samples were collected by WWTP employees or contractors using Method 1669. Field blanks were collected with each set of samples. During the course of the study, one field duplicate was collected from each of the seven participating WWTPs. Six of the seven facilities collected field measurements typically consisting of temperature, pH, conductivity, DO, and turbidity, although some facilities reported slightly different parameters.

A.3 Analytical Suite and Methods

Upon receipt by the analytical laboratory, Frontier Geosciences, Inc., samples for reactive Hg and dissolved Hg were filtered through a pre-cleaned 0.45μm filtration unit. The aliquot for reactive mercury was given directly to the analyst for analysis the day of receipt. The aliquot for dissolved Hg was preserved to 1-3% with a bromine monochloride (BrCl) solution. Water samples for total mercury and methyl mercury were preserved to 0.4% with concentrated hydrochloric acid. After sitting for at least 3 hours, samples were split for total mercury and methyl mercury. Samples for total mercury were oxidized with1-3% BrCl. Bottles for DOC, TSS, pH, sulfate and chloride were subcontracted to Test America for analysis.

Water samples for total and dissolved mercury determination were allowed to oxidize with BrCl at least overnight according to method FGS-012 prior to analysis. Water samples for methyl mercury determination were distilled according to method FGS-013 prior to analysis. There is no sample preparation for reactive Hg determination.

Total, dissolved and reactive mercury were analyzed in water samples by cold vapor atomic fluorescence spectrometry (CVAFS) according to method FGS-069, based on EPA Method 1631. For reactive Hg, the sample is treated exactly the same as samples for total and dissolved Hg except that no BrCl is used to oxidize the samples; only Hg that is reducible by stannous chloride is detected. Methylmercury was analyzed in distilled water samples by cold vapor gas chromatography atomic fluorescence spectrometry (CV-GC-AFS) according to method FGS-070, based on draft EPA Method 1630.

Due to foaming of the samples during the reactive Hg determination for the first set of samples, it was decided to analyze 50mL aliquots instead of the usual 100mL aliquot. The results were all below the MRL (maximum reporting limit) for a 50mL aliquot (1.0 ng/L) so an additional calibration standard was analyzed at half the MRL (0.5 ng/L for a 50mL aliquot). This standard recovered at 107%. Thus results are reported down to the MDL (maximum detection limit) (0.3 ng/L for a 50mL aliquot).

A.4 Data Review

Several quality related issues were addressed using the QA/QC samples collected during the study. The first was the issue of mercury in field blanks. As detailed below, the researchers feel this issue is mostly a result of the methods used to obtain the field blanks and the results do not reflect major concerns with the sampling strategy and approach.

From time to time, mercury was present in the field blanks above the reporting limit of 0.5 ng/L[31]. This occurred in almost every sampling round, with between 0 and 4 facilities having reportable values in any given sampling. Twenty-two samples out of 70 collected (about 31%) showed reportable mercury in the field blanks with levels ranging from 1.8% of total mercury measured in the primary sample to a maximum of 337%. The geometric mean of the percentage of field blank total mercury to mercury in the primary samples (when it was reportable) was about 35%. No single facility had consistently reportable values in the field blanks. The highest number of reportable field blanks out of seven sampling rounds at an individual facility was 5, the lowest, 0.

The data was reviewed carefully in an attempt to isolate any samples with reportable field blank levels that appeared to have anomalous speciation results. First, those samples with field blank levels less than 50% of the total mercury in the primary samples were identified along with samples having field blank levels greater than 50% of the total mercury in the primary sample. These subsets of samples from each facility were then examined with respect to speciation results (i.e. %MeHg, %R-Hg, %dissolved, %B-Hg) and compared to the speciation results from samples with no reportable mercury in the field blanks. In particular, the data were inspected to see if any of these samples had the highest or lowest levels of these results out of all the samples from that facility. This analysis revealed no particular patterns and none of the samples stood out as being anomalous. One sample with a reportable level in the field blank had the lowest %MeHg, %dissolved, and %R-Hg compared to the rest of the samples from that facility. On closer inspection however, this was deemed to be because this sample had the highest total

[31] Field blanks were analyzed for total mercury in an unfiltered sample.

mercury in the primary sample of all the samples from that facility and was probably not due to contamination of the sample, as the level of T-Hg in the field blank was only 6.1% of that of the primary sample result for total mercury.

There is also an important fact to consider when making judgments about the data based on the field blank results. That is, that the method for collecting field blanks did not mimic exactly the procedures for collecting the primary samples. Field blanks were made using "mercury-free" (< 0.5 ng/L) water from the analytical laboratory. This water was shipped to each facility in either 2.5L jugs on in 0.5L bottles. If the field blank water came in a jug, it was poured into a sample bottle in the field. If it came in a 0.5L bottle the bottle was simply opened to the atmosphere during sampling. Those facilities that used a peristaltic pump to collect samples pumped the water through tubing if it came in a 2.5L jug. The actual sampling involved either opening and immersing the bottle underwater to fill it after triple rinsing, or filling the bottle using a peristaltic pump. Thus, it is possible that mercury could have been introduced into a field blank and not into a primary sample, and vice versa. Because of these differences in collection procedures, there was no guarantee that the primary sample was contaminated based on a reportable result in the field blank. Based on these facts and examination of the data, all of the data were determined to be of adequate quality, and all the samples were retained for subsequent analysis.

The quality of the data was also examined using field duplicates. Relative percent differences (RPDs) were computed for each available analyte for each facility. RPDs were calculated for unfiltered T-Hg, filtered T-Hg, R-Hg, and MeHg, unless one of the parameters had a value less than the detection level in either the primary sample of duplicate. Five of 28 primary/duplicate pairs had RPDs >25%. Of these five, two occurred in the measurement of R-Hg, two for MeHg, and one for unfiltered T-Hg. The absolute values of these RPD values were 34% and 80%, 82% and 36%, and 77% respectively. Therefore, only three were widely outside of the 25% range normally regarded as acceptable. None of these data were eliminated or modified based on field duplicate results.

The data review revealed several other issues with the data. From time to time, (8 of 70 samples) filtered T-Hg was reported at values greater than unfiltered T-Hg. Due to the variability in the T-Hg measurement and, in general, the high percentages of total mercury in the dissolved phase (due to very low suspended solids concentrations) this was inevitable. Four of these eight occurrences were at one facility that had the second lowest average T-Hg and all TSS levels below detection. When this occurred, the dissolved level was reset to the unfiltered T-Hg level in any calculations involving the data.

Facility A had one sample from the July 22 event in which all mercury levels were about an order of magnitude higher than the rest of the samples from that facility. Other measured parameters (TSS, DOC) from this date were similar to the other events. The reason for these very high readings is unknown. However, the speciation of the sample was consistent with speciation of the remainder of the samples from the facility. Therefore, the sample was retained and used in subsequent calculations. This data point was, however, excluded from the averages in Table A-1 because in each instance the Grubbs Test identified it as an outlier.

Facility B had a filtered total Hg result on August 19 of 89 ng/L, nearly two orders of magnitude higher than the typical dissolved Hg result for this facility. This aliquot was obviously contaminated. Other results for the sample on this date for this facility were in their normal ranges. This result was eliminated from the statistical analysis of the data.

A.5 Results

Data from the study are summarized in Tables A-1 and A-2. Table A-1 contains the mercury data and Table A-2 contains the data for conventional water quality parameters which support various calculations and analysis.

Four mercury measurements were made at seven facilities over a 10-week period. These included unfiltered total mercury, filtered total mercury, reactive mercury, and methylmercury. During the last week of sampling, a single measurement of dissolved (filtered) methylmercury was made in each facility's effluent. Averages, standard deviations, and coefficients of variation were computed for each measured species.

Mean unfiltered total mercury ranged from 1.78 ng/L at Facility F to 13.4 ng/L at Facility D. With the exception of the outlier measured at Facility A on 7/22, the highest level recorded was 29.3 ng/L. With one exception, however, the average effluent concentration was less than 5 ng/L. CVs ranged from 34.4% to 94.8%. The average of all the CVs over all facilities was 51.4%.

Mean filtered total mercury ranged from 1.30 ng/L at Facility B to 3.41 ng/L at Facility C. Overall, the lowest recorded dissolved total mercury was 0.8 ng/L to and the highest was 5.27 ng/L, the data from 7/22 at Facility A and 8/19 at Facility B excepted. The average effluent concentration was less than 3.5 ng/L. CVs ranged from 25.3% to 65.8%. The average of the CVs over all facilities was 36.2%. It is reasonable that the dissolved Hg measurements are less variable as filtration of the sample removes the variability induced by mercury associated with solids.

Mean reactive mercury ranged from 0.25 ng/L at Facilities B and D to 0.84 ng/L at Facility E[32]. Overall, individual reactive mercury values ranged from <0.3 ng/L to 1.62 ng/L. Without the atypically high value measured at Facility A, average effluent concentration was less than 0.84 ng/L. CVs ranged from 34.3% to 41.3%. The average of all the CVs over all facilities was 37.6%. Since reactive Hg is measured on a filtered sample, the CVs for the reactive mercury and dissolved total mercury should be, and are, similar.

Mean methylmercury ranged from 0.04 ng/L at Facility G to 0.84 ng/L at Facility E[33]. Overall, the individual recorded methylmercury values ranged from <0.019 ng/L to 1.06 ng/L. Average effluent concentration was less than 0.84 ng/L. CVs ranged from 16.8% to 138%. The average of all the CVs over all facilities was 58.9% and was similar to that for total unfiltered Hg.

TSS concentrations were for the most part non-detectable at <4 mg/L, indicating the very efficient filtration of solids by these facilities. Facilities A, D, and E consistently had detectable levels, while Facilities B, C, F and G were almost always below 4 mg/L. The average of the values at facilities A, D, and E ranged from 5.3 at Facility E to 9.3 at Facility D. The overall CV for the detectable values was 32%.

---

[32] The mean values reported here include any data that were measured as <0.3 ng/L at ½ the detected level. Twenty-five (25) out of the 70 measured values (36%) were nondetectable. Standard deviations and CVs were not computed for an analyte if five or greater values were nondetectable at a given facility.

[33] The mean values reported here include any data hat were measured as <0.019 ng/L at ½ the detectable level. Ten (10) out of the 70 measured values (14%) were nondetectable. Standard deviations and CVs were not computed for an analyte if five or greater values were nondetectable at a given facility.

Table A-1. Mercury Data from the Reactive Mercury Study.

		Facility							
Parameter	**Date**	**A**	**B**	**C**	**D**	**E**	**F**	**G**	
Unfiltered	6/26/2008	5.08	8.56	6.03	29.3	10.6	2.88	3.1	
Total Hg	7/2/2008	8.25	2.7	4.85	18.4	5.1	3.16	3.67	
ng/L	7/9/2008	2.54	2.97	3.17	9.6	3.53	1.5	1.32	
	7/15/2008	7.12	1.4	3.94	12.5	4.21	1.45	1.12	
	7/22/2008	44.3	2.27	5.2	13.7	4.58	2.04	2.72	
	7/31/2008	6.6	1.25	4.64	12.3	4.31	1.67	1.51	
	8/6/2008	5.4	1.42	2.94	8.66	4	1.48	0.9	
	8/14/2008	3.3	1.3	3.1	10.6	4.73	1.3	1.06	
	8/19/2008	2.49	1.08	1.88	11.4	4.61	1.25	1.62	
	8/27/2008	2.39	1.02	2.54	7.35	3.49	1.05	1.01	
Mean		*4.80*	*2.40*	*3.83*	*13.38*	*4.92*	*1.78*	*1.80*	
Std Dev		*2.22*	*2.27*	*1.32*	*6.37*	*2.06*	*0.71*	*0.99*	*Overall CV*
CV (%)		*46.30*	*94.85*	*34.44*	*47.61*	*41.93*	*39.84*	*54.92*	*51.41*
Filtered	6/26/2008	2.11	1.97	5.27	2.51	4.04	1.44	1.89	
Total Hg	7/2/2008	2.34	1.77	4.35	3.48	4.66	2.38	1.95	
ng/L	7/9/2008	1.91	1.04	1.79	1.56	2.64	1.06	1.22	
	7/15/2008	1.59	1.2	3.62	3.03	2.67	1.3	0.8	
	7/22/2008	16.6	1.39	3.3	3.1	2.94	1.28	4.87	
	7/31/2008	3.35	1.47	4.43	3.7	3.4	1.48	1.57	
	8/6/2008	1.95	0.69	2.76	2.84	2.3	1.05	0.82	
	8/14/2008	1.55	1.32	5.01	4.19	4.26	1.25	2.43	
	8/19/2008	1.42	89.2	2.1	3.35	2.4	1.02	1.49	
	8/27/2008	1.3	0.87	1.43	2.25	2.61	0.86	1.09	
Mean		*1.95*	*1.30*	*3.41*	*3.00*	*3.19*	*1.31*	*1.81*	
Std Dev		*0.63*	*0.41*	*1.36*	*0.76*	*0.85*	*0.42*	*1.19*	*Overall CV*
CV (%)		*32.14*	*31.45*	*39.97*	*25.26*	*26.54*	*32.26*	*65.78*	*36.20*
Reactive Hg	6/26/2008	0.54	0.64	0.76	<0.3	0.77	0.55	0.77	
ng/L	7/2/2008	0.93	0.52	0.62	0.38	0.95	0.56	0.77	
	7/9/2008	0.64	<0.3	<0.3	0.4	1.14	0.39	0.67	
	7/15/2008	0.65	<0.3	0.35	1.13	0.59	0.47	<0.3	
	7/22/2008	7.53	<0.3	0.33	<0.3	0.89	<0.3	0.4	
	7/31/2008	1.14	<0.3	<0.3	<0.3	0.61	<0.3	<0.3	
	8/6/2008	0.59	<0.3	<0.3	<0.3	0.37	<0.3	0.35	
	8/14/2008	0.53	0.31	0.52	<0.3	0.86	0.35	0.82	
	8/19/2008	0.45	<0.3	<0.3	<0.3	0.64	<0.3	0.52	
	8/27/2008	0.37	<0.3	<0.3	0.61	1.62	<0.3	0.57	
Mean		*0.65*	*0.25*	*0.33*	*0.25*	*0.84*	*0.31*	*0.52*	
Std Dev		*0.24*	-	-	-	*0.35*	-	*0.18*	*Overall CV*
CV (%)		*37.24*	-	-	-	*41.32*	-	*34.31*	*37.62*
MeHg	6/26/2008	0.053	0.052	1.06	0.219	0.75	<0.019	0.032	
ng/L	7/2/2008	0.078	<0.019	0.398	0.297	0.591	<0.019	<0.019	
	7/9/2008	0.076	<0.019	<0.019	0.096	0.859	<0.019	<0.019	
	7/15/2008	0.053	0.063	0.111	0.281	0.714	<0.019	0.073	
	7/22/2008	0.497	0.104	0.155	0.339	0.807	0.064	0.061	
	7/31/2008	0.147	0.041	0.08	0.364	0.95	0.032	0.037	
	8/6/2008	0.11	0.02	0.055	0.325	0.949	0.045	0.031	
	8/14/2008	0.174	0.033	0.201	0.386	0.856	0.047	0.026	
	8/19/2008	0.089	0.022	0.152	0.38	0.845	0.061	0.04	
	8/27/2008	0.064	<0.019	0.083	0.024	1.1	0.07	0.034	
Mean		*0.09*	*0.04*	*0.23*	*0.27*	*0.84*	*0.04*	*0.04*	
Std Dev		*0.04*	*0.03*	*0.32*	*0.12*	*0.14*	*0.01*	*0.02*	*Overall CV*
CV (%)		*45.13*	*80.29*	*138.33*	*45.41*	*16.76*	*39.91*	*46.57*	*58.91*
Filtered MeHg *ng/L*	8/27/2008	0.058	<0.019	0.041	<0.019	0.962	<0.019	0.021	

Table A-2. Conventional Parameter Data from the Reactive Mercury Study.

Parameter	Date	A	B	C	Facility D	E	F	G	
TSS	6/26/2008	4	<4	<4	9	4.5	<4	<4	
mg/L	7/2/2008	8	<4	<4	12	5	<4	<4	
	7/9/2008	4	<4	<4	13	5	<4	<4	
	7/15/2008	10	<4	<4	6	5	<4	<4	
	7/22/2008	9.5	<4	<4	9	5.5	<4	<4	
	7/31/2008	4	<4	<4	8	5.5	<4	<4	
	8/6/2008	5.5	<4	<4	10	6	<4	<4	
	8/14/2008	5	<4	<4	11	6.5	<4	<4	
	8/19/2008	<2	<2	<4	5.5	2	<2	<2	
	8/27/2008	<4	<4	5	10	8	<4	<4	
Mean		*6.25*	*<4*	*2.30*	*9.35*	*5.30*	*<4*	*<4*	
Std Dev		*2.54*	-	-	*2.40*	*1.53*	-	-	*Overall CV*
CV (%)		*40.57*	-	-	*25.71*	*28.89*	-	-	*31.72*
DOC	6/26/2008	10	13	44	16	12	10	7.1	
mg/L	7/2/2008	9.7	11	47	19	13	13	6.5	
	7/9/2008	7.6	8.5	46	15	11	8.4	6.6	
	7/15/2008	10	11	45	18	12	8.7	7.9	
	7/22/2008	13	12	48	22	12	11	7.9	
	7/31/2008	8.1	12	61	23	11	7.8	6.9	
	8/6/2008	8.1	11	47	19	10	9	6.5	
	8/14/2008	8.3	10	15	15	9.1	8.8	7.2	
	8/19/2008	11	10	57	20	12	8.1	7.2	
	8/27/2008	6.8	8.5	59	18	15	11	8.8	
Mean		*9.26*	*10.70*	*46.90*	*18.50*	*11.71*	*9.58*	*7.26*	
Std Dev		*1.85*	*1.48*	*12.82*	*2.72*	*1.62*	*1.65*	*0.74*	*Overall CV*
CV (%)		*19.95*	*13.79*	*27.33*	*14.69*	*13.82*	*17.18*	*10.21*	*16.71*
Sulfate	6/26/2008	53.2	85.5	523	31.2	87.2	103	98.6	
mg/L	7/2/2008	65.7	88	380	27.6	97.4	104	104	
	Mean	*59.45*	*86.75*	*451.5*	*29.4*	*92.3*	*103.5*	*101.3*	
	Std Dev	*8.84*	*1.77*	*101.12*	*2.55*	*7.21*	*0.71*	*3.82*	*Overall CV*
	CV(%)	*14.87*	*2.04*	*22.40*	*8.66*	*7.81*	*0.68*	*3.77*	*8.60*
Chloride	6/26/2008	238	183	157	46.6	98.8	190	127	
mg/L	7/2/2008	348	227	178	51.8	95.2	184	124	
	Mean	*293*	*205*	*167.5*	*49.2*	*97*	*187*	*125.5*	
	Std Dev	*77.78*	*31.11*	*14.85*	*3.68*	*2.55*	*4.24*	*2.12*	*Overall CV*
	CV(%)	*26.55*	*15.18*	*8.87*	*7.47*	*2.62*	*2.27*	*1.69*	*9.24*

Mean DOC concentrations ranged from 7.3 mg/L to 46.9 mg/L. However, Facility C, with the highest DOC concentrations, receives the effluent from two paper mills and is therefore has a very different effluent than the rest of the plants. Excluding this facility, the high end of the DOC means was 18.5 mg/L at Facility D. CVs ranged from 10.2% to 27.3% with an overall average of 16.7%.

Sulfate and chloride concentrations were measured for the first two rounds of sampling. Mean sulfate concentrations ranged from 29.4 mg/L to 452 mg/L. The highest sulfate concentration was measured at Facility C, which is influenced by influent from the paper mills. Excluding this facility, the highest mean sulfate concentration was 103.5 mg/L. The average CV for the sulfate measurements was 8.6%. Mean chloride concentrations ranged from 49.2 mg/L to 187 mg/L. Interestingly, the facilities with both the lowest and the highest chloride concentrations are located in coastal areas. The average CV for chloride concentrations was 9.2%, similar to sulfate.

pH was measured by the laboratory at each facility both for the first two sampling rounds as well. However, it was determined that field measurements of pH being made by facility personnel were reasonable surrogates for laboratory pH. Therefore, laboratory pH measurements were suspended after the second round and field pH measurements were relied upon thereafter. The one exception was Facility E, which did not make field measurements. Therefore, the pH data for Facility E were all measured in the laboratory. Field measurements from all the facilities are shown in Table A-3[34].

A.6 Data Analysis

The primary objectives of the data analysis were to examine the relationships among the various species of mercury measured, to assess the level of bioavailable mercury species, and to examine the relationships between mercury species and other conventional water quality parameters. The speciation of the mercury present in these effluents is summarized in Figure A-1. Overall, it is clear that reactive Hg is greater than or similar to MeHg for all plants and that MeHg is <5% for all except Facility E. Facility G has a very high %R-Hg compared to the other plants, while Facility D has the lowest % R-Hg. Overall, there is little relationship between %R-Hg and %MeHg (Figure A-2).

Unfiltered methylmercury as a percent of total unfiltered mercury ranged from 2.0% (Facility B) to 19% (Facility E). Reactive mercury as a percent of total unfiltered ranged from 3.1% (Facility D) to 34% (Facility G). Bioavailable mercury, the sum of these two species, ranged from 5.3% (Facility D) to 38% (Facility E). There did not appear to be a strong relationship between %MeHg and %R-Hg, as evidenced by Figure A-2. The average total bioavailable mercury at the seven facilities was 21%. This percentage corresponds closely to the percentage of bioavailable mercury anticipated by the analysis in Phase.0 I, Chapter 5, which suggested that bioavailable Hg in wastewater effluents should on the order of 20% or less.

Figure A-3 shows the percentage of dissolved total mercury in these seven effluents. This percentage ranges from about 25% to over 86%. The primary reason the dissolved percentages are high is the very efficient removal of solids by these facilities. The logK_Ds for total mercury in these effluents are shown in Figure A-4. These K_Ds range from 4.7 to 5.5. As such, they are somewhat lower than the typical K_D measured for mercury in natural waters[35]. Regarding concentrations of total mercury in suspended solids, calculations show an overall mean of 0.46

[34] Not all facilities reported all field parameters requested and are thus missing from the table. Quality assurance of these values was the responsibility of the individual facilities collecting the information.

[35] The TSS MDL in these analyses was typically 4 mg/L. The MDL was used when calculating these K_Ds rather than ½ the MDL for the following reason. A plot of laboratory TSS versus field measured turbidity values showed a reasonable linear relationship between the detected TSS values and turbidity. However, the intercept of the linear regression was 3.85 mg/L TSS. Thus using the field measured turbidity values to correct the nondetectable TSS values would have given values very close to 4 or greater. Using a value of ½ the MDL may indeed have biased them upward. Therefore, it was decided to use the MDL for the K_D calculations. However, the reader is cautioned that these values may be biased due to the assumption used in their calculation.

Table A-3. Field Data Measured in the Reactive Mercury Study.

		Facility						
Parameter	Date	A	B	C	D	E	F	G
Conductivity	6/26/2008			1978	762	816		1051
uS / cm	7/2/2008			2080	802	878		1034
	7/9/2008	1322		2080	668	870		1042
	7/15/2008			1805	699	806		1016
	7/22/2008			1930	669	851		1150
	7/31/2008	1034		2130	801	910		1049
	8/6/2008	1151		1890	489	808		975
	8/14/2008	1181		2180	758	850		1050
	8/19/2008	1184		1970	720	850		1012
	8/27/2008	1173		2010	700	780		945
DO	6/26/2008		7.3	5.08	1.94	1.3		6.4
mg/L	7/2/2008	8.98	7.4	5.4	5.69	2.1		6.5
	7/9/2008	5.41	8.9	5	2.61	2.7		6.5
	7/15/2008	7	7.5	5.9	1.61	4.4		6.2
	7/22/2008	5.79	7.5	5.5	1.55	5.1		6.2
	7/31/2008	5.64	7.5	4.9	3.05	8.5		6.3
	8/6/2008	5.42	7.3	4.9	2.85	1.6		6.3
	8/14/2008	5.13	7.2	4.7	3.21	6.9		6.4
	8/19/2008	5.72	7.3	4.8	1.44	2.1		6.3
	8/27/2008	6.51	7.4	5.3	1.43	3		6.4
pH	6/26/2008	6.52	7.0	7.37	7.4	6.2	7.32	7.25
	7/2/2008	7.07	7.0	7.27	7.5	6.3	7.35	7.26
	7/9/2008	6.61	7.0	7.37	7.4	6.3	7.45	7.63
	7/15/2008	6.82	7.1	7.24	7.4	6.2	7.46	7.16
	7/22/2008	6.94	7.0	7.34	7.3	6.4	7.4	7.41
	7/31/2008	6.72	6.9	7.5	7.5	6.3	7.42	7.23
	8/6/2008	6.96	7.0	7.4	7.5	6.2	7.29	7.33
	8/14/2008	6.9	7.0	7.54	7.6	6.4	7.26	7.22
	8/19/2008	7.03	7.0	7.31	7.4	6.3	7.19	7.12
	8/27/2008	6.76	7.1	7.36	7.3	6.3	7.1	7.17
Temperature	6/26/2008	20	21.0	32.56	21.56	26.67		27
°C	7/2/2008	20.36	21.0	34.89	24.17	26.78		26.8
	7/9/2008	20.57	21.8	31.89	24.61	27.78		27
	7/15/2008	20.74	21.9	32.28	24.56	27.33		27.9
	7/22/2008	21.96	23.3	32.17	22.06	23.78		28.2
	7/31/2008	21.08	23.6	34	22.94	27.83		27.8
	8/6/2008	21.03	24.0	34.3	25.56	28.05		28.4
	8/14/2008	21.13	23.5	34.78	26	28.22		29
	8/19/2008	21.29	23.8	35.28	24.56	27.89		28.4
	8/27/2008	20.68	23.1	35.06	23.06	28		28.4
Turbidity	6/26/2008		0.2	2.5	12.3	5.2		
NTU	7/2/2008		0.3	5.5	13.3	3.2		
	7/9/2008	2.7	0.7	8	8.97	3.6		
	7/15/2008		0.2	5.8	10.3	6.2		
	7/22/2008	6.8	0.2	5.7	9.33	6.1		
	7/31/2008	0.6	0.2	8.8	7.9	5		
	8/6/2008	2.8	0.2	5.1	5.2	5.5		
	8/14/2008	3.7	0.2	2.3	6.78	7.5		
	8/19/2008	2.2	0.3	6.1	11.3	5.8		
	8/27/2008	0.6	0.1	6.58	4.85	7.2		
Flow	6/26/2008	34.05	128	34.16				
MGD	7/2/2008	38.79	97	40.77				
	7/9/2008	37.91	221	30.31				
	7/15/2008	32.54	109	36.91				
	7/22/2008	34.32	94	34.66				
	7/31/2008	40.17	98	34.77				
	8/6/2008	38.92	88	33.83				
	8/14/2008	34.99	91	30.29				
	8/19/2008	35.51	86	32.98				
	8/27/2008	32.83	82	34.35				

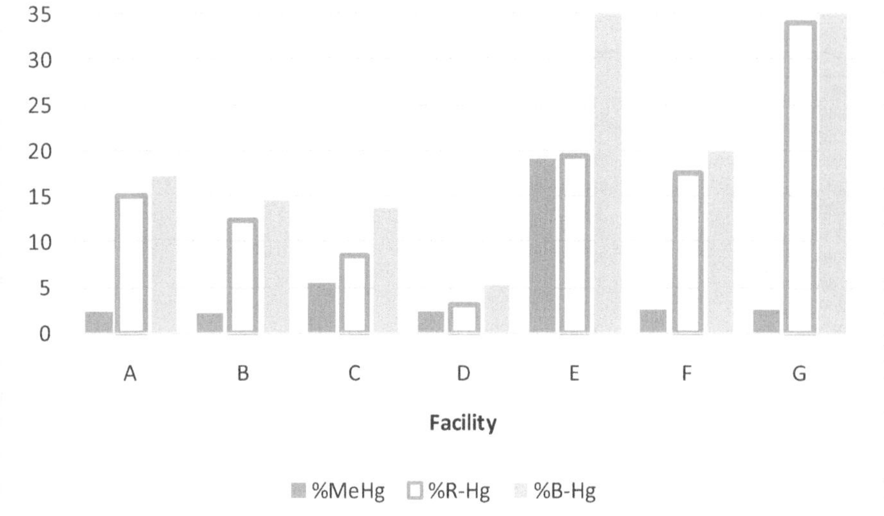

Figure A-1. Relationships Among Bioavailable Mercury Species in Seven Wastewater Effluents.

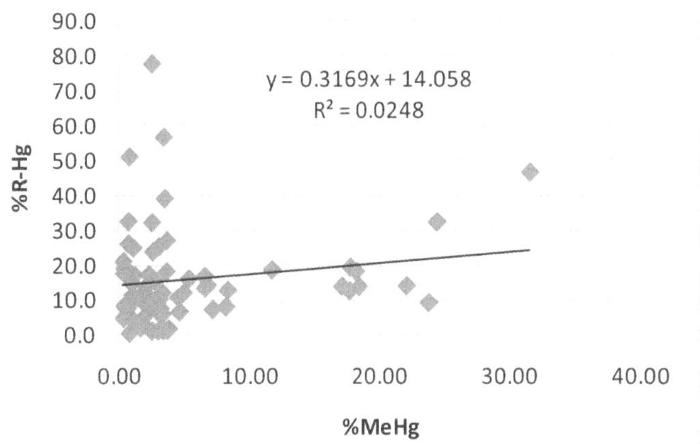

Figure A-2. Relationship Between %MeHg and %R-Hg in Seven Wastewater Effluents.

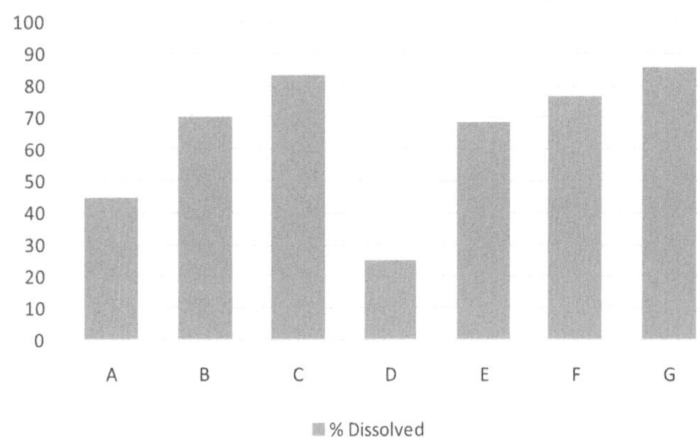

Figure A-3. Percent Dissolved Mercury in Seven Wastewater Effluents.

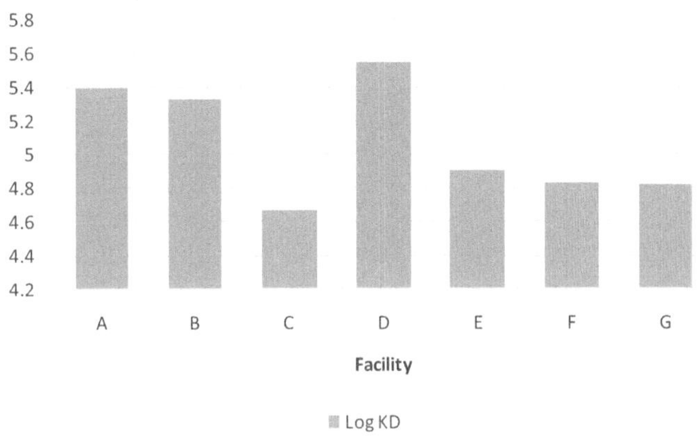

Figure A-4. LogK$_D$s for Total Mercury Measured in Seven Wastewater Effluents.

mg/kg[36]. Means for individual facilities ranged from 0.122 to 1.18 mg/kg. The highest observed individual concentration was 2.98 mg/kg and the minimum observed individual concentration was 0.012 mg/kg.

In general, the researchers observed no significant relationships among the mercury species and the conventional parameters. There was a weak relationship between chloride levels and %R-Hg as evidenced in Figure A-5. A weak relationship was also observed between DO levels in the effluents and %MeHg, as shown in Figure A-6.

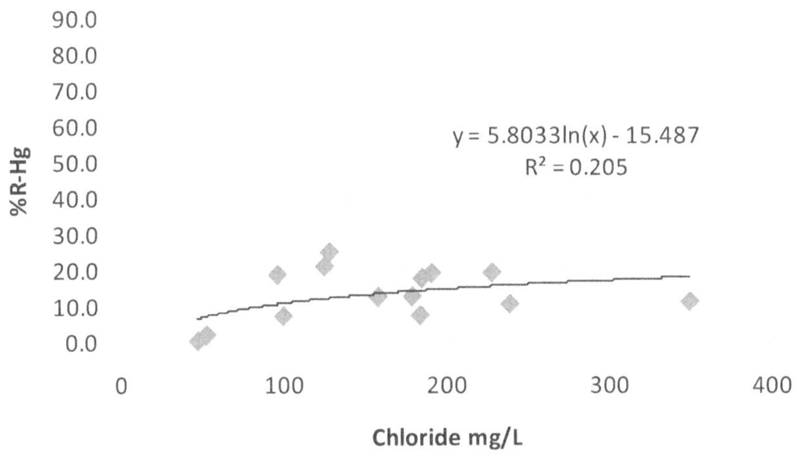

Figure A-5. Relationship Between Chloride and %R-Hg in Seven Wastewater Effluents.

[36] Values are calculated as total unfiltered mercury in the sample minus total filtered mercury divided by the TSS concentration.

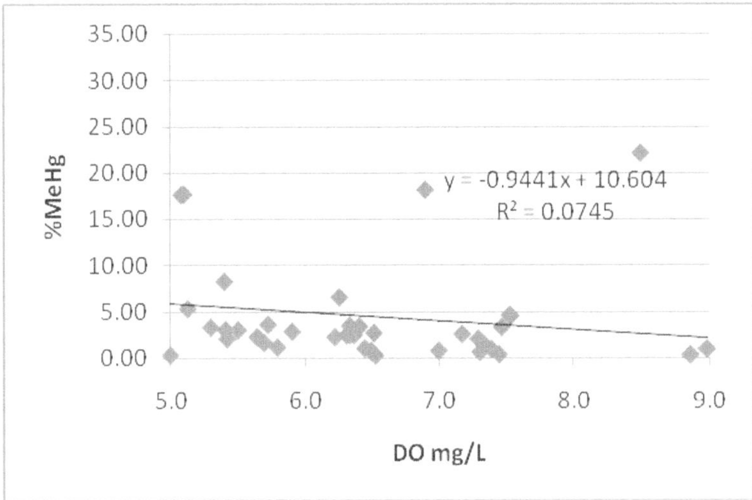

Figure A-6. Relationship Between DO and %MeHg in Seven Wastewater Effluents.

Perhaps the most significant finding of this research was the confirmation of the theorized relationship between the fraction of reactive to dissolved mercury (expressed as a percent) with DOC. This relationship was parameterized as a declining exponential function in Phase I, Chapter 5.0, based on speciation model results. Using the measurements in this study, this plot is reconstructed as Figure A-7.

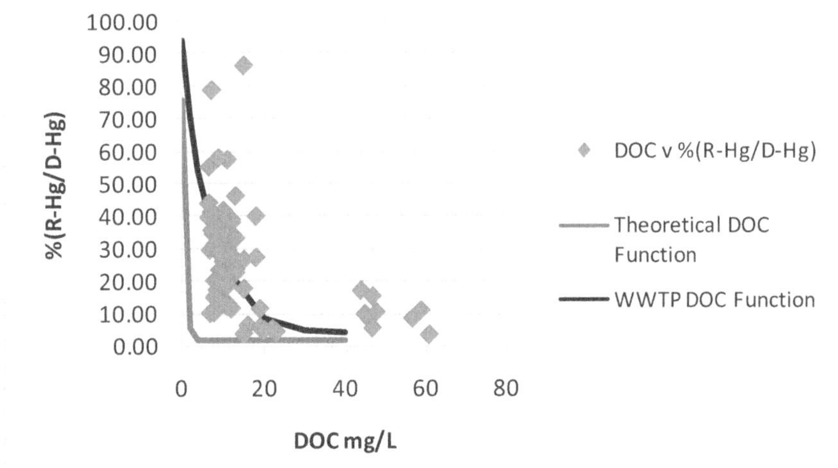

Figure A-7. Relationship Between DOC and Reactive Mercury as a Percent of Dissolved Inorganic Mercury.

From this figure it is apparent that the levels of reactive mercury in these seven wastewater effluents are higher than would be predicted based on the DOC characteristics and binding constants reported for natural waters. Therefore a new function was developed based on the observed data for this study for wastewater effluents.

The plot reveals a cluster of data points in the 40 to 60 mg/L DOC range. These points are all from one of the treatment plants in this study, Facility C, which receives a significant inflow from two paper mills. As a result, the DOC levels for this effluent are much higher than the others and indeed much higher than for most natural waters, even those with relatively high ambient DOC levels. When the data were plotted on a semi-log plot, it was obvious that these data did not fall onto the same line as the remainder of the data. Therefore, it was decided that these ten data points would be excluded when formulating the general relationship for wastewater treatment plants to be used in the Bioavailability Tool and screening assessments.

The resulting regression equation for the remaining data is:

$$\%(R\text{-}Hg/D\text{-}Hg) = 4 + 90.1e^{-0.147(DOC)} \qquad \text{Equation A-1}$$

which has an R^2 of 0.2839.

While there is an apparent relationship between DOC and reactive mercury, it by no means explains the major proportion of the variance between the two parameters. This means that there are other factors that may be as important or of greater importance than DOC in predicting reactive mercury levels.

While no other obvious relationships were observed among the data, there are a few interesting notes to be made concerning the data. In Figure A-7, there is another cluster of data points at a DOC level of about 20 mg/L having %(R-Hg/D-Hg) levels of less than about 10%. These data are also all from one treatment plant, Facility D, which is shown in Figure A-1 to have by far the lowest bioavailable mercury levels of the seven plants. Facility D also has the following characteristics, when compared to the other treatment plants:

- outside of Facility C, the highest average DOC levels of the six plants,
- the highest average effluent pH,
- the highest average TSS levels,
- the highest average partition coefficient (K_D) between mercury and suspended solids,
- the lowest average dissolved oxygen effluent levels,
- the lowest sulfate levels, and
- the lowest chloride levels.

Among the facilities that measured conductivity of the effluent, it was also the lowest, and of those that reported turbidity, it had the highest levels. While Facility D had the lowest % reactive mercury and % MeHg levels, it is also noted that it had the highest mean unfiltered total mercury levels, which tends to make these percentages lower.

Facilities E and G tended to have the highest percentages of bioavailable mercury. In terms of the relationship of this observation to other measured parameters:

- Facility E had the lowest average effluent pH, and
- Facility G had the lowest average DOC level, among the lowest TSS levels, and the highest % dissolved Hg level.

In the foregoing discussion of bioavailability, it should not be overlooked that the three facilities with the best solids removal also had the lowest effluent concentration of both total mercury and methylmercury. Also note that the three facilities with the highest ratios of methyl to reactive mercury had the three highest DOC levels, while those four plants with the lowest methyl to reactive ratios had the four lowest DOC levels. This perhaps indicates a tendency for plants dealing with higher DOC concentrations to have higher methylmercury levels. This is by no means an indication that this is due to methylation within the plant, as higher DOC levels may promote methylation in the influent as well. The DOC may be just acting as a "carrier" by helping keep the MeHg in solution, or preventing its degradation by photochemistry or other processes. Alternatively, the relationship may just indicate that %R-Hg is inversely related to DOC; thus the high DOC plants have a relatively low %R-Hg, and the methyl to reactive ratio is relatively elevated. There is not sufficient information to assess which is the most likely reason for these trends.

It is worthwhile to compare the results of this study to other similar studies that have been done. While there have been few studies that have measured reactive Hg in sewage treatment plant effluents or other waters, the studies of Lamborg et al (2003, 2004) are worth examining. In these studies, these authors used reactive Hg titrations to determine the binding constants for Hg to DOC in a variety of waters. The approach of these authors was used to compare the results between studies. Considering the equilibrium between Hg and DOC, the conditional stability constant (K') for the reaction:

$$Hg^{2+} + DOC = HgDOC$$

can be written as:

$$K' = [HgDOC]/[Hg^{2+}][DOC] \qquad \text{Equation A-2}$$

where all concentrations are in molar units. As discussed by Lamborg et al. (2003), it is possible to rewrite this relationship in terms of reactive Hg if the ratio of $[R\text{-}Hg]/[Hg^{2+}]$ is known. Lamborg et al. (2003) calculated this ratio for a variety of conditions, including the range of pH and Cl concentrations in the effluents used in this study. Based on their work, a value for the ratio of 10^{13} is a reasonable average. Then:

$$K' = 10^{13}[HgDOC]/[\underline{R\text{-}Hg}][DOC]$$

Now, given the discussions above and elsewhere in the report, [HgDOC] is taken to be the unreactive fraction of the dissolved Hg and the above equation can be rewritten as:

$$K' = 10^{13}(1\text{-}f\underline{R\text{-}Hg})/(f\underline{R\text{-}Hg})[DOC] \qquad \text{Equation A-3}$$

where fR-Hg is the fraction of the dissolved Hg that is reactive. The relationship obtained between fR-Hg and DOC (Figure A-7) can then be used to calculate K' as long as the concentration of "binding ligands" in the DOC can be determined. To do this, the estimation of the number of binding ligands per mole of DOC determined by Lamborg et al. (2004) was used (5×10^{-6}). This is similar to the value used in the initial calculations in the Phase I report (Section 5.4.5) of 8.3×10^{-6} binding sites per mol DOC.

Using these relationships, the value of logK' can be determined using the relationship in Equation A-1 over a range of DOC concentrations. The average value determined was logK' = 21.9. This value compares well with, but is lower, than the value estimated by Lamborg et al. (2004) for the sewage treatment plant effluent they measured from Groton, CT (logK' = 23). Alternatively, the actual DOC and reactive Hg concentrations for the individual facilities can be used to calculate the logK' values assuming all have a similar number of binding ligands per mol of DOC. The results of this calculation are shown in Table A-4.

Table A-4. Calculated Binding Constants for Hg to Effluent Organic Matter for Each Facility.

Facility	fR-Hg /Diss.Hg	DOC (mg/L)	logK'
A	0.344	9.26	21.7
B	0.188	10.7	22.0
C	0.101	46.9	21.7
D	0.129	18.5	21.9
E	0.278	11.71	21.7
F	0.231	9.58	21.9
G	0.390	7.26	21.7

The calculated binding constants are similar for all facilities (the variability is 0.3 log units) and to the average value estimated using the R-Hg and DOC regression equation. This is not surprising given the relatively good fit of the equation to the data. The results suggest that there is overall little difference between treatment facilities in the capacity of the DOC to bind Hg and that the relatively fraction of reactive Hg is driven primarily by the amount of DOC in solution. In systems where there is more DOC, the reactive Hg fraction is lower. All facilities have relatively similar DOC except Facility C, and these all have similar R-Hg fractions. Facility C has a much lower R-Hg fraction. However, systems with higher DOC are likely to have a higher dissolved concentration such that the overall concentration of R-Hg may not be lower overall.

Overall, the consistency of these results with the previous work of Lamborg et al. (2003; 2004) and the relationships between fR-Hg and DOC, and their theoretical basis, suggest that the measurement of R-Hg is providing a suitable measure of the fraction of Hg in the dissolved fraction that is bioavailable or reactive on a kinetically fast timescale. Thus, it seems reasonable to use the R-Hg measurement as a surrogate for bioavailability of the inorganic Hg pool in WWTP's and in their receiving waters as well.

APPENDIX B

USGS DATA AND CALCULATED VALUES FOR BIOAVAILABLE INORGANIC MERCURY AND RATE CONSTANTS FOR NET METHYLMERCURY FORMATION AND DESTRUCTION

Data Sources: Krabbenhoft et al., 1999; Brumbaugh et al., 2001

Code	Name	Type	% Wetland	MeHg ng/L	Total Hg ng/L	TOC mg/L	pH	SO4 mg/L	%MeHg	MeHgsed ng/g	Total Sed ng/g	%LOI	%MeHgsed	kdemeth per day	HgD ng/L	IB-Hg ng/L	[MeHg]/[IB-Hg]	kmeth per day
ACAD	Bayou Lacassine	Ag.	--	0.46	1.61	11.1	6.7	3.03	28.57	0.26	88.1	10.06	0.295	0.04	0.64	0.01	35.71	1.431
ACAD	Calcasieu River @ Kinder	A/F	--	0.15	3.68	5.2	6.8	3.7	4.08	0.05	1.9	0.26	2.632	0.09	1.47	0.03	5.01	0.462
ACAD	Bayou Boeuf	Urb.	--	0.05	1.66	8.6	8.9	21.06	3.01	0.1	53.5	5.93	0.187	0.05	0.66	0.01	3.76	0.198
ACAD	Mermentau R. @ Mermentau	Urb.	--	0.09	1.11	8.9	7.5	3.6	8.11	0.06	157	7.87	0.038	0.05	0.44	0.01	10.13	0.515
ACAD	Turtle Bayou nr B. Penchant	Bkg.	--	0.03	1.03	8.6	7.4	29.07	2.91	0.41	83.6	32.46	0.490	0.05	0.41	0.01	3.64	0.192
ALMN	Youghiogheny R.	Mine	2	0.02	1.16	2.1	7.6	94	1.72	0.52	96.78	4.01	0.537	0.29	0.46	0.02	0.82	0.237
ALMN	Clarion River	Mine	1	0.09	8.09	4.2	7.4	39	1.11	1.08	76.5	2.81	1.412	0.12	3.24	0.07	1.30	0.153
ALMN	Dunkard Crk. @ Shann.	Mine	0	0.04	0.64	2.6	7.5	195	6.25	0.15	12.8	4.26	1.172	0.22	0.26	0.01	4.38	0.945
ALMN	Tenmile Crk @ Amity	Bkg.	0	0.1	2.48	3.9	7.7	31	4.03	0.38	20.7	3.04	1.836	0.13	0.99	0.02	4.52	0.584
ALMN	Allegheny R. @ New Kensing.	Urb.	1	0.01	0.85	1.8	7.8	86	1.18	0.27	56.51	4.1	0.478	0.37	0.34	0.02	0.41	0.152
COOK	SF Campbell Creek	Bkg.	0	0.02	2.5	1.6	7.8	9.3	0.80	0.67	200	2.83	0.335	0.45	1.00	0.09	0.23	0.100
COOK	Chester Creek	Urb.	0	0.02	2.96	3.8	8	28	0.68	0.38	109.9	3.24	0.346	0.13	1.18	0.03	0.75	0.099
COOK	Costello Creek	Bkg.	0	0.02	4.97	0.7	8.1	41	0.40	0.04	169.1	3.45	0.024	1.00	1.99	0.56	0.04	0.036
COOK	Deshka River	Bkg.	39	--	--	8.4	6.8	0.2		5.1	21	4.07	24.286	0.05				
COOK	Johnson R. abv Lateral Glacier	Bkg.	0	0.02	9.78	0.7	7.7	6.1	0.20	0.01	50.4	0.79	0.020	1.00	3.91	1.10	0.02	0.018
DELR	Manataway Creek	Urb.	0	0.06	1.37	3.3	8.7	17	4.38	1.08	62.8	8.85	1.720	0.16	0.55	0.01	4.28	0.679
DELR	Tulpehocken Ck	Ag.	1	0.09	2.14	2.7	7.9	28	4.21	0.57	45.8	6.46	1.245	0.21	0.86	0.03	3.14	0.644
DELR	Little Neshaminy Ck	Ag.	0	0.1	4.08	4.9	7.8	39	2.45	0.38	40.1	3.04	0.948	0.10	1.63	0.03	2.99	0.295
DELR	Raccoon Ck. @ Swedesboro	Ag	5	0.05	1.11	2.9	7.3	21	4.50	0.2	33.9	4.49	0.590	0.19	0.44	0.01	3.75	0.700
DELR	Hay Creek nr Birdsboro, PA	Ag.	1	0.04	0.77	1.4	7.8	24	5.19	1.64	36.4	5.45	4.505	0.57	0.31	0.03	1.15	0.653
GRSL	Cub River nr. Richmond, UT	Ag.	0	0.03	2.6	2.1	8.5	9.9	1.15	0.14	11.3	1.8	1.239	0.29	1.04	0.05	0.55	0.159
GRSL	Jordan River @ Salt L. City, UT	Urb.	2	0.03	4.8	3.3	7.5	217.6	0.63	3.2	116.3	3.97	2.752	0.16	1.92	0.05	0.61	0.097
GRSL	Weber River nr. Coalville, UT	Mine	13	0.1	21.76	3.1	8.5	14	0.46	4.02	1041	7.13	0.386	0.17	8.70	0.24	0.42	0.072
LINJ	Swan River	Urb..	0	0.06	2.13	2.4	6	11	2.82	10.85	161.4	34.69	6.722	0.24	0.85	0.03	1.71	0.413
LINJ	Muddy Run	Ag.	30	0.06	2.25	3.9	7.3	10	2.67	6.03	252.3	43.48	2.390	0.13	0.90	0.02	2.99	0.386
LINJ	Passaic River, Millington, NJ	Urb..	37	0.24	2.72	5.5	6.9	15	8.82	2.67	89.8	15.2	2.973	0.09	1.09	0.02	10.91	0.944
LINJ	Great Egg @ Sicklerville	A/F	19	0.35	12.26	3.4	5.8	8.4	2.85	5.36	31.8	8.96	16.855	0.15	4.90	0.12	2.88	0.440
LTEN	Buffalo River, Flatwoods, TN	Bkg.	--	0.01	1.45	0.9	7.2	6.1	0.69	0.64	47.7	2.88	1.342	1.00	0.58	0.12	0.08	0.081
LTEN	Indian Creek nr Madison, AL	Urb..	--	0.03	3.4	1.6	7.2	7.2	0.88	1.14	66.4	6.18	1.717	0.45	1.36	0.12	0.25	0.111
LTEN	Seqwatchie R. @ Whitnell, TN	A/F	--	0.01	1.38	1.4	7.6	6.7	0.72	0.18	10.4	2.32	1.731	0.57	0.55	0.06	0.16	0.091
MIAM	Holes Creek	Urb..	0	0.05	1.1	2.5	7.8	37.23	4.55	0.26	10.5	1.02	2.476	0.23	0.44	0.02	2.98	0.677
MIAM	Mad R., Hwy 41, Springfield	Ag.	0	0.08	0.79	2.1	8	61.5	10.13	1.54	30.8	2.27	5.000	0.29	0.32	0.02	4.79	1.393
MIAM	Great Miami R., Hamilton, OH	Urb..	1	0.19	3	5	8.4	73.5	6.33	0.67	87.2	2.36	0.768	0.10	1.20	0.02	7.74	0.747
MIAM	Whitewater R. @ Nulltown, IN	Ag.	0	0.03	0.81	1.9	7.7	33.67	3.70	0.43	11.7	1.51	3.675	0.34	0.32	0.02	1.44	0.487
MIAM	Little Miami R. @ Milford, OH	Urb.	0	0.05	1.57	3.5	8.3	49.3	3.18	0.45	9.2	1.12	4.891	0.15	0.63	0.02	3.30	0.486
MIAM	EF L. Miami R., @ Williamsburg	Ag.	0	0.07	2.09	4.4	8.3	44.9	3.35	0.24	13.3	1.03	1.805	0.11	0.84	0.02	3.97	0.444
MIAM	Stillwater R. on Springfield Rd	Ag.	1	0.05	2.1	3.3	7.6	56.87	2.38	0.17	34	1.04	0.500	0.16	0.84	0.02	2.33	0.369
MIAM	Great Miami R.	Ag.	1	0.05	2.52	3.4	7.9	64.2	1.98	0.34	31.4	1.41	1.083	0.15	1.01	0.02	2.00	0.306
MOBL	Chickasaw Creek	A/F	--	0.21	2.27	5.5	6.3		9.25	0.65	11.1	2.11	5.856	0.09	0.91	0.02	11.44	0.989
MOBL	Satilpa Ck.nr Coffeeville, AL	A/F	--	0.07	2.09	3.8	7.5		3.35	0.18	11.8	1.06	1.525	0.13	0.84	0.02	3.70	0.493
MOBL	Cahaba Valley Creek	Urb..	--	0.02	0.97	1.7	6.7		2.06	0.31	34.8	3.54	0.891	0.40	0.39	0.03	0.65	0.262
MOBL	Shades Ck. @ Homewood, AL	Urb..	--	0.04	1.31	2.6	8.2		3.05	0.17	15.2	1.42	1.118	0.22	0.52	0.02	2.14	0.462
MOBL	Alabama River @ Clairborne	Ag.	--	0.04	1.78	4.4	7.8		2.25	0.15	19.5	2.38	0.769	0.11	0.71	0.02	2.66	0.298
MOBL	Tombigbee R. @ Coffeeville	Urb..	--	0.04	2.74	4.3	7.9		1.46	0.06	26.3	2.81	0.228	0.11	1.10	0.02	1.72	0.197
MOBL	Coosa River @ Rome	Ag.	--	0.04	4.68	2.4	7.5		0.85	0.15	33.2	1.74	0.452	0.24	1.87	0.08	0.52	0.125
NECB	Stillwater River, Sterling, MA	Bkg.	2	0.25	0.53	2.1	6.6	10.9	47.17	7.02	72.9		9.630	0.29	0.21	0.01	22.31	6.489
NECB	Ipswich R. nr S. Middleton, MA	Urb..	27	0.44	2.72	7	6.6	15.6	16.18	9.91	380		2.608	0.07	1.09	0.02	20.20	1.333
NECB	Neponset R. @ Norwood, MA	Urb..	16	0.28	4.4	6.6	6.6	7.9	6.36	7.93	2477		0.320	0.07	1.76	0.04	7.94	0.559
NECB	Saugus R. @ Saugus Iron.	Urb..	34	0.11	2.79	4.4	7	19.5	3.94	4.41	309.2		1.426	0.11	1.12	0.02	4.67	0.523
NECB	Aberjona River	Urb..	2	0.08	9.11	4	6.5	24.4	0.88	7.14	1488		0.480	0.13	3.64	0.08	1.00	0.125
NROK	Clark Fork @ Turah Bridge	Mine	0	0.09	5.57	2	8.6	39.6	1.62	3.75	337.7	3.27	1.110	0.31	2.23	0.13	0.70	0.217
NROK	Clark Fork@ St. Regis, MT	Ag.	1	0.02	1.53	1.5	8.1	12.8	1.31	3.39	41.4	1.85	8.188	0.50	0.61	0.06	0.33	0.163
NROK	Flathead River @ Perma, MT	Ag.	1	0.01	1.14	1.7	8.1	2.7	0.88	0.18	19.2	1.62	0.938	0.40	0.46	0.04	0.28	0.111
NROK	MF Flathead R. nr W. Glacier.	Bkg.	0	0.01	1.63	0.7	8.5	4.5	0.61	2.1	24	3.53	8.750	1.00	0.65	0.18	0.05	0.055
NROK	S. Fork Coeur d'Alene	Mine	0	0.01	8.91	0.6	7	61	0.11	8.21	4517	5.73	0.182	1.00	3.56	1.15	0.01	0.009
NVBR	Carson @ Dresslerville	Mine	0	0.16	3.42	1.5	8.3	24	4.68	2.73	66.2	5.53	4.124	0.50	1.37	0.14	1.17	0.584
NVBR	Carson @ Markleeville	Bkg.	0	0.08	4.74	1.3	8.31		1.69	0.55	45.3	5.69	1.214	0.65	1.90	0.24	0.33	0.217
NVBR	Carson @ Deer Run Rd.	Ag.	3	0.68	31.08	6.9	8.1	52	2.19	1.21	78.4	1.73	1.543	0.07	12.43	0.25	2.73	0.183
NVBR	Carson @ Tarzan Rd	Ag.	4	1.34	204.57	5.5	8.4	145	0.66	1.34	778.3	0.82	0.172	0.09	81.83	1.65	0.81	0.070
NVBR	Carson @ Fort Churchill	Mine	2	5.12	1106	4.7	8.2	77	0.46	4.2	4130	3.33	0.102	0.10	442.40	9.16	0.56	0.058
OAHU	Nuuanu Reservoir	Ref	--	0.1	24.27	2.5	7.8		0.41	0.46	291.2	23.97	0.158	0.23	9.71	0.37	0.27	0.061
OAHU	S. Fork Lake Wilson	Urb..	--	0.12	1.93	3.3	6.8		6.22	1.18	300.2	17.02	0.393	0.16	0.77	0.02	6.08	0.964
OAHU	Kawainui Canal	Bkg.	--	0.02	1	3.1	7		2.00	0.64	106.2	41.79	0.603	0.17	0.40	0.01	1.82	0.312
OAHU	Waikele Stream	Ag.	--	0.01	1.27	0.8	7.2		0.79	3.55	186	15.95	1.909	1.00	0.51	0.12	0.08	0.080
OAHU	Ala Wai Canal	Urb..	--	0.01	1.17	0.6	8		0.85	0.34	255.7	21.62	0.133	1.00	0.47	0.15	0.07	0.066
SACR	Cottonwood Creek	Mine	0	0.03	1.02	1.5	8.5	10.8	2.94	0.36	26.3	2.61	1.369	0.50	0.41	0.04	0.74	0.367
SACR	Sacramento Slough	Ag.	2	0.15	10.19	3.2	7.9	7.9	1.47	2.84	128.5	7.07	2.210	0.16	4.08	0.11	1.39	0.229
SACR	Bear River @ Hwy 70	Mine	0	0.24	17.82	3.5	7.4	5.6	1.35	0.55	176.8	2.89	0.311	0.15	7.13	0.17	1.39	0.206
SACR	Putah Creek @ Davis	Mine	0	0.05	4.1	2.3	8	28	1.22	0.27	275.6	3.86	0.098	0.26	1.64	0.07	0.69	0.175
SACR	Colusa Basin Drain	Ag.	2	0.08	6.9	5.9	7.9	55.3	1.16	0.52	53.3	6.85	0.976	0.08	2.76	0.06	1.44	0.115
SANA	Santa Ana R. @ MWD Cross.	Urb..	--	0.05	2.28	2.3	8.1	87.6	2.19	3.42	27.3	5.81	12.527	0.26	0.91	0.04	1.23	0.315
SANA	S. Fork, Santa Ana River	Bkg.	0	0.02	0.61	0.6	7.7	1.4	3.28	3.65	28.8	17.22	12.674	1.00	0.24	0.08	0.25	0.254
SANA	Mill Ck. @ Chino-Cor. Rd	Urb..	--	0.08	2.99	6.9	10	53.8	2.68	1.29	30.3	7.75	4.257	0.07	1.20	0.02	3.34	0.224
SANA	Santa Ana R. @ Hamner Rd	Urb..	--	0.03	3.71	3	8.2	94.4	0.81	1.7	24.3	5.02	6.996	0.18	1.48	0.04	0.71	0.126
SANA	Santa Ana R. blw Prado Dam	Urb..	0	0.14	14.67	4.8	7.8	105	0.95	1.47	45.6	5.77	3.224	0.10	5.87	0.12	1.16	0.117
SANT	Edisto River, Givhans	A/F	36	1.36	8.18	4.6	6.6	4	16.63	5.8	70.1	21.31	8.274	0.11	3.27	0.07	19.97	2.121
SANT	Saluda River, Silverstreet, SC	Bkg.	0	0.09	1.05	2.9	7.1	5.5	8.57	0.7	23.9	6.67	2.929	0.19	0.42	0.01	7.13	1.333
SANT	NF Edisto River nr Branchville	A/F	16	1.48	9.41	3.8	6.4	2.9	15.73	2.5	6.1	15.46	40.984	0.13	3.76	0.09	17.36	2.314
SANT	NF Edisto R. nr Fairview Cross	A/F	16	0.32	3.77	4.8	6.1	1.1	8.49	6.8	86.6	12.04	7.852	0.10	1.51	0.03	10.30	1.041
SANT	SF Edisto River, Canaan	A/F	20	0.4	7.27	5.9	6.4	2.1	5.50	3.1	69.4	31.03	4.467	0.08	2.91	0.06	6.84	0.546
SANT	SF Edisto River @ Springfield	A/F	16	0.41		3	6.3			2.7	41.5	28.09	6.506	0.18				
SOFL	Everglds. Natl. Pk., site TS7	Bkg.	50	0.2	2.38	12	6.8	0.5	8.40	5.8	145	90	4.000	0.04	0.95	0.02	10.50	0.388
SOFL	WCA 3, site 3A15	Bkg.	50	0.5	1.9	16	7.1	0.5	26.32	7.8	288	92	2.708	0.03	0.76	0.02	32.89	0.897
SOFL	WCA 2, site 2BS	Bkg.	50	0.45	2.1	18	6.9	10.2	21.43	3.9	234	90	1.667	0.02	0.84	0.02	26.79	0.646
SOFL	WCA 2, site U3	Bkg.	50	0.61	3.5	24	7.5	25.6	17.43	2.7	194	91	1.392	0.02	1.40	0.03	21.79	0.391
TRIN	Lake Livingston	Urb.	1	0.02	1.34	6			1.49	0.2	45.2	7.59	0.442	0.08	0.54	0.01	1.86	0.145
TRIN	White Rock Ck. Dallas, TX	Urb.	0	0.04	1.65	4.9			2.42	0.14	8.6		1.628	0.10	0.66	0.01	2.95	0.292
TRIN	White Rock Lake, Dallas	Urb.	0	0.02	1.24	5.6	7.9		1.61	0.52	55.4	6.24	0.939	0.08	0.50	0.01	2.00	0.169
TRIN	Trinity River nr Crockett, TX	Urb.	0	0.02	6.5	5.8	7.5		0.31	0.28	30.6	4.42	0.915	0.08	2.60	0.05	0.38	0.031
TRIN	Clear Crk. @ Sanger	Bkg.	0	--	--		8	55		0.23	5.9	2.18	3.898					
UCOL	Red Mountain Creek	Mine	0	0.02	1.68	0.4	3.3	484	1.19	0.13	107.2	6.4	0.121	1.00	0.67	0.29	0.07	0.070
UCOL	Colorado River @ Baker Gulch	Bkg.	13	0.05	0.57	1.3	7.8	6.2	8.77	2.84	27.4	6.18	10.365	0.65	0.23	0.03	1.72	1.128
UCOL	Snake River @ Peru Creek	Mine	0	0.02	0.48	1.3	6.7	46	4.17	0.28	56.2	7.45	0.498	0.65	0.19	0.02	0.82	0.536
UCOL	French Gulch nr Breckenridge	Mine	0	0.02	0.64	1.4	7.8	63	3.13	0.21	113.2	5.76	0.186	0.57	0.26	0.03	0.69	0.393
UCOL	Dry Creek @ Begonia Road	Ag.	1	0.15	6.05	3.9	8.2	467	2.48	1.68	37.4	3.82	4.492	0.13	2.42	0.05	2.78	0.359
UIRB	Pitner Ditch nr LaCrosse, IN	Ag.	0	0.03	0.27	3.1	7.8	89	11.11	0.09	9.1	1.66	0.989	0.17	0.11	0.00	10.11	1.734
UIRB	Nippersink Ck. bv Wonder L.	Ag.	0	0.04	1.42	3.5	7.8	64	2.82	0.08	8.7	2.3	0.920	0.15	0.57	0.01	2.92	0.430
UIRB	Mukwanago R. @ Mukwanago	Ref	3	0.06	1.48	6.9	8		4.05	0.11	31.6	1.15	0.348	0.07	0.59	0.01	5.06	0.339
UIRB	Salt Creek @ W. Springs, IL	Urb..	0	0.13	9.26	6.3	7.3	89	1.40	1.1	46.8	3.29	2.350	0.07	3.70	0.07	1.75	0.130
UIRB	Des Plaines River	Ag.	15	0.1	4.18	11.4	7.5	94	2.39	3.58	11.6	1.4	30.862	0.04	1.67	0.03	2.99	0.117
YELL	Yellowstone R. near Sidney	Ag.	1	0.15	4.07	2.9	8.6	110	3.69	0.45	18.7	2.14	2.406	0.19	1.63	0.05	3.07	0.573
YELL	Tongue River	Ag.	4	0.06	1.87	3.4	8.5	94	3.21	3.05	27.7	8.23	11.011	0.15	0.75	0.02	3.23	0.494
YELL	Shoshone River	Ag.	0	0.13	5.31	4.2	8.1	240	2.45	0.53	11.1	1.83	4.775	0.12	2.12	0.05	2.85	0.337
YELL	Bighorn Lake @ Hwy14A	Ag.	1	0.1	2.48	3.8	8.3	190	4.03	0.59	33	5.31	1.788	0.13	0.99	0.02	4.45	0.593
YELL	Bighorn River nr. Kane	Ag.	1	0.13	3.89	3.7	8.4	170	3.34	0.6	16.3	2.09	3.681	0.14	1.56	0.04	3.62	0.498
	Average		7.24	0.20	16.79	4.19	7.57	54.73	4.64	1.89	211.49	10.57	3.47	0.26	6.72	0.19	4.40	0.50
	Stdev		13.67	0.55	110.82	3.57	0.83	82.95	6.70	2.40	651.50	18.66	5.96	0.27	44.33	0.93	6.73	0.74

APPENDIX C

SUMMARY INFORMATION FROM THE STUDY OF MARYLAND RESERVOIRS AND THE CALCULATED PARAMETERS FOR EVALUATING METHYLATION AND DEMETHYLATION

Data Source: Gilmour et al., 2007

Reservoir	Depth	SurfArea	Cap	WatArea	SurftoWat	Flow	ResTime	HgDep	HgUNF	HgD	MeHgUNF	MeHgD	%MeHg	pH	TSS	log Kd Hg water	log Kd MeHg water	DOC	Cl	NO3	SO4	BottDO	kdemeth	IB-Hg	MeHg/ IB-Hg	k meth
Clopper	12	0.36	2000000	7	19	0.17	0.38	2.81	5.63	4.93	0.21	0.12	2.46	7.4	4.17	4.532115	5.254925	7.12	130.19	0.37	4.5	0.48	0.02	0.10	2.127744	0.032726
DeepCreek	25	18	1.1E+08	163	9	8.66	0.4	0.97	1.13	0.45	0.29	0.27	60.39	6.84	2.32	5.813808	4.504178	2.66	7.67	0.05	12.91	0.44	0.02	0.02	18.76644	0.344575
Duckett	22.6	3.13	21000000	343	110	2.38	0.28	0.24	31.51	31.5	0.11	0.06	0.2	6.84	9.79	1.510907	4.930036	2.77	61.37		6.26	0.77	0.02	1.01	0.108497	0.002036
Lariat	9.1	0.39	1900000	7	18	0.09	0.71	0.37	2.28	1.78	0.61	0.4	22.7	6.49	7.75	4.559248	4.830858	4.14	20.7	0.53	9.3	0.17	0.02	0.04	15.86245	0.367733
Liberty	44	12.57	1.6E+08	424	34	5.46	0.93	0.17	1.3	0.85	0.14	0.11	13.13	7.06	2.32	5.358306	5.070241	10.57	14.91	1.96	5.28	0.48	0.01	0.02	8.235245	0.090227
LochRaven	23.2	9.71	90000000	789	81	8.6	0.33	0.65	1.83	1.95	0.34	0.12	6.3	7.58	27.9		4.817637	6.79	49.37	1.77	3.87	0.15	0.01	0.04	8.704124	0.11257
PineyRunLake	9.8	0.49	1700000	31	63	0.65	0.08	0.16	1.5	0.84	0.15	0.26	31.12	6.98	1.74	5.654715		4.31	24.18	0.81	7.49	2.25	0.02	0.02	8.404385	0.182121
PineyFrostburg	54.5	1.21	9560000	28	23	0.36	0.84	0.16	1.52	0.74	0.2	0.17	23.22	7.2	5.65	5.270814	4.494624	2.89	27.96	0.58	5.99	0	0.01	0.02	8.948492	0.120256
PrettyBoy	30	6.07	74000000	206	34	2.9	0.81	0.13	2.31	1.36	0.13	0.07	5.5	7.3	7.61	4.9628	5.051669	10.44	23.78	1.89	2.61	0.17	0.01	0.03	4.779378	0.054586
RockyGap	25	0.85	6600000	23	27	0.34	0.62	0.1	0.66	0.44	0.06	0.05	12.12	7.51	1.99	5.400117	5.002177	3.77	3.29	0.04	10.27	3.99	0.02	0.01	5.987704	0.092034
Savage	46.1	1.46	25000000	270	185	4.17	0.19	0.3	0.79	0.43	0.07	0.07	17.11	7.22	3.92	5.329548		2.03	10.7	0.58	11.49	3.82	0.02	0.02	2.88468	0.047971
StMarysLake	6.4	1.01	3900000	22	22	0.26	0.48	0.33	3.72	1.25	1.53	1.04	83.77	4.84	14.56	5.132626	4.510001	9.06	9	6.51	3.95	0.25	0.02	0.03	61.19662	1.086647
Triadelphia	15.8	3.24	23000000	205	63	2.35	0.31	0.23	3.11	0.89	0.13	0.07	7.79	6.71	21.74	5.059703	4.595794	3.38			5.55	0.04	0.02	0.02	5.856631	0.115488
Tuckahoe	2.7	0.35	32000	223	640	2.78	0.000365	0.09	1.69	1.03	0.12	0.09	9	7.47	2.5	5.408767	5.124939	8.47	15.76	2.82	5.57	5.44	0.03	0.02	5.824473	0.173431

APPENDIX D

EVALUATING THE NORMALITY OF A DATASET

There are two popular ways to test for normality of the data (or logarithms of the data); the Chi-square test and the Kolmogorov-Smirnov test (Haan, 1977). For grouped data involving larger datasets, the Chi-square (X^2) test can be used to test for normality. For smaller datasets, the Kolmogorov-Smirnov (K-S) test may be preferable. The reader is referred to the above reference for a more theoretical discussion of these tests.

Figures D-1(a) and (b) show relative frequency plots (or histograms) of largemouth bass data collected by the Department of Energy (DOE) Savannah River Site from the Savannah River. A frequency histogram is constructed by setting up class intervals for the observations and counting the number of occurrences in each interval. There are 208 fish tissue values in this dataset covering a three year period from 1996 to 1998. Notice that the histogram of the untransformed data appears to be positively skewed (Figure D-1(a) as compared to the normal distribution with mean and variance calculated from the data. Figure D-1(b) shows the natural logarithms of the data. This histogram appears to be a better "fit" to the lognormal distribution.

Table D-1 shows the calculations required to perform a Chi-Square test for both the original and log-transformed data. Focusing first on the upper portion of the table, the number of intervals for which the frequencies are to be determined are provided in Column 1. The number of intervals can be determined in a number of ways but the most basic if the take the range and divide it into a manageable number of intervals, about 10. These intervals are set up so that the number of occurrences in any given interval "n" are counted as the number of occurrences less than the entry in the nth row, minus the number of occurrences in the $(n-1)^{th}$ row (Column 2). For example, the number of occurrences in interval number 2 are those fish tissues values less than 0.4 mg kg^{-1} but greater than 0.2 mg kg^{-1}, as shown in Column 3. In column 4, the cumulative normal distribution function (CDF) is computed at the upper end of the class interval. In column 5, the difference in the cumulative probability between the nth and $(n-1)^{th}$ cell interval is multiplied by the total number of observations (208) to arrive at the expected interval frequency per the normal distribution. The difference between the observed (column 3) and

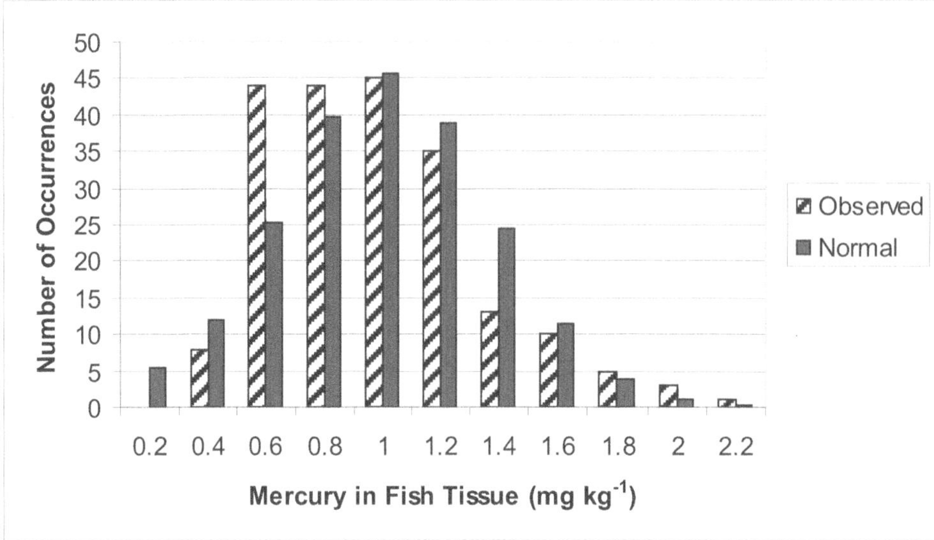

Figure D-1(a). Frequency Histogram of Mercury in Largemouth Bass in the Savannah River.
Source: EPRI, 2006. Copyright Electric Power Research Institute, 2006. Reprinted with permission.

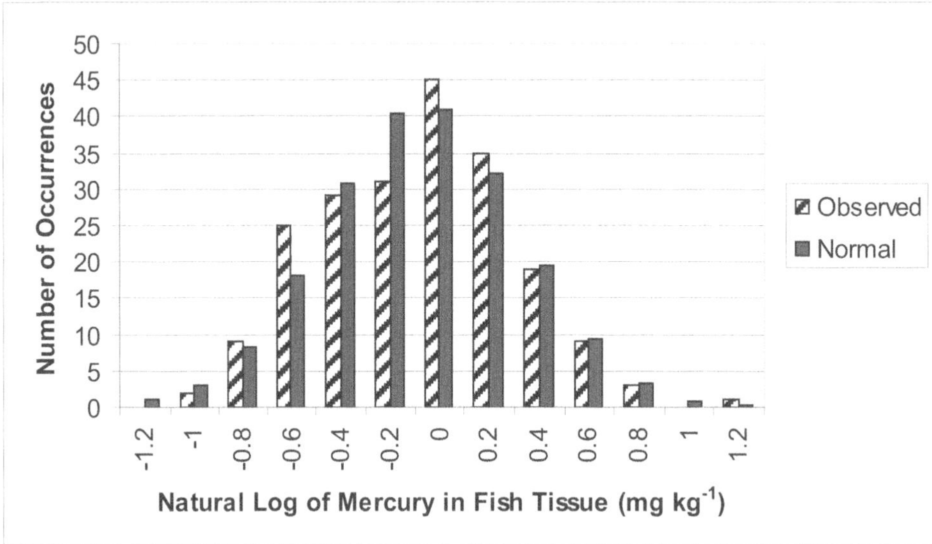

Figure D-1(b). Frequency Histogram of the Natural Logarithms of Mercury in Largemouth Bass in the Savannah River.
Source: EPRI, 2006. Copyright Electric Power Research Institute, 2006. Reprinted with permission.

Table D-1. Chi-Square Test for Goodness-of-fit for Largemouth Bass Data from the Savannah River and Tributaries, Original and Log-transformed Data.
Source: EPRI, 2006. Copyright Electric Power Research Institute, 2006. Reprinted with permission.

(1) Interval Number	(2) Interval Upper Limit (mg kg^{-1})	(3) Number of Occurrences	(4) Cumulative Normal Distribution	(5) Probability Density Function	(6) $(O-E)^2/E$
Original Data					
1	0.2	0	0.026	5.496	5.496
2	0.4	8	0.084	12.001	1.334
3	0.6	44	0.206	25.398	13.625
4	0.8	44	0.397	39.667	0.473
5	1	45	0.617	45.725	0.012
6	1.2	35	0.804	38.904	0.392
7	1.4	13	0.921	24.430	5.348
8	1.6	10	0.976	11.322	0.154
9	1.8	5	0.994	3.871	0.329
10	2	3	0.999	0.976	4.194
11	2.2	1	1.000	0.182	3.687
				$\Sigma((O-E)^2/E)$ =	**35.04**
		Total = 208		**X^2 (0.9,8)=**	**13.4**
Log-Transformed Data					
1	-1.2	0	0.005	1.017	1.017
2	-1	2	0.019	2.955	0.308
3	-0.8	9	0.059	8.327	0.054
4	-0.6	25	0.147	18.179	2.559
5	-0.4	29	0.294	30.745	0.099
6	-0.2	31	0.488	40.283	2.139
7	0	45	0.685	40.892	0.413
8	0.2	35	0.839	32.162	0.250
9	0.4	19	0.933	19.597	0.018
10	0.6	9	0.978	9.251	0.007
11	0.8	3	0.994	3.383	0.043
12	1	0	0.999	0.958	0.958
13	1.2	1	1.000	0.210	2.970
				$\Sigma((O-E)^2/E)$ =	**10.84**
		Total = 208		**X^2 (0.9,10) =**	**16.0**

expected (column 5) cell frequencies are then squared and divided by the expected interval frequency (column 6). The sum of the computations in column 6 over all intervals is the X^2 value for the test. The critical value is obtained in a look up table of the X^2 statistic. For the X^2 test, the critical value is found by entering the table with $1-\alpha$ and degrees of freedom equal to the number of intervals (k) minus the number of parameters (p) estimated for the normal distribution (μ, σ) minus 1, or

$$d.f. = k-p-1 \qquad \text{Equation D-1}$$

When the data are log-transformed (lower portion of Table D-1), the X^2 test shows that the computed statistic is less than the critical value. So the null hypothesis that the data are log-normally distributed cannot be rejected. In this case, then, the data can be assumed to be log-normally distributed and the geometric mean is a more appropriate statistic to represent the central tendency in the data.

When there are enough data that they can be organized into a frequency histogram, the X^2 test can be used to determine goodness-of-fit. However, when the data set is small, the Kolmogorov-Smirnoff test offers an attractive alternative.

Table D-2 shows a set of methylmercury data in fish tissue from Liberty Reservoir, MD (column 1). To perform the K-S test on log-transformed data, the log values (column 2) are ranked (column 3). Using the ranks, a probability plotting position is assigned (column 4). A number of methods are available for assigning a plotting position. The Weibull plotting position formula (m/n+1), where m is the rank of the observation and 'n' is the total number of observations, is used here. Values of the cumulative normal distribution for the mth fish tissue value with a mean and standard deviation computed from the data are next computed (column 5). The absolute values of the difference between the plotting position and cumulative normal distribution is then calculated (Column 6), and the maximum deviation is determined (0.084). The value of the K-S test statistic is found by table lookup for the significance level and degrees of freedom (in this case equal to the number of data points). The value of the statistic (0.295) is then compared to the maximum deviation between the sample distribution and the normal distribution. If the maximum deviation is less than the value of the K-S statistic, then the hypothesis that the data are drawn from a log-normal distribution cannot be rejected, as is the case in this example.

Table D-2. Example Application of the Kolmogorov-Smirnov One Sample Test for Normality.
Source: EPRI, 2006. Copyright Electric Power Research Institute, 2006. Reprinted with permission.

1 MeHg in Fish Tissue (mg kg^{-1})	2 Ln (MeHg in Fish Tissue) (mg kg^{-1})	3 Rank	4 Plotting Position S(x)	5 Cumulative Normal Distribution F(x)	6 ABS[F(x) – S(x)]
0.314	-1.158	11	0.647	0.621	0.026
0.282	-1.266	9	0.529	0.551	0.021
0.298	-1.211	10	0.588	0.587	0.001
0.219	-1.519	6	0.353	0.383	0.030
0.126	-2.071	2	0.118	0.109	0.008
0.202	-1.599	5	0.294	0.332	0.038
0.503	-0.687	14	0.824	0.865	0.041
0.154	-1.871	3	0.176	0.186	0.010
0.198	-1.619	4	0.235	0.320	0.084
0.226	-1.487	8	0.471	0.403	0.068
0.224	-1.496	7	0.412	0.397	0.015
0.549	-0.600	15	0.882	0.894	0.012
0.709	-0.344	16	0.941	0.954	0.012
0.355	-1.036	12	0.706	0.697	0.009
0.449	-0.801	13	0.765	0.819	0.054
0.068	-2.696	1	0.059	0.011	0.048
Mean	-1.341			**Maximum**	0.084
Standard Deviation	0.593			**K-S(0.1,16)**	**0.295**

APPENDIX E

TESTS FOR DATA OUTLIERS

Identification of outliers should first be done by visual inspection of the data. However, as with any decisions concerning data, it is preferable to have some statistical verification of one's notions to help decide whether a particular data point should or should not be used in subsequent analyses. The Walsh Test is used to verify whether the observation was a statistical outlier for larger numbers of observations. Walsh's test is a nonparametric test, meaning that its application does not depend on the normality of the data distribution. Walsh's Test is recommended when there are greater than 50 values in the dataset ($N > 50$). If there are $n > 220$ data points, the significance level (α) of the test is at least 0.05 and if $N > 60$, then the α is at least 0.10.

The Walsh Test uses the statistic:

$$W = x_r - (1+a)x_{r+1} + ax_k \qquad \text{Equation E-1}$$

for identifying small outliers, and

$$W = x_{n+1-r} - (1+a)x_{n-r} + ax_{n+1-k} \qquad \text{Equation E-2}$$

for identifying large outliers where 'r' is the possible number of outliers and 'N' is the number of observations in the sample.

The parameter 'a' for the Walsh test is computed as follows:

$$a = \{1 + b * SQRT[(c - b_2)/(c - 1)]\} / (c - b_2 - 1) \qquad \text{Equation E-3}$$

in which

$$c = [SQRT(2N)] \qquad \text{Equation E-4}$$

where [ ] indicates rounding the value to the next largest possible integer,

$$k = r + c \qquad \text{Equation E-5}$$

and,

$$b_2 = 1/\alpha \qquad \text{Equation E-6}$$

where α is the significance level.

For small outliers, if the result (W) is < 0, then the value is confirmed to be an outlier at the chosen significance level. For large outliers, if the result (W) is > 0, then the value is confirmed to be an outlier at the chosen significance level.

The Grubbs Test can be used for smaller sample sizes (N < 50) (U.S. EPA, 2000). The Grubbs Test focuses on the standardized distance between the suspected outlier and the mean given by:

$$G = (x - x_{mean}) / s_x \qquad \text{Equation E-7}$$

for values higher than the mean, and

$$G - (x_{mean} - x) / s_x \qquad \text{Equation E-8}$$

for values less than the mean.

In order to use the Grubbs Test, it should be confirmed that the data are normally distributed. Note that the test assumes that the values, excluding the suspected outlier, are normally distributed so the test for normality should exclude the suspect data point. Critical values for the Grubbs Test can be found in U.S. EPA (2000).

APPENDIX F

SELECTION OF BIOINDICATOR ORGANISMS

This appendix will review various organisms that have been used in biomonitoring applications and provide recommendations on their in the specific application to estimation of bioavailability in effluents and receiving waters. A summary of the conclusions reached here is provided in Table 3-3 in the main text. In all cases, literature references are provided where available and appropriate.

F.1 Microorganisms

The use of phytoplankton to examine the relative bioavailability of inorganic Hg and MeHg for different waters and the impact of dissolved constituents has been widely used for the examination of metal uptake and accumulation (Campbell, 1996; Fisher and Reinfelder, 1995; Morel and Hering, 1993) at the base of the food chain. Such studies have shown the impact of water column speciation on the uptake of metals and have been specifically used in the examination of Hg and MeHg accumulation by phytoplankton (Lawson and Mason, 1998a; Mason et al., 1996; Morel et al., 1998; Pickhardt and Fisher, 2007). Early studies with phytoplankton suggested that accumulation was mainly related to the uptake of neutrally-charged complexes, and more recently, the same conclusion was reached for methylating bacteria (Benoit et al., 2003; Benoit et al., 1999a). For example, Mason et al. (1996) demonstrated passive uptake of neutral inorganic chloride complexes of both inorganic Hg and MeHg ($HgCl_2$ and CHgCl) by the estuarine diatom, *Thalassiosira weisflogii.* It was also shown that these complexes were hydrophobic (had a relatively large octanol-water partition coefficients (K_{ow})) compared to the neutral hydroxide complexes ($Hg(OH)_2$, CH_3HgOH) and the charged complexes in solution. These experiments were conducted in the absence of DOC. Further studies with both diatoms and sulfate-reducing bacteria similarly showed that neutral complexes with sulfide (e.g. HgS and CH_3HgSH) and with small organic thiols were also efficiently taken up and had measurable K_{ow} values (Benoit et al., 2001; Lawson and Mason, 1998). While recent studies have confirmed the concept of passive uptake, there do appear to be other, albeit less rapid, uptake processes (Lin and Jay, 2007; Najera et al., 2005). Many studies have also examined the role of DOC and other factors in controlling uptake into phytoplankton such that this is now relatively well understood (Mason, 2000; Pickhardt and Fisher, 2007).

Recommendation - As noted in Table 2.3, it is not recommended that these approaches are used in most situations although it is possible that there use could provide important information in some instances.

F.2 Bioreporters

These organisms are generally bacteria, such as *E. coli*, that have been modified so that they have a measurable response when exposed to inorganic Hg (Golding et al., 2002; Golding et al., 2008; Scott, 2001). Also, given the way the organisms are modified and constructed, it is only Hg inside the cell that causes a response. Therefore these genetically modified organisms show a specific response to the accumulation of Hg. The bioreporter is constructed using the *mer* gene which is involved in Hg detoxification in some bacteria and by combining this gene to the *lux* gene, which promotes the release of light when turned on (as in a firefly). Thus, these bioreporters give off a light signal in response to the presence of Hg inside the cell. The results obtained from studies over a pH range, using the anaerobic organism, are shown in Figure F-1 below (Golding et al., 2008). This indicates uptake of the Hg into the cell. Note that in the phytoplankton studies discussed above it is possible that the Hg and MeHg is attached to the cell surface and not truly assimilated by the organisms. Current available technology and bioreporter design is only for inorganic Hg, and most of the organisms that have been used can only be used in oxic systems. Thus, there is a difficulty in extrapolating from the results obtained to what may occur in a methylating (reduced) environment. The recent report of using such an organism under anaerobic conditions (Golding et al., 2008) gives some promise for this approach in the future. Overall, the application and results are still relatively experimental and it is not clear in all situations what complexes of inorganic Hg is inducing a response from the bioreporter (Golding et al., 2002; Golding et al., 2008). However, overall, the response of the organisms is directly related to the uptake rate across the cell membrane, even if this appears to be by both passive and active processes.

As noted, there are both positive aspects and negatives related to this approach. As accumulation of inorganic Hg by the bioreporters involves both passive and facilitated transport, the results are clearly comparable to those obtained using the phytoplankton culture approaches described above and the bioreporters in essence only represent a different measurement approach (detection system). In some studies there is evidence of an impact of growth media on speciation, raising similar concerns as with the phytoplankton culture medium, as the bioreporter medium needs to be relatively specialized and contains relatively high concentrations of some potential complexing agents. If the medium impacts Hg speciation, then this clearly is a disadvantage. Also, while the bioreporters are very sensitive, they may not respond adequately if Hg levels are low. However, as with the phytoplankton assays, it is possible to spike the solutions to enhance the response as long as the added Hg is near background concentrations and does not overwhelm the binding of the *in situ* ligands. This, however, also raises an issue in terms of the timescale of response in terms of partitioning of the added spike with the *in situ* ligands and how this may impact bioavailability. Equilibration of an added spike with inorganic ligands (e.g. Cl, sulfate) is relatively instantaneous while equilibration with natural organic ligands can take two or more hours to reach equilibrium (e.g. Miller et al., 2007). Therefore, if such spike additions are used, it is essential to allow for equilibration of the spike with ligands in solution prior to the start of the experiment.

Many investigators use two hour or greater for the pre-equilibration time and this appears to be sufficient in most circumstances although up to 24 hours has been used in some studies. In some cases, this can be done with a small aliquot of the medium prior to the spiking of the assay solution.

Recommendation - While the bioreporter approach has promise, it is still in a relatively experimental stage and there appear still to be difficulties in interpreting the results of the studies that have been done to date. Thus, it is concluded that this approach does not satisfy the needs of the proposed applications and it is recommended that this approach not be used routinely - although there may be situations where its use may be appropriate.

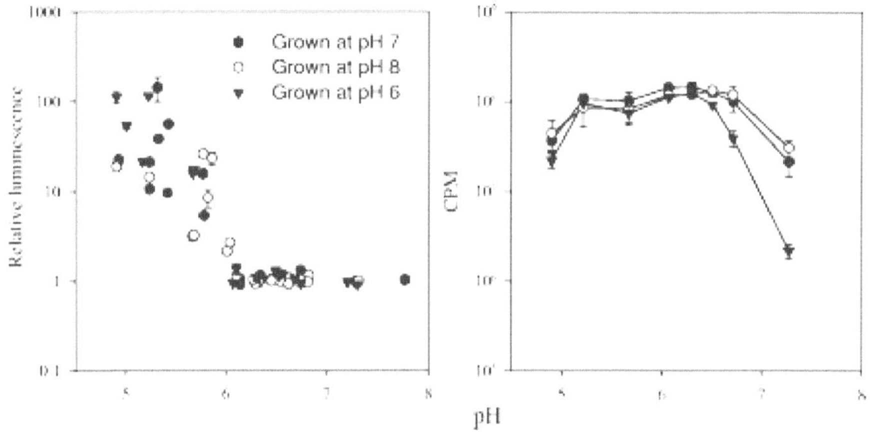

Figure F-1. Effect of Varying pH on the Hg(II)-induced Right response from *E. coli* HMS174(pRB28) Preconditioned in pH 6 to 8 Growth Medium (Left) and on Constitutive Light Production from *E. coli* HMS174(pRB27) (Right)[37].
Source: Golding et al. (2008). Copyright American Society for Microbiology, 2007. Reprinted with permission.

F.3 Biosentinel Organisms

The use of biosentinel organisms relies on the assumption that the concentration in the organism is a direct reflection of the bioavailable concentration of the contaminant in the medium, be that water or sediment. This is clearly a reasonable assumption if the organism has relatively defined feeding habits and does not acquire the contaminant to a significant degree from many sources. For example, an organism living at the sediment/water interface may acquire a contaminant from: 1) the overlying waters, through uptake of dissolved species or from consumption of particulate material; 2) the sediment pore water; and 3) from ingestion of sediment (Lawrence and Mason, 2001). A further complication could occur if there is substantial sediment resuspension, and thus particles in the water may be "sediment-derived", but this can also result in a more direct link and relationship between surface sediment concentrations and water column biota concentrations. For example, Lawrence and Mason (2001) were able to demonstrate through modeling and laboratory exposure experiments that uptake from particle ingestion was the most important route for the amphipod, *Leptochirus*. It is also well established that uptake of Hg and MeHg from food is greater than direct uptake from water for most pelagic organism (Wiener et al., 2007). There have been many applications of bioindicator use and one of the longest standing monitoring programs for contaminants uses this approach, the NOAA Mussel Watch Program (NOAA, 2008).

The response of biosentinels to changes in concentration is overall a function of size and growth rate. Zooplankton will respond in days to weeks to changes in water/food concentration

---

[37] Hg(II) was added to a final concentration of 1 ng $liter^{-1}$ in the bioreporter assays, and light production was measured following 120 min of incubation. Error bars for the *mer-lux* bioreporter assays represent the standard deviations of duplicate samples. Error bars for the constitutive *lux* assays represent the standard errors of three independent assays (*n*= 6).

as they grow quickly and have a short lifetime (Wiener et al., 2007; Mason et al., 2005). Filters feeding organisms, such as mussels and clams, will take a month to a few months to reach steady state, depending on their size and growth rate. Small fish respond to longer term (seasonal to yearly) changes. Large fish respond on an even slower basis. Birds can show seasonal effects, as blood can be used as a short-term measure. Larger mammals would respond slowly (Wolfe et al., 2007).

Epi-benthic invertebrates such as amphipods, crayfish, and other larger invertebrates can be used to assess overall water column/ecosystem changes, and are found in higher concentrations and as larger organisms in estuarine and coastal waters (Wiener et al., 2007). Filter feeding organisms have advantages in that, even though they made reside in or on the sediment surface, they feed on suspended matter in the water column. Thus, they potentially reflect a pelagic rather than a benthic signal.

Based on the issues and focus of this study, it is concluded that filter feeding organisms and small fish are the likely best candidates for biological monitoring of bioavailability in effluents and receiving waters using transplanted organisms. However, the other approaches will be discussed in more or less detail, depending on their likelihood to be used as a biological indicator. Overall, small resident fish, which often do not move large distances, are the best "biosentinels", in the sense that they are in situ organisms that reflect their environment (Slotton et al., 2007). Filter feeders are likely to be suitable organisms for transfer or caged studies as they are typically more robust and hardy and can be transferred between locations with less mortality.

Biosentinels are good indicators of MeHg bioavailability and bioaccumulation as they magnify the signal in the water and integrate it over time. Overall, trophic transfer factors between predator and prey are in the region of 80% for MeHg, and <20% for inorganic Hg (Fisher and Reinfelder, 1995; Mason, 2000). Thus, organisms high in the food chain have most of their Hg as MeHg while the lower trophic organisms have some lower fraction of their Hg as MeHg, and this is especially true for bivalves and other herbivorous/detrital organisms. Thus, these lower food chain organisms may reflect the relative bioavailability of inorganic Hg for uptake into the food chain, but they do not provide any indication or information about the bioavailability of inorganic Hg for methylation. Clearly, however, with the strategic deployment or sampling of sentinel organisms it may be possible to ascertain this, especially if the biological monitoring is coupled with some water column sampling.

Given this background, some specific comments about each class of organism will be given. As noted above, there will be a focus on the organisms deemed most useful for this study.

F.3.1 Phytoplankton (Seston)

Phytoplankton and other microorganisms respond very rapidly to changes in concentration in MeHg or Hg concentrations in water, within minutes or hours (Mason et al. 1996). Depending on their growth rate, they will come to steady-state with their environment in a few days. Sampling waterbodies for *in situ* plankton however is not simple because of the presence of inorganic, detrital matter or dead algae, and the differences in algal size results in an overlap in size range with zooplankton. Therefore it is extremely difficult to get an environmental sample of seston that represents a true algal signal (Krabbenhoft et al., 2007). This is the major reason why such a measure is not recommended for this study. Furthermore as, algal density and growth rate can have a strong influence on their concentrations due to growth and biomass dilution effects (Pickhardt et al., 2002), it is not easy to compare across sites unless some information about system productivity and dynamics is available. Such information is not easily obtained and will vary on a short timescale. Species composition and size will also vary widely making comparisons difficult.

Recommendation - As phytoplankton are difficult to sample cleanly in field and separate from detritus/abiotic particles, and the signal is difficult to interpret, they are more useful in estimating bioavailability directly as discussed above. They are not recommended as a field bioavailability indicator species.

F.3.2 Zooplankton and Other Small Herbivores

These organisms are widely distributed, common, and important in pelagic food webs. However, there are only a few species that are truly cosmopolitan and can be used for monitoring purposes (Wiener et al., 2007). For freshwaters, daphnia are often used and in marine environments, genera such as *Arcatia* are abundant. The life cycle of zooplankton is short that populations will show a rapid response to changes in concentration in the water column, within hours or days (Herrin et al., 1998; Paterson et al., 1998; Tsui and Wang, 2004). Zooplankton can accumulate significant quantities of MeHg from food (typically algae) and/or water (Monson and Brezonik, 1999; Peech Cherewyk, 2002; Tsui and Wang, 2004; Lawrence and Mason, 2001) which is a potential confounding factor. An example of uptake from water directly, and the impact of DOC on uptake is shown in Figure F-2 (Lawrence and Mason, 2001). It is evident that while the presence of DOC enhanced the dissolved concentration somewhat by helping keep the Hg and MeHg in solution (rather than being adsorbed to the beaker walls), the accumulation factors decreased as the DOC increased. In addition, the bioconcentration factor for MeHg was much higher than that for inorganic Hg, likely reflecting the stronger binding of inorganic Hg to DOC.

However, even if zooplankton accumulate both Hg and MeHg from water and food, the dominant uptake route is food, and the overall integrated signal will still be reflective of the water column, except if food sources between sites are mixed. This is a complication if concentrations in food vary widely. Zooplankton are readily sampled but samples should be checked and processed to remove phytoplankton and detritus, and it is recommended for studies of Hg that the organisms are "picked" individually under a light microscope or visually for larger species. The zooplankton can be separated from much of other seston using a large mesh filter (>50 μm typically used) and a single bulk sample from a plankton net can be used to characterize the open-water community of zooplankton at a particular time. The processing of samples can require substantial effort if picking of individual species or genera is done (which would provide the most robust signal). In terms of their being representative, these organisms are relatively local, although as they are planktonic with relatively restricted capabilities for movement, they are as mobile as their environment. Thus, zooplankton may not be the species of choice in a tidal reaches, fast flowing systems or similar locations. They are good candidate species for large water bodies, such as lakes and estuaries. Their use is restricted to the sampling of the *in situ* population as they are relatively fragile organisms, not really suitable for *in situ* deployment, and they are difficult to cage without mortality in the field.

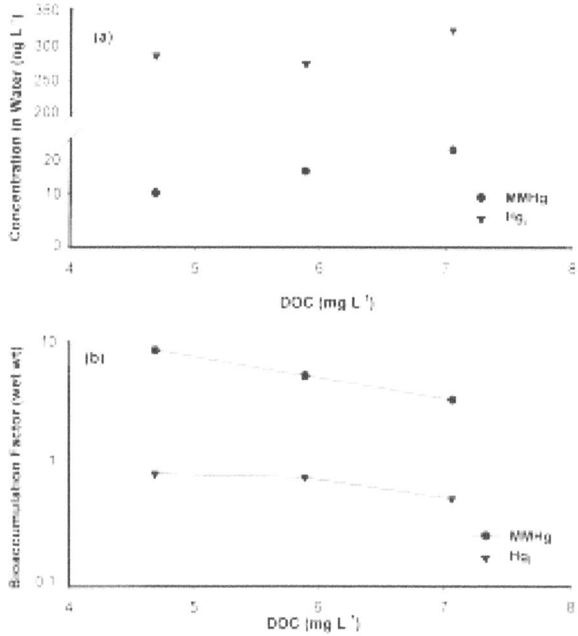

Figure F-2. The Impact of Dissolved Organic Matter on the Bioaccumulation of Hg and Methylmercury (MMHg in Figure) from Water. Source: Lawrence and Mason (2001). Copyright Elsevier, 2001. Reprinted with permission.

Specific applications of zooplankton in monitoring bioavailable MeHg are in the literature. For instance, they have been used to show the short-term increase in water column MeHg in surface waters of lakes which have seasonally anoxic bottom waters with elevated MeHg concentrations, and which mix in the fall. These mixing events result in the input of MeHg to the surface waters, and a concomitant increase in zooplankton concentration has been documented (Herrin et al., 1998; Slotton et al., 1995). Also, in the METAALICUS study in Canada (Figure F-3), where an Hg isotope was added to a lake, the zooplankton were the first to show a biological response (an increase in the isotopic MeHg) (Harris et al., 2007b). These studies confirm the usefulness of pelagic invertebrates as indicators of bioavailable Hg and MeHg and of their rapid response to changes in water column concentration.

Recommendation - As zooplankton obtain most of their Hg and MeHg from food, laboratory-based exposure studies using water collected from the sites, and exposure of phytoplankton and then subsequent feeding experiments to zooplankton are difficult to implement and interpret. Therefore, the use of zooplankton is not recommended in general for the type of studies discussed in this report, although there is the potential for their use in some specific applications. For example, they may be useful to examine situations where changes are occurring on a relatively short timescale (days to weeks) where other biosentinels may not be suitable.

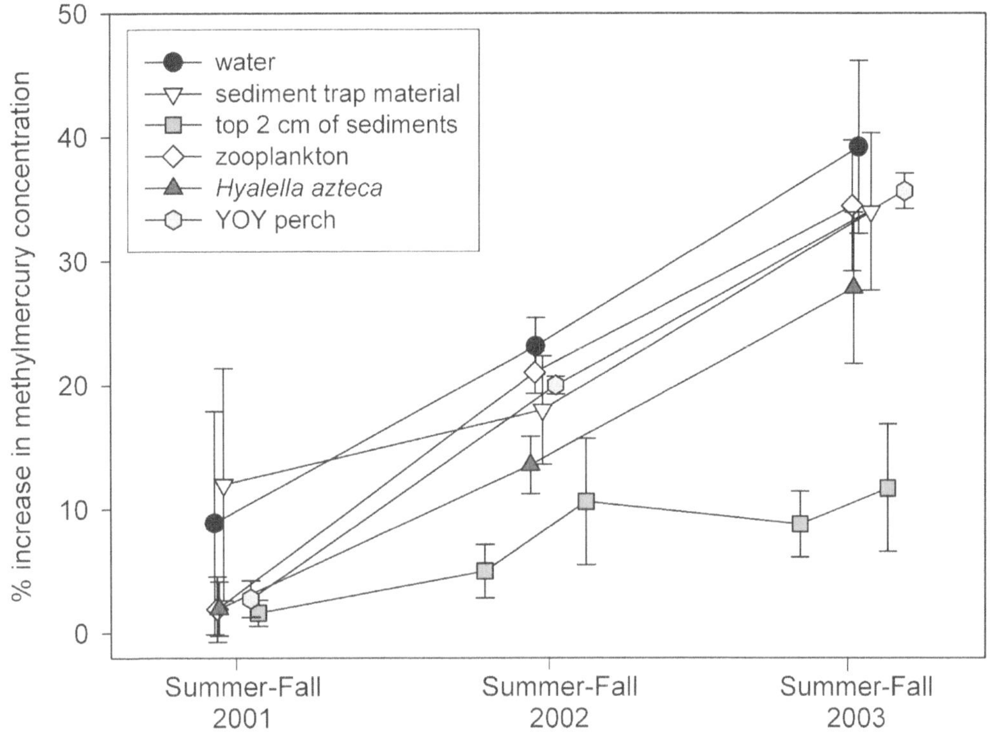

Figure F-3. Percentage Increase of Methylmerucry Concentrations in Water, Sediment trap Material, Top 2 cm of Sediments, and Biota Resulting from Additions of Lake Spike Mercury to Lake 658 over a 3-Year Period.
Source: Harris et al. (2007b)[38]. Copyright 2007, National Academy of Sciences, U.S.A. Reprinted with permission.

F.3.3 Benthic and Other Invertebrates

Benthic invertebrates are macroscopic animals that live at or near the sediment/water interface. Some benthic invertebrates, particularly mussels, readily accumulate metals, prompting their use as biological indicators of mercury contamination (NOAA, 2008). Overall, benthic invertebrates have been widely used in biological monitoring of freshwater and marine habitats (Resh and McElvary, 1993; Weigel et al., 2003), providing a foundation for their use in trend monitoring programs. They have relatively short life spans (≤1 year except for large species such as crayfish, crabs) (Wiener et al., 2007). However, in many cases, little is known about species uptake dynamics or how quickly they respond to changes in loading or concentration. The larger organisms are often relatively well-distributed, especially in the case of riverine and saline species, and therefore these organisms have been extensively used for comparative monitoring. However, they do occupy a variety of different environmental niches and some may not be suitable monitors for this application. Most of the benthic invertebrates are

[38] The graph shows the rate of change in the percentage of the total MeHg in an organism that is the added methylmercury isotope spike over time. Dosing of the lake with the isotope spike (as inorganic Hg) began in summer 2001 and isotope was added each year through the summer season. Chironomidae and Chaoborus are insect larvae living in sediments. The rapid response of the invertebrate species and the slow response of the large fish (whitefish and pike) is evident from the graph.

either herbivores or detritivores and are therefore relatively well-related to water column or sediment concentrations. In terms of their use for estimation of mercury bioavailability, this may be the major complication in their wide use, as their contact or close proximity to sediments, and their feeding on surface or deeper sedimentary material, could result in their not being a good indicator of water column changes in bioavailability. Clearly there is a need to take care in choosing the species to use in any specific study. Filter feeding organisms, even though they may reside on or at the sediment interface are the preferred benthic organisms, and these are dealt with separately and in detail below.

Recommendation - The use of infaunal or deposit-feeding organisms is recommended if other choices or species for biological monitoring are not available.

F.3.4 Filter Feeding Bivalves

In the literature, and in recent Hg studies in freshwater environments, there has been a focus on the use of the invasive Asian clam, *Corbicula*, which has become resident in many locations throughout the freshwaters of the U.S. (LWA, 2007; Paller et al., 2004; Simon et al., 2000). The use of mussels (particularly *Mytilus* species) for monitoring in estuarine and coastal waters has a long history (NOAA, 2008). Other species of mussel, as well as clams or oysters, can be used in brackish water or other locations where mussels are not widespread. These have been detailed in a number of publications (Gunther et al., 1999; Paller et al., 2004; Wang et al., 2005; Zhong and Wang, 2006).

It is also possible to use filter feeders suspended in the water column. Filter feeders are also preferred as their food source (suspended particulate matter) is more defined and can be characterized. In California and elsewhere, the invasive clam, *Corbicula*, appears to be a useful monitor and has been used in a number of studies (e.g. LWA, 2007). Clearly, a non-native organism should not be used in an environment where it is not already established and therefore, while *Corbicula* may be a suitable candidate, it should only be used in environments where it is already well established and there are no efforts for its eradication. Bivalves are typically robust organisms, they can be readily transplanted, and they assimilate Hg and MeHg rapidly (achieving steady-state concentrations within a month), based on studies in the Sacramento River and in other locations. These same studies however showed that these bivalves readily eliminate MeHg and therefore their body burdens at steady state may not be as high as another organism which is less able to depurate MeHg. In estuarine systems, there is a wider variety of benthic invertebrates such as bivalves (mussels, oysters, clams) and crustaceans (crabs, lobster, shrimp), with some larger species, and there have been extensively used for monitoring and analysis of mercury in shellfish and macro crustaceans, which could be useful (Wiener et al., 2007).

Recommendation - These organisms are recommended for most situations.

F.3.5 Small (Prey or Forage) Fish

Small fish have been used extensively in fish surveys in the San Francisco Bay and delta region, for example (Slotton et al., 2007). The three species that have been used extensively in monitoring for the California Bay-Delta Authority (CBDA) are Mississippi silverside, juvenile black bass (primarily largemouth and spotted bass), and prickly sculpin. In lakes and other freshwater habitats, young of the year (YOY) yellow perch and other prey fish can be useful indicators (Chen et al., 2005).

As shown above in Figure F-3, the YOY yellow perch and shiner responded to the Hg spike much more rapidly than the larger fish and both gave a similar response curve. In estuarine environments, *Fundulus* (mummichogs) are ubiquitous resident species in intertidal habitats with relatively stationary life styles and it has been shown that their metal burdens are good indicators of local metal exposure. They are common prey food for larger more transient fish species. In addition to being common in the field, they are an extremely hardy species, easily reared in the laboratory, and could be used in caged studies. *Fundulus* can withstand rapid and dramatic changes in their external environment (i.e., salinity, temperature, oxygen content) allowing evaluations across the range of natural conditions that encompass the limits of most other marine organisms.

Recommendation - These organisms are recommended for most situations.

F.3.6 Large (Piscivorous) Fish

Piscivorous fish are present in most surface waters, and they require moderate sampling effort as samples cannot likely be collected without techniques such as boats, electro-shocking, nets or fishing (Wiener et al., 2007). They are the primary pathway for dietary exposure of humans, as noted above, and so are relevant in this sense. However, it is somewhat difficult to relate the tissue concentration to a specific bioavailable concentration in the water and so their measurement provides a relative estimate of bioavailability. However, their concentration has been shown to correlate with water column MeHg (Sveinsdottir and Mason, 2005). If used, analyses of the filets or axial muscle for total mercury are typically sufficient as most of the Hg is as MeHg. Non-migratory piscivorous fish would obviously be the preferred candidates because of the stronger site fidelity although most large fish move relatively large distances unless constrained to a particular waterbody. In coastal waters, many species are migratory, which is a significant complication in the interpretation of their tissue burden.

In many states, there are legal limits pertaining to fish collection and capture and it is therefore necessary to obtain the required permits for the scientific collection of fish, when sampling resident fish (and other biota whose collection is regulated). The application for such permits should be submitted several weeks before sampling begins, and should be received prior to sampling.

Recommendation - Because of the slow response of predatory fish to changes in water concentration, they are not the preferred organism type for this application.

F.3.7 Birds, Reptiles, and Other Organisms

These organisms have been used extensively in monitoring programs (Wolfe et al., 2007) but again are not ideally suited for the current application. Many birds and mammals are migratory and this is a complication, as noted above. To use these species successfully, there is a need to choose species with good site fidelity, even if only on a seasonal basis. Blood is a short-term indicator and can be used to assess accumulation into migratory birds, while feathers, cuticles, etc. give an indication of longer term trends. Amphibians may be useful candidate species but have not been examined in detail in terms of Hg exposure. Some reptiles (e.g. turtles) could be useful but they tend to be difficult to catch, and their capture can be restricted.

Recommendation – While there are many organisms that could be suitable in some applications, as a class they not generally useful for this particular application.

REFERENCES

Babiarz, C.L.; Hurley, J.P.; Benoit, J.M.; Shafer, M.M.; Anders, A.W.; and Webb D.A. 1998. *Biogeochem.* **41**: 237-257.

Baker, R.F.; Blanchfield, P.J.; Paterson, M.J.; Flett, R.J.; and Wesson, L. 2004. *Trans. Am. Fish. Soc.* **133:** 568-576.

Balcom, P.H.; Fitzgerald, W.F.; Vandal, G.M.; Lamborg, C.H.; Rolfhus, K.R.; Langer, C.S.; and Hammerschmidt, C.R. 2004. *Marine Chem.* **90**: 53-74.

Balcom, P.H.; Hammerschmidt, C.R.; Fitzgerald, W.F.; Lamborg, C.H.; and O'Connor, J.S. 2008. *Marine Chem.* **109**: 1-17.

Balogh, S.J.; Huang, Y.; Offerman, H.J.; Meyer, M.L.; and Johnson, D.K. 2002. *Environ. Sci. Technol.* **36**: 1665-1670.

Balogh, S.J.; Nollet, Y.H.; and Swain, E.B. 2004. *Environ. Sci. Technol.* **38**: 4921-4927.

Bekvar, N.; Salazar, S.; Salazar, M.; and Finkelstein, K. 2000. *Can. J. Fish. Aquat. Sci.* **57**(5): 1103-1112.

Benoit, J.M.; Gilmour, C.C.; Heyes, A.; Mason, R.P.; and Miller, C.L., 2003. *Geochemical and biological controls over methylmercury production and degradation in aquatic ecosystems, Biogeochemistry of Environmentally Important Trace Elements.* ACS Symposium Series, pp. 262-297.

Benoit, J.M.; Gilmour, C.C.; and Mason, R.P. 2001. *Appl. Environ. Microbiol.* **67:** 51-58.

Benoit, J.M.; Gilmour, C.C.; Mason, R.P.; and Heyes, A. 1999a. *Environ. Sci. Technol.* **33:** 1780-1780.

Benoit, J.M.; Mason, R.P.; and Gilmour, C.C. 1999b. *Environ. Toxicol. Chem.* **18:** 2138-2141.

Bloom, N.S. 1989. *Can. J. Fish. Aquat. Sci.* **46:** 1131-1140.

Bloom, N.S. 1992. *Can. J. Fish. Aquat. Sci.* **49:** 1010-1017.

Bloom, N.S. and Crecelius, E.A. 1983. *Marine Chem.* **14:** 49-59.

Bodaly, R.A.; Rudd, J.W.M.; Fudge, R.J.P.; and Kelly, C.A. 1993. *Can. J. Fish. Aquat. Sci.* **50:** 980-987.

Brumbaugh, W.G.; Krabbenhoft, D.P.; Helsel, D.R.; Wiener, J.G.; and Echols, K.R. 2001. *A national pilot study of mercury contamination of aquatic ecosystems along multiple gradients: Bioaccumulation in fish.* Biological Science Report. USGS/BRD/BSR-2001-0009.

Campbell, P. and Tessier, A. 1996. *Ecotoxicology of metals in the aquatic environment: Geochemical aspects.* CRC Press, pp. 11-58.

Chen, C.Y.; Stemberger, R.S.; Kamman, N.C.; Mayes, B.M.; and Folt, C.L. 2005. *Ecotoxicol.* **14:** 135-147.

Chow, V.T. 1954. *Proc. Am. Soc. Civil Engrs.* **80**: 1-25.

Choy, E.S.; Hodson, P.V.; Campbell, L.M.; Fowlie, A.R.; and Ridal, J. 2008. *Arch Environ Contam. Toxicol.* **54:** 473-481.

Conaway, C.H.; Squire, S.; Mason, R.P.; and Flegal, A.R. 2003. *Marine Chem.* **80**: 199-225.

Domagalski, J. 1998. *J. Geochem. Explor.* **64:** 277-291.

Domagalski, J. 2001. *Appl. Geochem.* **16:** 1677-1691.

Dyer, K.R. 1973. Estuaries: A Physical Introduction. John Wiley and Sons, New York.

Eckley, C.S. and Hintelmann, H. 2005. *Sci. Total Environ.* **368:** 111-125.

Eckley, C.S.; Watras, C.J.; Hintelmann, H.; Morrison, K.; Kent, A.D.; and Regnell, O. 2005. *Can. J. Fish. Aquat. Sci.* **62:** 400-411.

EPRI. 2003. *Implementation of United States Environmental Protection Agency's methylmercury criterion in fish tissue.* Electric Power Research Institute, Palo Alto, CA. Report No. 1005520.

EPRI 2006. *Mercury TMDLs – significance to the power industry and guidance for individual power plants.* Electric Power Research Institute, Palo Alto, CA. Technical Report #1010102.

Fischer, H.B.; List, E.J.; Koh, R.C.Y.; Imberger, J.; and Brooks, N.H. 1979. *Mixing in Inland and Coastal Waters*. Academic Press, New York.

Fisher, N. and Reinfelder, J. 1995. *The trophic transfer of metals in marine systems.* In: A.a.T. Tessier, D. R. (Editor), Metal Speciation and Bioavailability in Aquatic Systems. Wiley & Sons, NY, pp. 1-65.

Fitzgerald, W.F.; Lamborg, C.H.; and Hammerschmidt, C.R. 2007. *Chem. Rev.* **107**b 641-662.

Foe, C.; Davis, J., et al. 2003. *Conceptual Model and Working Hypotheses of Mercury Bioaccumulation in the Bay-Delta Ecosystem and its Tributaries.* California Bay-Delta Authority (CALFED), Sacramento, CA.

Ganguli, P.M.; Mason, R.P.; Abu-Saba, K.E.; Anderson, R.S.; and Flegal, A.R. 2000. *Environ. Sci. Technol.* **34**: 4773-4779.

Golding, G.R. et al. 2002. *Limnol. Oceanog.* **47:** 967-975.

Golding, G.R.; Sparling, R.; and Kelly, C.A. 2008. *Appl. Environ. Microbiol.* **74:** 667-675.

Grigal, D.F. 2002. *Environ. Rev.* **10**: 1-39.

Guentzel, J.L.; Powell, R.T.; Landing, W.M.; and Mason, R.P. 1996. *Marine Chem.* **55**: 177-188.

Gunther, A.J. et al. 1999. *Marine Pollut. Bull.* **38:** 170-181.

Haan, C.T. 1977. *Statistical methods in hydrology.* The Iowa State University Press, Ames, Iowa.

Hammerschmidt, C.R. and Fitzgerald, W.F. 2006. *Environ Sci. Technol.* 40: 1212-1216.

Hammerschmidt, C.R.; Lamborg, C.H.; and Fitzgerald, W.F. 2006. *Atmos. Environ.* **41:** 1663-1668.

Han, S.; Gill, G.A.; Lehman, R.D.; and Choe, K.Y. 2006. *Marine Chem.* **98**: 156-166.

Harris, R.C. and Bodaly, R.A. 1998. *Biogeochemistry*, **40:** 175-187.

Harris, R. et al., 2007a. *Introduction.* In: Harris, R.; Krabbenhoft, D.P.; Mason, R.; Murray, M.W.; Reash, R.; Saltman, T. (Eds.), Ecosystem Response to Mercury Contamination: Indicators of Change. CRC press, Boca Raton, pp. 1-11.

Harris, R.C. et al., 2007b. *Whole-ecosystem study shows rapid fish-mercury response to changes in mercury deposition.* Proceedings of the National Academy of Sciences of the United States of America, 104(42): 16586-16591.

Herrin, R.T.; Lathrop, R.C.; Gorski, P.R.; and Andren, A.W. 1998. *Limnol. Oceanog.* **43:** 1476-1486.

Heyes, A.; Mason, R.P.; Kim, E.H.; and Sunderland, E. 2004. *Marine Chem.* **102**: 134-147.

Hintelmann, H.; Keppel-Jones, K.; and Evans, R.D. 2000. *Environ. Toxicol. Chem.* **19:** 2204-2211.

Hollweg, T.; Mason, R.P.; and Gilmour, C.C. *Mercury methylation and cycling in the Chesapeake Bay and continental shelf.* Paper in final stages of preparation for publication.

Hopkinson, C.S. and Vallino, J.J. 1995. *Estuaries.* **18**: 598-621.

Horvat, M. 1996. *Mercury analysis and speciation in environmental samples.* In: W. Baeyens, R. Ebinghaus and O. Vasiliev (Editors), Global and Regional Mercury Cycles: Sources, Fluxes and Mass Balances. Kluwer Academic Publishers, Dordrecht, pp. 1-31.

Hurley, J.P.; Benoit, J.M.; Babiarz, C.L.; Shafer, M.M.; Andren, A.W.; Sullivan, J.R.; Hammond, R.; and Webb, D.A. 1995. *Environ. Sci. Technol.* **29:**1867-1875.

Kim, E.H.; Mason, R.P.; Porter, E.T.; and Soulen, H.L. 2006. *Marine Chem.* **102:** 300-315.

Krabbenhoft, D.; Engstrom, D.; Gilmour, C.; Harris, R.; Hurley, J.; and Mason, R. 2007. *Monitoring and evaluating trends in sediment and water indicators.* pp. 47-86. In: Harris, R; Krabbenhoft, D.P.; Mason, R.; Murray, M.W.; Reash, R.; Saltman, T. (Eds.), Ecosystem Responses to Mercury Contamination: Indicators of Change, SETAC, CRC Press, Boca Raton.

Krabbenhoft, D.P.; Wiener, J.G.; Brumbaugh, W.G.; Olson M.L.; DeWild, J.F.; and Sabin, T.J. 1999. *A national pilot study of mercury contamination of aquatic ecosystems along multiple gradients*. In: Geological Survey Toxic Substances Hydrology Program,

Proceedings of the Technical Meeting, Charleston, South Carolina, March 8-12, 1999, Volume 2, Contamination of Hydrologic Systems and Related Ecosystems: U.S Geological Survey Water-Resources Investigations Report 99-4018B, 16 p.

Krabbenhoft, D.P.; Olson, M.L.; De Wild, J.F.; Clow, D.W.; Striegl, R.S.; Hornblaser, M.M.; and Van Metre, P. 2002. *Water, Air, Soil Pollut. Focus*. **2**: 233-249.

Lamborg, C.H.; Fitzgerald, W.F.; Skoog, A.; and Visscher, P.T. 2004. *Marine Chem.* **90**: 151-163.

Lamborg, C.H.; Tseng, C.M.; Fitzgerald, W.F.; Balcom, P.H.; and Hammerschmidt, C.R. 2003. *Environ. Sci. Technol.* **37**: 3316-3322.

Lawrence, A.L. and Mason, R.P. 2001. *Environ. Pollut.* **111:** 217-231.

Lawson, N.M. and Mason, R.P. 1998a. *Biogeochem.* **40:** 235-247.

Lin, C.C. and Jay, J.A. 2007. *Environ. Sci. and Technol.* **41:** 6691-6697.

LWA. 2007. Localized mercury bioaccumulation study, Larry Walker Associates, Davis, CA.

Marvin-Dipasquale, M. and Oremland, R.S. 1998. *Environ. Sci. Technol.* **32:** 2556-2563.

Marvin-DiPasquale, M.; Agee, J.; Bouse, R.; and Jaffe, B. 2003. *Environ. Geol.* **43:** 260-267.

Mason, R.P. 2000. *The bioaccumulation of mercury, methylmercury and other toxic trace metals into pelagic and benthic organisms.* In: Newman, M.C; Hale, R.C. (Eds.), Coastal and Estuarine Risk Assessment. CRC Press, Boca Raton, pp. 127-149.

Mason, R. 2002. *An investigation of the influence of water quality on the mercury, methylmercury, arsenic, selenium and cadmium concentrations in fish of representative Maryland streams, Chesapeake Bay and Watershed programs: Monitoring and Non-Tidal Assessment.* Annapolis, MD.

Mason, R.P. and Gill, G.A. 2005. Mercury in the Marine Environment. In *Mercury: Sources, Measurements, Cycles and Effects*, Parsons, M.B.; Percival, J.B. (Eds.), Mineralogical Association of Canada Short Course Series. **34**: 179-216.

Mason, R.P.; Laporte, J.M.; and Andres, S. 2000. *Arch. Environ. Contam. Toxicol.* **38:** 283-297.

Mason, R.P.; Lawson, N.M.; and Sullivan, K.A. 1997. *Atmos. Environ.* **31**: 3541-3550.

Mason, R.P.; Reinfelder, J.R.; and Morel, F.M.M. 1996. *Environ. Sci. and Technol.* **30:** 1835-1845.

Mason, R.P. and Sullivan, K.A. 1997. *Environ. Sci. Technol.* **31:** 942-947.

Mason, R.P, et al. 2005. *Environ. Sci. Technol.* **39:** 14A-22A.

Matilainen, T. and Verta, M. 1995. *Can. J. Fish. Aquat. Sci.* **52:** 1597-1608.

Mauro, J.; Guimaraes, J.; Hintelmann, H,; Watras, C.; Haack, E.; and Coelho-Souza, S. 2002. *Analytical Bioanalytical Chem.* **374:** 983-989.

Meybeck, M. 1982. *American Journal of Science.* **282**: 401-450.

Miller, C.L.; Mason, R.P.; Gilmour, C.C.; and Heyes, A. 2007. *Environ. Toxicol. Chem.* **26**: 624-633.

Mills, W.B.; Porcella, D.B.; Ungs, M.J.; Gherini, S.A.; Summers, K.V.; Mok, L.; Rupp, G.L.; Bowie, G.L.; and Haith, D.A. 1985. *Water quality assessment: A screening procedure for toxic and conventional pollutants in surface and groundwater (Revised 1985). Part I.* EPA/600/6-85/002a. Unites States Environmental Protection Agency, Office of Research and Development, Environmental Research Laboratory, Athens, GA.

Monperrous, M.; Tessier, E.; Amouroux, D.; Leynaert, A.; Huonnic, P.; and Donard, O.F.X. 2007. *Marine Chem.* **107:** 49-63.

Morel, F.; Kraepiel, A.; and Amyot, M. 1998. *Annu. Rev. Ecol. Syst.* **29:** 543-566.

Morel, F.M.M. and Hering, J.G. 1993. *Principals and applications of aquatic chemistry.* Wiley and Sons, NY.

NACWA. 2007. An examination of mercury levels at clean water agencies 2003-2006. National Association of Clean Water Agencies. Draft Report July 31, 2007.

Naimo, T.J,; Weiner, J.G.; Cope, W.G.; and Bloom, N.S. 2000. Can. J. Fish. Aquatic Sci. **57**: 1092-1102.

Najera, I.; Lin, C.C.; Kohbodi, G.A.; and Jay, J.A. 2005. *Environ. Sci. Technol.* **39:** 3116-3120.

NEORSD. 2006. Cuyahoga River Mercury Sampling Report. Northeast Ohio Regional Sewer District.

NOAA 2008. Mussel Watch Website, http://ccma.nos.noaa.gov/about/coast/nsandt/welcome.html.

ODEQ. 2004. DRAFT Willamette Basin TMDL. Chapter 3. Mercury TMDL. Oregon Department of Environmental Quality. Salem, Oregon.

Orihel, D.M. et al. 2006. *Environ. Sci. Technol.* **40:** 5992-6000.

Paller, M.H.; Jagoe, C.H.; Bennett, H.; Brant, H.A.; and Bowers, J.A. 2004. *Sci. Total Environ.* **325:** 209-219.

Parmer, A.; Maidrand, M.; Fry, V.; Abu-Saba, K.; and Whalin, L. 2005. Methylmercury fate and transport study. Conference Proceedings WEFTEC, 2005.

Perez-Fuentetaja, A.; Dillon, P.J.; Yan, N.D.; McQueen, D.J. 1999. *Aquatic Ecology* **33**: 127-133.

Pickhardt, P.C. and Fisher, N.S. 2007. *Environ. Sci. Technol.* **41:** 125-131.

Pickhardt, P.C.; Folt, C.L.; Chen, C.Y.; Klaue, B.; and Blum, J.D. 2002. *Algal blooms reduce the uptake of toxic methylmercury in freshwater food webs.* Proceedings of the National Academy of Sciences of the United States of America, 99(7): 4419-4423.

Pickhardt, P.C.; Folt, C.L.; Chen, C.Y.; Klaue, B.; and Blum, J.D. 2005. *Sci. Total Environ.* **339:** 89-101.

Resh, V.H. and McElvary, E.P. 1993. *Contemporary quantitative approaches to biomonitoring using benthic invertebrates.* In: Rosenberg, D.M.; Resh, V.H. (Eds.), Freshwater Biomonitoring and Benthic Macroinvertebrates. Chapman & Hall, NY, pp. pp 159-194.

Rolfhus, K.R. and Fitzgerald, W.F. 2001. *Geochim. Cosmochim. Acta* **65**: 407-418.

Rolfhus. K.R.; Lamborg, C.H.; Fitzgerald, W.F.; and Balcom, P.H. 2003. Evidence for enhanced mercury reactivity in response to estuarine mixing. JGR-Oceans **108** (C11) Article #: 3353.

Rolfhus, K.R.; Sakamoto, H.E.; Cleckner, L.B.; Storr, R.W.; Manolopoulos, H.; and Hurley, J.P. *Environ. Sci. Technol.* **37:** 865-872.

Saltman, T.; Harris, R.; Murray, M.; and Reash, R. 2007. *An Integrated Framework for Ecological Mercury Assessments.* In: Harris, R; Krabbenhoft, D.P.; Mason, R.; Murray, M.W.; Reash, R.; Saltman, T. (Eds.), Ecosystem Response to Mercury Contamination: Indicators of Change. CRC Press, Boca Raton.

Schmitt, C.J. and Brumbaugh, W.G. 2007. *Arch. Environ. Contam. Toxicol.* **53:** 84-95.

Scott, K.J. 2001. *Arctic* **54:** 92-95.

Sellers, P.; Kelly, C.A.; and Rudd, J.W.M. *Nature* **380**: 694-697.

Simon, O.; Ribeyre, F.; and Boudou, A. 2000. *Arch. Environ. Contam. Toxicol.* **38:** 317-326.

Slotton, D.G.; Ayers, S.M.; and Weyand, R.D. 2007. CBDA Biosentinel mercury monitoring program: Second year draft data report.

Slotton, D.G.; Reuter, J.E.; and Goldman, C.R. 1995. *Water Air Soil Pollut.* **80:** 841-850.

Stordal, M.C. and Gill, G.A. 1995. *Water Air Soil Pollut.* **80:** 725-734.

Sunderland, E.M. and Mason, R.P. 2007. *Global Biogeochem. Cycles* **21**: Article # GB4022.

Sunderland, E.M.; Krabbenhoft, D.P.; Moreau, J.W.; Strode, S.A.; and Landing, W.M. *Mercury sources, distribution and bioavailability in the North Pacific Ocean: Insights from data and models.* Accepted for publication in *Marine Chemistry.*

Sveinsdottir, A.Y. and Mason, R.P. 2005. *Arch. Environ. Contam. Toxicol.* **49:** 528-545.

Taylor, J.R. 1982. *An introduction to error analysis.* University Science Books, Mill Valley, CA.

Tetra Tech, Inc. 2005. *Guadalupe River watershed mercury TMDL project: Final conceptual model report.* Prepared for the San Francisco Regional Water Quality Control Board, Oakland, CA. May 20, 2005.

Thomas, M.A.; Conaway, C.H.; Steding, D.J.; Marvin-DiPasquale, M.; Abu-Saba, K.E.; and Flegal, A.R. 2002. *Geochemistry: Exploration, Environment, Analysis.* **2**: 211-217.

U.S. EPA. 1985. *Initial Mixing Characteristics of Municipal Ocean Discharges: Volume 1, Procedures and Applications.* EPA 600385073a. US Environmental Protection Agency, Washington, D.C.

U.S. EPA. 1992. *Technical Guidance Manual for Performing Wasteload Allocations, Book III: Estuaries – Part 3: Use of Mixing Zone Models in Estuarine Waste Load Allocations.* In: U.S. EPA (Editor). U.S. Environmental Protection Agency, Washington, D.C.

U.S. EPA. 1995. *Method 1631: Mercury in Water by Oxidation, Purge and Trap, and Cold Vapor atomic Fluorescence Spectrometry.* EPA 821-R-95-027, US Environmental Protection Agency, Washington, D.C.

U.S. EPA. 1998. *EPA Methods and Guidance for Analysis of Water*: CD ROM. Government Institutes, Washington, D.C.

U.S. EPA. 2000. *Guidance for data quality assessment. Practical methods for data analysis.* EPA/600/R-96/084. United States Environmental Protection Agency, Office of Environmental Information. Washington, D.C. July.

USGS. 2002. *Processing of water samples.* Chapter A5, Book 9 Handbook for water resources investigations, National field manual for the collection of water quality data. Wilde F.D.; Radtke D.B.; Gibs J.; Iwatsubo R.T. (Eds.) United States Geological Survey, Reston, VA.

Vandal, G.M.; Fitzgerald, W.F.; Rolfhus, K.R.; Lamborg, C.H.; Langer C.S.; and Balcom, P.H. 2001. *Sources and cycling of mercury and methylmercury in Long Island Sound.* Final Report to the Connecticut Department of Environmental Protection Long Island Sound Program, September 2001.

Van Veen, V.; Burton, N.; Comber, S.; and Gardner, M. 2002. *Environ. Toxicol. Chem.* **21:** 275–280.

Waldron, M.C.; Coleman, J.A.; and Breault, A.F. 2000. *Can. J. Fish. Aquat. Sci.* **57**: 1080–1091

Wang, Y.W.; Liang, L.N.; Shi, J.B.; and Jiang, G.B. 2005. *Environ. Pollut.* **135:** 457-467.

Watras, C.J. and Bloom, N.S. 1992. *Limnol. Oceanogr.*, **37:** 1313.

Weigel, B.; et al., 2003. *Freshwat. Biol.* **48:** 1440-1461.

Wiener, J.; et al. 2007. *Monitoring and evaluating trends in methylmercury accumulation in biota.* In: Harris, R; Krabbenhoft, D.P.; Mason, R.; Murray, M.W.; Reash, R.; Saltman, T. (Eds.), Ecosystem Response to Mercury Contamination: Indicators of Change. CRC Press, Boca Raton, pp. 87-122.

Wiener, J.; Krabbenhoft, D.; Heinz, G.; and Scheuhammer, A.M. 2003. *Ecotoxicology of mercury.* In: R.B. Hoffman D.J., Burton G.A., and Cairns J. (Editor), Handbook of Ecotoxicology. CRC Press, Boca Raton, pp. 409-463.

Whalin, L.; Kim, E.; and Mason, R. 2007. *Marine Chem.* **107**: 278-294.

Wolfe, M.; et al. 2007. *Wildlife indicators.* In: Harris, R; Krabbenhoft, D.P.; Mason, R.; Murray, M.W.; Reash, R.; Saltman, T. (Eds.), Ecosystem Response to Mercury Contamination: Indicators of Change. CRC Press, Boca Raton, pp. 123-190.

Zhong, H. and Wang, W.X. 2006. *Environ. Sci. Technol.* **40:** 3794-3799.

WERF SUBSCRIBERS

WASTEWATER UTILITY

Alabama
Montgomery Water Works & Sanitary Sewer Board

Alaska
Anchorage Water & Wastewater Utility

Arizona
Avondale, City of
Glendale, City of, Utilities Department
Mesa, City of
Peoria, City of
Phoenix Water Services Dept.
Pima County Wastewater Management
Safford, City of
Tempe, City of

Arkansas
Little Rock Wastewater Utility

California
Central Contra Costa Sanitary District
Corona, City of
Crestline Sanitation District
Delta Diablo Sanitation District
Dublin San Ramon Services District
East Bay Dischargers Authority
East Bay Municipal Utility District
El Dorado Irrigation District
Fairfield-Suisun Sewer District
Fresno Department of Public Utilities
Inland Empire Utilities Agency
Irvine Ranch Water District
Las Gallinas Valley Sanitary District
Las Virgenes Municipal Water District
Livermore, City of
Los Angeles, City of
Los Angeles County, Sanitation Districts of
Napa Sanitation District
Novato Sanitary District
Orange County Sanitation District
Palo Alto, City of
Riverside, City of
Sacramento Regional County Sanitation District
San Diego Metropolitan Wastewater Department, City of
San Francisco, City & County of
San Jose, City of
Santa Barbara, City of
Santa Cruz, City of
Santa Rosa, City of
South Bayside System Authority
South Coast Water District
South Orange County Wastewater Authority
Stege Sanitary District
Sunnyvale, City of
Union Sanitary District
West Valley Sanitation District

Colorado
Aurora, City of
Boulder, City of
Greeley, City of
Littleton/Englewood Water Pollution Control Plant
Metro Wastewater Reclamation District, Denver

Connecticut
Greater New Haven WPCA
Stamford, City of

District of Columbia
District of Columbia Water & Sewer Authority

Florida
Broward, County of
Fort Lauderdale, City of
Jacksonville Electric Authority (JEA)
Miami-Dade Water & Sewer Authority
Orange County Utilities Department
Pinellas, County of
Reedy Creek Improvement District
Seminole County Environmental Services
St. Petersburg, City of
Tallahassee, City of
Tampa, City of
Toho Water Authority
West Palm Beach, City of

Georgia
Atlanta Department of Watershed Management
Augusta, City of
Clayton County Water Authority
Cobb County Water System
Columbus Water Works
Fulton County
Gwinnett County Department of Public Utilities
Savannah, City of

Hawaii
Honolulu, City & County of

Idaho
Boise, City of

Illinois
Decatur, Sanitary District of
Greater Peoria Sanitary District
Kankakee River Metropolitan Agency
Metropolitan Water Reclamation District of Greater Chicago
Wheaton Sanitary District

Iowa
Ames, City of
Cedar Rapids Wastewater Facility
Des Moines, City of
Iowa City

Kansas
Johnson County Wastewater
Lenexa, City of
Unified Government of Wyandotte County/ Kansas City, City of

Kentucky
Louisville & Jefferson County Metropolitan Sewer District

Louisiana
Sewerage & Water Board of New Orleans

Maine
Bangor, City of
Portland Water District

Maryland
Anne Arundel County Bureau of Utility Operations
Howard County Bureau of Utilities
Washington Suburban Sanitary Commission

Massachusetts
Boston Water & Sewer Commission
Massachusetts Water Resources Authority (MWRA)
Upper Blackstone Water Pollution Abatement District

Michigan
Ann Arbor, City of
Detroit, City of
Holland Board of Public Works
Saginaw, City of
Wayne County Department of Environment
Wyoming, City of

Minnesota
Rochester, City of
Western Lake Superior Sanitary District

Missouri
Independence, City of
Kansas City Missouri Water Services Department
Little Blue Valley Sewer District
Metropolitan St. Louis Sewer District

Nebraska
Lincoln Public Works and Utilities Department

Nevada
Henderson, City of
Las Vegas, City of
Reno, City of

New Jersey
Bergen County Utilities Authority
Ocean County Utilities Authority
Passaic Valley Sewerage Commissioners

New York
New York City Department of Environmental Protection

North Carolina
Charlotte/Mecklenburg Utilities
Durham, City of
Metropolitan Sewerage District of Buncombe County
Orange Water & Sewer Authority
University of North Carolina, Chapel Hill

Ohio
Akron, City of
Butler County Department of Environmental Services
Columbus, City of
Metropolitan Sewer District of Greater Cincinnati
Montgomery, County of
Northeast Ohio Regional Sewer District
Summit, County of

Oklahoma
Oklahoma City Water & Wastewater Utility Department
Tulsa, City of

Oregon
Albany, City of
Clean Water Services
Eugene, City of
Gresham, City of
Portland, City of Bureau of Environmental Services
Water Environment Services

Pennsylvania
Hemlock Municipal Sewer Cooperative (HMSC)
Philadelphia, City of
University Area Joint Authority

South Carolina
Charleston Water System
Mount Pleasant Waterworks & Sewer Commission
Spartanburg Water

Tennessee
Cleveland Utilities
Murfreesboro Water & Sewer Department
Nashville Metro Water Services

Texas
Austin, City of
Dallas Water Utilities
Denton, City of
El Paso Water Utilities

WERF SUBSCRIBERS

Fort Worth, City of
Houston, City of
San Antonio Water System
Trinity River Authority

Utah
Salt Lake City Corporation

Virginia
Alexandria Sanitation Authority
Arlington, County of
Fairfax County
Hampton Roads Sanitation District
Hanover, County of
Henrico, County of
Hopewell Regional Wastewater Treatment Facility
Loudoun Water
Prince William County Service Authority
Richmond, City of
Rivanna Water & Sewer Authority

Washington
Everett, City of
King County Department of Natural Resources
Seattle Public Utilities
Sunnyside, Port of
Yakima, City of

Wisconsin
Green Bay Metro Sewerage District
Kenosha Water Utility
Madison Metropolitan Sewerage District
Milwaukee Metropolitan Sewerage District
Racine, City of
Sheboygan Regional Wastewater Treatment
Wausau Water Works

Australia
ACTEW (Ecowise)
South Australian Water Corporation
South East Water Limited
Sydney Water Corporation
Water Corporation of Western Australia

Canada
Edmonton, City of/Edmonton Waste Management Centre of Excellence
Lethbridge, City of
Regina, City of, Saskatchewan
Toronto, City of, Ontario
Winnipeg, City of, Manitoba

New Zealand
Watercare Services Limited

United Kingdom
Anglian Water Services, Ltd.

STORMWATER UTILITY

California
Fresno Metropolitan Flood Control District
Los Angeles, City of, Department of Public Works
Monterey, City of
San Francisco, City & County of
Santa Rosa, City of
Sunnyvale, City of

Colorado
Aurora, City of
Boulder, City of

Florida
Orlando, City of

Iowa
Cedar Rapids Wastewater Facility
Des Moines, City of

Kansas
Overland Park, City of

Kentucky
Louisville & Jefferson County Metropolitan Sewer District

Maine
Portland Water District

North Carolina
Charlotte, City of, Stormwater Services

Pennsylvania
Philadelphia, City of

Tennessee
Chattanooga Stormwater Management

Texas
Harris County Flood Control District, Texas

Washington
Bellevue Utilities Department
Seattle Public Utilities

STATE

Connecticut Department of Environmental Protection
Kansas Department of Health & Environment
New England Interstate Water Pollution Control Commission (NEIWPCC)
Ohio River Valley Sanitation Commission
Urban Drainage & Flood Control District, CO

CORPORATE

ADS Environmental Services
Advanced Data Mining International
AECOM
Alan Plummer & Associates
Alpine Technology Inc.
Aqua-Aerobic Systems Inc.
Aquateam–Norwegian Water Technology Centre A/S
ARCADIS
Associated Engineering
Bernardin Lochmueller & Associates
Black & Veatch
Blue Water Technologies, Inc.
Brown & Caldwell
Burgess & Niple, Ltd.
Burns & McDonnell
CABE Associates Inc.
The Cadmus Group
Camp Dresser & McKee Inc.
Carollo Engineers Inc.
Carpenter Environmental Associates Inc.
CET Engineering Services
CH2M HILL
CRA Infrastructure & Engineering
CONTECH Stormwater Solutions
D&B/Guarino Engineers, LLC
Damon S. Williams Associates, LLC
Ecovation
EMA Inc.
Environmental Operating Solutions, Inc.
Environ/The ADVENT Group, Inc.
Fay, Spofford, & Thorndike Inc.
Freese & Nichols, Inc.
ftn Associates Inc.
Gannett Fleming Inc.
Garden & Associates, Ltd.
Geosyntec Consultants
GHD
Greeley and Hansen LLC
Hazen & Sawyer, P.C.
HDR Engineering Inc.
HNTB Corporation
Hydromantis Inc.
HydroQual Inc.
Infilco Degremont Inc.
Jacques Whitford NAWE, Inc.
Jason Consultants LLC Inc.
Jordan, Jones, & Goulding Inc.
KCI Technologies Inc.
Kelly & Weaver, P.C.
Kennedy/Jenks Consultants
KMK Consultants
Larry Walker Associates
Limno-Tech Inc.
The Low Impact Development Center Inc.
Malcolm Pirnie Inc.
Material Matters
McKim & Creed
MPR Engineering Corporation, Inc.
MWH
NTL Alaska, Inc.
O'Brien & Gere Engineers Inc.
Odor & Corrosion Technology Consultants Inc.
Paradigm Environmental Technologies, Inc.
Parametrix Inc.
Parsons
Post, Buckley, Schuh & Jernigan
Praxair, Inc.
Ring Industrial Group
RMC Water & Environment
Ross & Associates Ltd.
SAIC
Siemens Water Technologies
The Soap & Detergent Association
Smith & Loveless, Inc.
Southeast Environmental Engineering, LLC
Stearns & Wheler, LLC
Stone Environmental Inc.
Stratus Consulting Inc.
Synagro Technologies Inc.
Tetra Tech Inc.
Trojan Technologies Inc.
Trussell Technologies, Inc.
Uni-Bell PVC Pipe Association
URS Corporation
Westin Engineering Inc.
Wright Water Engineers
Zoeller Pump Company

INDUSTRY

American Electric Power
American Water
Anglian Water Services, Ltd.
Chevron Energy Technology
The Coca-Cola Company
Dow Chemical Company
DuPont Company
Eastman Chemical Company
Eli Lilly & Company
Johnson & Johnson
Merck & Company Inc.
Procter & Gamble Company
Suez Environnment
United Utilities North West (UUNW)
United Water Services LLC

Note: List as of 2/21/09

IWA Publishing's authorised EU representative for General Product Safety Regulations is Diane D'Arras, 15 rue Duret, 75116 Paris, France, e-mail: safety@iwap.co.uk.

Printed and bound by CPI Group (UK) Ltd, Croydon, CR0 4YY

12/05/2026

02108865-0003

OHIM (Office for Harmonization in the Internal Market - Trademarks and Designs)	http://oami.europa.eu/en/default.htm
Republic of Korea	http://www.kipo.go.kr/kpo2/user.tdf?a=user.english.html.HtmlApp&c=50100&catmenu=ek50100 (Patents)
Romania	http://193.230.133.4/cgi-bin/invsearch (Patents)
Russian Federation	http://www.fips.ru/ensite/ (Patents & Trademarks)
Singapore	http://www.epatents.gov.sg/PE/ (Patents)
Slovakia	http://www.indprop.gov.sk/eskipo.php?lang=en&idd=2&idd2=
Slovenia	http://www2.uil-sipo.si/dse.htm
Spain	http://invenes.oepm.es (Patents)
Switzerland	http://www.swissreg.ch (Patents & Trademarks)
Thailand	http://www1.ipic.moc.go.th/dipdcd/owa/p5query1.main?LANG=1 (Patents)
United Kingdom	hhttp://www.ipo.gov.uk/patent.htm (Patents)
United States of America	http://www.uspto.gov/patft/index.html (Patents) http://tess2.uspto.gov/bin/gate.exe?f=login&p_lang=english&p_d=trmk (Trademarks)

4

Copyright: Essential Concepts for Authors and Innovators

Copyright is a legal entitlement granted to creators of literary, dramatic, musical, and artistic works, as well as producers of cinematographic films and sound recordings. This legal right encompasses various privileges, including the reproduction, adaptation, translation, and communication to the public of the work. The composition of these rights may vary slightly, depending on the nature of the work. Along with Intellectual Property (IP), Copyright is a legal concept that safeguards the rights of creators and inventors. Copyright confers upon creators the exclusive legal right to reproduce, distribute, and exhibit their work, while IP comprises other legal rights such as patents, trademarks, and trade secrets. Copyright applies to a diverse range of creative works, including literary, dramatic, musical, and artistic works, as well as sound recordings and cinematographic films. Piracy is a worldwide issue that adversely affects various content industries, such as software, television, recording, and movies.

Copyright and IP laws play a crucial role in promoting creativity and innovation by safeguarding the rights of creators and inventors. These laws grant exclusive rights to creators and inventors, allowing them to financially benefit from their work. This encourages investment in research and development, leading to the creation of new ideas, products, and services. Without copyright and IP protection, creators and inventors may be discouraged from investing their time and resources into creating new works, resulting in a decline in innovation and economic growth. These laws also protect consumers by ensuring that they receive original, high-quality products and services instead of counterfeit or low-quality imitations.

Historical Background

Copyright has a long history dating back to ancient civilizations such as Greece and Rome. However, the modern concept of copyright protection began to emerge with the invention of printing in the 15th century, which made it possible for literary works to be duplicated by mechanical processes instead of being copied by hand.In the early days, copyright was granted as a form of privilege by authorities and kings, entitling beneficiaries exclusive rights of reproduction and distribution for a limited period. In England, the Statute of Anne, which was passed in 1710, is considered to be the first copyright statute. This statute granted authors the exclusive right to publish their works for a period of 14 years, which could be renewed for another 14 years if the author was still alive.

In the 18th century, there was a dispute over the relationship between copyright subsisting in common law and copyright under the Statute of Anne. This was finally settled by the House of Lords in 1774, which ruled that at common law, the author had the sole right of printing and publishing his book, but that once a book was published, the rights in it were exclusively regulated by the Statute. This common law right in unpublished works lasted until the Copyright Act of 1911, which abolished the Statute of Anne.

The history of copyright protection in India dates back to the colonial era. The first copyright law was introduced in 1847 and was based on the British Copyright Act of 1842. Over time, the law has been amended several times to keep pace with changing technology and the times. The current copyright law in India is the Copyright Act of 1957, which has been amended multiple times, with the most recent amendment taking place in 2012. The law provides protection for various types of creative works, including literary, artistic, musical, and dramatic works, as well as for performers' rights and the rights of producers of sound recordings. The objective of the law is to promote creativity, protect copyright owners' interests, and ensure that the public has access to a broad range of creative works.

The Copyright Act of 1957 came into effect in January 1958, and it has been amended five times since then, in 1983, 1984, 1992, 1994, 1999, and 2012. The most significant amendment was the Copyright (Amendment) Act, 2012, which was introduced to bring the Act in line with two WIPO internet treaties, protect the music and film industry, address concerns of the physically disabled, and enforce copyright protection. The amendments introduced new penalties for circumvention of technological protection measures, rights management information, and liability of internet service providers, as well as introduced statutory licenses for cover versions and broadcasting organizations. The amendments also ensured authors and music composers' right to receive royalties, performers' exclusive economic and moral rights, equal membership rights for authors and other right owners in copyright societies, and an exception for physically disabled individuals to access copyrighted works.

Types of Works Protected by Copyright: Copyright protection typically covers works of literature, music, art, architecture, computer software, and other original intellectual creations. These works must be original and fixed in a tangible form to be eligible for copyright protection.

Work: Work encompasses literary, dramatic, musical, artistic, cinematographic films, and sound recordings. A "work of joint authorship" refers to a collaboration of two or more authors, where the contributions are not distinct from one another. India offers copyright protection for original works in literary, dramatic, musical, and artistic categories, as well as for cinematographic films and sound recordings. Additionally, artistic works can include paintings, sculptures, drawings, engravings, photographs, works of architecture, and artistic craftsmanship. A musical work is defined as a composition of music that can be notated graphically, but does not include any accompanying words or actions intended for singing, speaking, or performing with the music.

Nature and Scope of Copyright Protection

Copyright protection is a legal concept that grants creators of original works exclusive rights to control the use and distribution of their works. The nature and scope of copyright protection varies depending on the jurisdiction, but generally, it applies to various types of creative works.

Duration of Copyright Protection

In India, the duration of copyright protection varies depending on the type of work. For literary, dramatic, musical, or artistic works, the copyright lasts for the lifetime of the author plus 60 years from the year of the author's death. In the case of joint authorship, the copyright lasts for 60 years from the year of the death of the last surviving author. For cinematograph films, the duration is 60 years from the date of publication. In the case of sound recordings, the duration is 60 years from the year in which the recording was first published. After the expiration of the copyright term, the work enters the public domain and can be used freely without the need for permission or payment of royalties.

Exclusive Rights of Copyright Holders

- Copyright holders have exclusive rights over their works.
- They can control the use and distribution of their works.
- Exclusive rights include reproduction, creating derivative works, distributing copies, and publicly displaying or performing the work.
- Anyone who wants to use a copyrighted work must obtain permission from the copyright holder, either through a license or by obtaining explicit permission.
- Using a copyrighted work without permission or outside the scope of the license constitutes copyright infringement.

- Infringement of copyright can lead to legal action, including civil lawsuits and criminal charges.

Copyright Infringement

Copyright infringement refers to the unauthorized use of copyrighted material without the permission of the copyright owner. Copyright infringement occurs when someone exercises one of the exclusive rights of the copyright owner without permission. These rights include the right to reproduce, distribute, display, perform, and create derivative works.

- **Types of Copyright Infringement:** There are different types of copyright infringement, such as direct infringement, contributory infringement, vicarious infringement, and inducement of infringement.

a. **Direct infringement:** This occurs when someone intentionally or knowingly violates the exclusive rights of the copyright owner.

b. **Contributory infringement:** This happens when someone intentionally contributes to the infringement committed by another person.

c. **Vicarious infringement:** This occurs when someone has the right and ability to control the infringing activity and receives direct financial benefit from it.

d. **Inducement of infringement:** This happens when someone encourages, induces, or causes another person to infringe on a copyright.

Each type of infringement may have different legal consequences, and the copyright owner may take legal action against the infringer to protect their exclusive rights.

Remedies for Copyright Infringement

Copyright holders can seek remedies for copyright infringement, such as injunctive relief, damages, and an account of profits. Injunctive relief involves obtaining a court order to stop the infringing activity. Damages can be awarded to compensate for the harm caused by the infringement, including lost profits and any other losses suffered by the copyright holder. An account of profits requires the infringer to account for and pay over to the copyright holder any profits made as a result of the infringement. Additionally, criminal penalties can be imposed for willful infringement.

Fair Use and Public Domain

- **Fair Use:** Fair use is a legal principle that permits the limited use of copyrighted material without obtaining permission from the copyright owner. This principle is based on the idea that certain uses of copyrighted works can benefit the public, such as criticism, commentary, news reporting, teaching, scholarship, or research. Fair use is determined on a case-by-case basis, and several factors are taken into consideration, including the purpose and character of the use, the nature of the copyrighted work, the amount used, and the effect on the potential market for the copyrighted work.Some examples of fair use in India include using a limited amount of copyrighted material in a news report, using a quote from a copyrighted work in a book review, or using a portion of a copyrighted work for educational purposes.

- **Public Domain Works:** Public domain works are those that are no longer protected by copyright, or that were never protected by copyright in the first place. Anyone can use these works without obtaining permission from the copyright owner. Public domain works in India include works where the copyright has expired, works created by the Indian government, and works that have been dedicated to the public domain by the copyright owner.

International Copyright Law

International copyright law refers to the set of rules and regulations that govern the protection of copyright and intellectual property rights across different countries. Here are some important aspects of international copyright law:

a. **Berne Convention for the Protection of Literary and Artistic Works:** The Berne Convention is an international treaty that was established in 1886 to provide a framework for the protection of literary and artistic works across different countries. It provides a minimum standard of protection for copyright holders and ensures that copyright works created in one country are protected in all other countries that are members of the convention. As of 2023, there are over 170 member countries of the Berne Convention.

b. **World Intellectual Property Organization (WIPO):** WIPO is a specialized agency of the United Nations that is responsible for promoting and protecting intellectual property rights worldwide. It provides a forum for member states to develop and harmonize intellectual property policies and laws, and also administers a number of international treaties related to intellectual property.

c. **International Copyright Treaties:** There are several international treaties that deal with copyright protection, including the Berne Convention, the Universal Copyright Convention, the WIPO Copyright Treaty, and the Agreement on Trade-Related Aspects of Intellectual Property Rights

(TRIPS Agreement). These treaties provide a framework for the protection of copyright across different countries and aim to harmonize copyright laws across borders.

International copyright law plays a crucial role in protecting the rights of copyright holders and ensuring that their works are recognized and protected across different countries. It also promotes cross-border collaboration and innovation by providing a level playing field for creators and inventors to share their works and ideas with a global audience.

Registration of Copyright

Copyright registration is not necessary for acquiring copyright. Copyright comes into existence as soon as a work is created, and no formality is required. However, registering a copyright provides prima facie evidence in a court of law with reference to disputes relating to ownership of copyright. The procedure for registration is set out in Chapter VI of the Copyright Rules, 1956.

To register a work, an application must be made on Form IV, which includes a Statement of Particulars and a Statement of Further Particulars, as prescribed in the first schedule to the Rules. Separate applications should be made for each work, and the application should be accompanied by the requisite fee prescribed in the second schedule to the Rules. The applications should be signed by the applicant or the advocate in whose favor a Vakalatnama or Power of Attorney has been executed, and each and every column of the Statement of Particulars and Statement of Further Particulars should be replied specifically.

Both published and unpublished works can be registered. For works published before January 21, 1958, three copies of the published work may be sent along with the application for registration. If the work to be registered is unpublished, a copy of the manuscript has to be sent along with the application for affixing the stamp of the Copyright Office in proof of the work having been registered. In case two copies of the manuscript are sent, one copy of the same duly stamped will be returned, while the other will be retained, as far as possible, in the Copyright Office for record and will be kept confidential. If a work has been registered as unpublished and subsequently it is published, the applicant may apply for changes in particulars entered in the Register of Copyright in Form V with the prescribed fee.

The Copyright Office, which provides registration facilities for all types of works, is headed by a Registrar of Copyrights and is located at B.2/W.3, C.R. Barracks, Kasturba Gandhi Marg, New Delhi- 110 003, Tel: 338 4387. The Act and Rules can be obtained from the Manager of Publications, Publication Branch, Civil Lines, Delhi or his authorized dealers on payment.

What is Plagiarism

Plagiarism is the act of using someone else's work or ideas and passing them off as your own without giving proper credit to the original source. This can include copying text, images, music, or any other creative work without permission or attribution. Plagiarism is considered a serious offense in academia, journalism, and other fields that value originality and intellectual property rights. It can have consequences ranging from failing grades or loss of employment to legal action and damaged reputation. Proper citation and attribution of sources are necessary to avoid plagiarism and give credit where it is due.

There are several disadvantages of plagiarism, including:

i. **Legal consequences:** Plagiarism is considered a violation of intellectual property rights and can result in legal action, including lawsuits and fines.
ii. **Academic consequences:** Plagiarism in academic settings can lead to severe consequences, including failing grades, suspension, or expulsion from school or university.
iii. **Loss of credibility:** Plagiarism damages the credibility and integrity of the person who committed it, making it difficult for them to gain trust from peers, colleagues, or the public.
iv. **Damage to reputation:** Plagiarism can damage the reputation of an individual or organization, leading to negative publicity, loss of business or job opportunities, and a tarnished image.
v. **Ethical considerations:** Plagiarism is unethical, as it involves taking credit for someone else's work without their permission or acknowledgment. It is considered a breach of academic and professional ethics and can lead to a loss of trust and respect from others.
vi. **Stifling creativity:** Plagiarism can discourage creativity and innovation, as individuals may be less inclined to generate new ideas if they can simply copy the work of others without consequences.

Plagiarism has far-reaching negative consequences, affecting the legal, academic, professional, ethical, and creative aspects of an individual or organization's life.

There are several methods that can help reduce plagiarism:

i. **Paraphrasing:** This involves rewording someone else's ideas in your own words. However, it is important to ensure that the meaning of the original text is not changed.
ii. **Citing sources:** Whenever you use someone else's work or ideas, it is important to give them proper credit. This involves citing your sources using the appropriate citation style.

iii. **Quoting:** If you need to use someone else's exact words, you can do so by using quotation marks and citing the source.

iv. **Using plagiarism checker tools:** There are several online tools available that can help you check for plagiarism in your work. These tools can compare your work with other published material and highlight any similarities.

v. **Time management:** Proper time management can also help reduce plagiarism. By managing your time effectively, you can avoid the temptation to copy and paste information from other sources.

vi. **Learning proper research skills:** Knowing how to conduct proper research and taking good notes can also help reduce plagiarism. By understanding the material and being able to summarize it in your own words, you are less likely to copy and paste someone else's work.

There are several plagiarism reducing tools available, including:

i. **Turnitin:** This is a widely used tool that checks for plagiarism by comparing submitted work to its database of internet sources, academic journals, and previously submitted student papers.

ii. **Grammarly:** This tool not only checks for grammar and spelling errors, but also includes a plagiarism checker that compares text to online sources and academic databases.

iii. **Copyscape**: This is a tool designed specifically to check for online plagiarism. It allows users to enter a website URL or block of text to see if it appears elsewhere on the internet.

iv. **Plagiarism Checker:** This is a free online tool that compares text to internet sources and academic databases.

v. **Dupli Checker:** This is another free online tool that compares text to online sources and provides a percentage of originality.

vi. **Quetext:** This is a plagiarism checker that is available for both free and paid versions. It compares text to internet sources and academic databases, and provides a detailed report of any potential plagiarism.

vii. **Unicheck:** This is a cloud-based plagiarism checker that is available for both individuals and institutions. It compares text to online sources and academic databases, and provides a detailed report of any potential plagiarism.

These tools can be helpful in reducing plagiarism, but it is important to note that they are not foolproof and should be used in conjunction with proper citation practices and academic integrity.

Conclusion

In conclusion, copyright and intellectual property protection play a critical role in promoting creativity, innovation, and economic growth. By granting exclusive rights to creators and inventors, these legal concepts encourage investment in research and development, leading to the creation of new ideas, products, and services. Additionally, copyright and IP laws protect consumers by ensuring that they receive high-quality, original products and services. However, with the rapid advancement of technology, the future of copyright and IP protection is uncertain. New challenges arise with the ease of copying and distributing digital content, and it is important for lawmakers to adapt to these changes to ensure that the rights of creators and innovators are protected.

In the coming years, it is likely that copyright and IP laws will continue to evolve to keep up with technological advancements. It will be crucial for lawmakers to balance the interests of creators and innovators with the public interest in access to information and content. Additionally, international cooperation and agreements will be essential to ensure that copyright and IP protection is effective in a globalized world. As technology continues to develop, it is important to ensure that copyright and IP laws remain relevant and effective in protecting the rights of creators and innovators.

5

Industrial Designs, Integrated Circuits Trademarks, and Trade Secrets: Essential Concepts for Innovators and Entrepreneurs

Intellectual Property (IP) refers to intangible assets that are created through human creativity and innovation. These assets can be legally protected and provide exclusive rights to the creators and owners for a certain period of time.

Importance of IP for Innovators and Entrepreneurs

a) Protecting intellectual property (IP) is essential for fostering innovation and creativity, as it encourages inventors, creators, and entrepreneurs to invest their time, energy, and resources into developing new products, services, and technologies without fear of having their ideas stolen or copied.

   i. For Example, if a software company develops a new program that can revolutionize how people interact with technology, it's essential to protect the IP of the product by patenting and copyright. This protection ensures that no other company can steal the idea or copy the software, providing an incentive for the software company to invest in the product's development.

   ii. Similarly, if a fashion designer creates a unique design for a dress, they can protect their IP by trademarking the design. This protection provides incentivizes the designer to invest their time and resources into developing new designs, knowing that their creations will be legally protected from copycats.

b) Having intellectual property (IP) protection grants a competitive advantage in the marketplace as it enables creators and owners to have exclusive rights to use, sell, and license their creations without the fear of infringement by others.

   i. For instance, if a pharmaceutical company invests millions of dollars in research and development to create a new medication, obtaining a patent for the drug gives them exclusive rights to manufacture, distribute, and sell the product for a certain period. This protection allows the company to recoup their investment and make a profit without competition from other companies producing a similar drug.

ii. Similarly, if a software developer creates a new application and secures a copyright, it can license the product to other companies and individuals for a fee, generating revenue and expanding their market share. This protection ensures that no one else can use or distribute the software without permission, giving the developer a competitive advantage in the marketplace.

c) Intellectual property (IP) protection allows innovators and entrepreneurs to attract investors, secure funding, and expand their businesses. This is because IP protection assures potential investors that their investment will be protected, and their return on investment will not be undermined by competitors.

d) Protecting intellectual property (IP) helps businesses to establish a strong brand and reputation by preventing others from using similar trademarks, logos, or designs that could confuse customers or dilute the business's identity.

i. For instance, if a clothing company designs a unique logo and obtains a trademark, they can prevent other companies from using a similar logo to market their products. This protection ensures that the clothing company's logo is associated with their brand only, helping to build brand recognition and customer loyalty.

ii. Similarly, if a tech company creates a new design for a mobile device and patents it, they can prevent other companies from using a similar design, which could confuse customers and harm the company's reputation. This protection ensures that the tech company's design is recognized as unique and innovative, enhancing the company's reputation and increasing its market share.

Industrial Designs

Industrial designs are the decorative or visual elements of a product that set it apart from others in the market. They refer to the appearance of an article resulting from specific features of shape, configuration, pattern, or ornamentation. These features are intended to make the product visually appealing, unique, and easily recognizable. The industrial design recognizes the creation of new and original features such as shape, configuration, surface pattern, ornamentation, and composition of lines or colors applied to articles that are judged solely by the eye in the finished state.

In India, the registration and protection of industrial designs are governed by the Designs Act of 2000 and the corresponding Designs Rules of 2001, which were enforced on May 11, 2001, replacing the earlier Act of 1911. The Design Rules of 2001 were subsequently amended by Designs (Amendment) Rules in 2008 and 2014. The most recent amendment to the Designs Rules came into effect on December 30, 2014, which introduced a new category of applicant known as a small entity, in addition to natural persons and other than small entities. Industrial design rights are critical for businesses that rely on the visual appeal of their products to attract customers and differentiate themselves from competitors in the market.

Benefits of Design Registration

i. **Legal protection:** Registration provides legal protection to the design and prevents others from using or copying it without permission.

ii. **Exclusive rights:** The registered owner gets exclusive rights to use and license the design, providing a competitive advantage in the market.

iii. **Commercial value:** A registered design adds commercial value to the product or article and can be used as a valuable asset in negotiations with investors or buyers.

iv. **Deterrent to infringement**: Registration acts as a deterrent to potential infringers as it provides the owner with the legal means to take action against infringement.

v. **International protection:** Registration in one country can be used as a basis for seeking protection in other countries, through international agreements such as the Hague Agreement.

vi. **Renewal:** The design registration can be renewed for additional periods of time, ensuring ongoing protection of the design.

vii. **Licensing opportunities:** The registered owner can license or sell the design to third parties, generating revenue and increasing exposure to the design.

Essential Requirements for Registration

i. The design should be new or original.

ii. It should not have been made available to the public in tangible form, by use, or in any other way before the filing date, or priority date where applicable, of the registration application.

iii. The design should be significantly distinguishable from known designs or combinations of known designs.

iv. It should not contain scandalous or obscene material.
v. The design should not be a mere mechanical contrivance.
vi. It should be applied to an article and be visually appealing.
vii. The design should not be contrary to public order or morality.
viii. It should not be prejudicial to the security of India.

Types of Industrial Designs: Industrial designs can be classified into three main categories:

- **Two-dimensional designs:** These are designs that are applied to surfaces of products, such as wallpapers or textiles.
- **Three-dimensional designs:** These refer to designs that are applied to three-dimensional products, such as furniture or vehicles.
- **Designs applied to specific articles:** These are designs that are specifically created for a particular article, such as a computer mouse or a phone case.

Registration Process and Requirements: To obtain protection for an industrial design, the owner must register it with the relevant intellectual property office. The application must include a clear representation of the design, a description of the article to which it is applied, and the application fee. The registration process varies by country, but generally takes several months to complete.

Enforcement of Industrial Design Rights: Owners of industrial designs can enforce their rights against unauthorized use or infringement by taking legal action against infringers. The owner can also grant licenses to others to use the design in exchange for a fee or royalty.

Examples of Industrial Designs

There are several examples from various industries, such as fashion, automotive, and electronics, to illustrate the importance of industrial design protection. The case studies provides real-life examples of the benefits and challenges of industrial design protection, including the role of intellectual property in promoting innovation and creativity, protecting investments, and enhancing brand value.

i. The unique shape of a Coca-Cola bottle
ii. The design of Apple's iPhone
iii. The pattern on a Louis Vuitton handbag
iv. The distinctive shape of a Lamborghini sports car
v. The design of a Philips electric shaver
vi. The ornamental design of a Swarovski crystal necklace

vii. The appearance of a Herman Miller Eames chair

viii. The design of a Dyson vacuum cleaner

xi. The shape and design of a Rolex watch

x. The aesthetic features of a Bang & Olufsen speaker.

Integrated Circuits

Integrated Circuits (ICs) are electronic devices that contain multiple components, such as transistors and capacitors, on a single semiconductor chip. An IC is a miniature electronic device that contains multiple components, such as transistors, resistors, and capacitors, on a single semiconductor chip. ICs are widely used in various electronic devices, including computers, smartphones, and televisions.

Semiconductor Integrated Circuits Layout-Design Registry (SICLDR) in India

The Semiconductor Integrated Circuits Layout-Design Registry (SICLDR) is responsible for accepting and processing applications for the registration of intellectual property rights related to layout-designs of integrated circuits. The jurisdiction of the SICLDR covers the entire country of India. The office operates under the guidelines established by the Semiconductor Integrated Circuits Layout Design (SICLD) Act of 2000 and the Semiconductor Integrated Circuits Layout-Design (SICLD) Rules of 2001. Upon examination of the layout-designs of integrated circuits, the SICLDR issues a registration certificate to the original layout-designs of the semiconductor integrated circuits.

Criteria for Protection: To qualify for protection, an IC must meet the following criteria:

i. **Originality**: The design must be new and original, meaning that it has not been made public before.

ii. **Functionality:** The design must have a specific function or purpose.

iii. **Significant Investment:** The design must represent a significant investment of time, effort, and resources.

Types of Integrated Circuits: ICs can be classified into two main categories:

i. **Analog ICs:** These are ICs that process continuous signals, such as sound or temperature, and are widely used in communication and audio equipment.

ii. **Digital ICs:** These are ICs that process discrete signals, such as binary code, and are widely used in computers, microprocessors, and digital signal processing.

Registration Process and Requirements: To obtain protection for an IC, the owner must register it with the relevant intellectual property office. The application must include a clear description of the design and the application fee. The registration process varies by country, but generally takes several months to complete.

Enforcement of Integrated Circuit Rights: Owners of ICs can enforce their rights against unauthorized use or infringement by taking legal action against infringers. The owner can also grant licenses to others to use the design in exchange for a fee or royalty.

Several case studies from various industries, such as telecommunications, computing, and consumer electronics, to illustrate the importance of IC protection.

Trademarks

Trademarks are an essential aspect of branding and business identity. A trademark is a symbol, word, phrase, or design that identifies and distinguishes the source of a product and service. A company can have serval trademark for their products. A trademark helps consumers to identify and choose between products/services based on their reputation and quality

Legislation in India

In 1940, the Trademarks Registry was established in India, and it is currently responsible for administering the Trademarks Act, 1999 and its associated rules. The registry serves as a valuable resource and information center, and helps facilitate matters relating to trademarks throughout the country. The Head Office of the Trademarks Registry is located in Mumbai, with branch offices in Ahmedabad, Chennai, Delhi, and Kolkata. An International Registration wing has been established in the Head Office in Mumbai to handle functions related to international applications and registrations under the Madrid Protocol. The Controller General of Patents, Designs, and Trademarks is in charge of the Trademarks Registry offices and acts as the Registrar of Trade Marks.Currently, all Trademarks Registry functions are performed using an automated Trademarks System. The Central Server of the TMR is located at the Intellectual Property Office (IPO) Building in Delhi, and a Disaster Recovery server is located in IPO Mumbai.

Objectives and functions of Trademarks Act, 1999 and Registry

Objectives of the Trademarks Act, 1999

1. To register trademarks applied in the country.
2. To provide better protection for goods and services.
3. To prevent fraudulent use of the trademark.

Functions of the Trademarks Registry

1 Register trademarks that meet the requirements under the Trade Marks Act and Rules.
2. Maintain the Register of trademarks.
3. Act as an office of origin for Indian entrepreneurs for international registration of their trademarks under the Madrid Protocol.
4. Act as an office of the designated Contracting party for international registrations in which India has been designated for trademark protection.
5. Offer preliminary advice on registrability.
6. Conduct trademark searches and issue certificates under Section 45(1) of the Copyright Act, 1957.
7. Provide public information and guidance on trademarks.
8. Provide information to various government agencies.
9. Maintain a top-class intellectual property (IP) library.
10. Produce an annual statistical report and an official Trade Marks Journal in electronic form.
11. Submit an Annual Report to Parliament.

Types of Trademarks: Trademarks can be categorized into different types based on their characteristics, such as:

i. **Word Marks:** These are trademarks that consist of a word or combination of words, such as "Apple" or "Coca-Cola."
ii. **Design Marks:** These are trademarks that consist of a graphic or pictorial design, such as the Nike "swoosh" logo or the Starbucks mermaid.
iii. **Color Marks**: These are trademarks that consist of a specific color or combination of colors, such as the Tiffany blue color.
iv. **Sound Marks:** These are trademarks that consist of a specific sound or musical jingle, such as the NBC chimes.

v. **Scent Marks:** These are trademarks that consist of a specific scent or fragrance, such as the scent of a perfume.

vi. **3D Marks:** These are trademarks that consist of a three-dimensional shape or design, such as the shape of a Coca-Cola bottle.

Duration of the Registration: The duration of trademark registration is 10 years, and it can be renewed periodically in accordance with the specified provisions.

Registration Process and Requirements: To register a trademark, the owner must file an application in Head Office (Mumbai)or in branch offices (Ahmedabad, Chennai, Delhi, and Kolkata). The application must include a clear description of the trademark and the application fee. The registration process varies by country, but generally takes several months to complete. The trademark must also meet certain requirements, such as being distinctive and not confusingly similar to existing trademarks.

Enforcement of Trademark Rights: Trademark owners can enforce their rights against unauthorized use or infringement by taking legal action against infringers. This can include filing a lawsuit for damages, obtaining an injunction to prevent further use of the trademark, or pursuing criminal charges in some cases. The owner can also grant licenses to others to use the trademark in exchange for a fee or royalty.

Infringement of Trademark

Trademark infringement happens when someone uses a trademark that is identical or similar to a registered trademark for the same or similar goods or services. This can lead to the sale of counterfeit products to customers who believe they are buying genuine products, a practice also known as "passing off". If customers receive substandard products without knowing about the "passing off" practice, they may mistakenly believe that the manufacturer is producing low-quality products and choose another trademark in the future. These imitation products in the market are commonly referred to as counterfeits.

Trademark Advantages and Disadvantages

Advantages

i. **Exclusive Rights:** Trademark registration gives the owner exclusive rights to use the mark in connection with their goods or services, preventing others from using the same or a similar mark.

ii. **Brand Recognition:** A registered trademark can help build brand recognition and create goodwill among customers, helping to distinguish your products or services from those of your competitors.

iii. **Legal Protection:** Trademark registration provides legal protection against infringement and allows the owner to take legal action against anyone who uses their mark without permission.

iv. **Asset Value:** A trademark can be a valuable asset for a business and can even be licensed or sold to others for a profit.

v. **International Protection:** Trademark registration can be used to protect your mark in other countries, providing a basis for international expansion.

Disadvantages

i. **Cost:** The cost of registering a trademark can be high, and ongoing maintenance and renewal costs can add up over time.

ii. **Time:** The trademark registration process can be lengthy and time-consuming, taking months or even years to complete.

iii. **Limitations:** Trademarks only protect specific goods or services, and may not provide complete protection against similar marks used for different goods or services.

iv. **Enforcement:** While trademarks provide legal protection, enforcement can be difficult and costly, especially when dealing with infringers located in other countries.

v. **Limited Duration:** Trademarks have a limited duration and must be renewed periodically, adding to ongoing costs and administrative requirements.

Trade Secrets

Trade secrets are confidential information that provides a competitive advantage to businesses. Trade secrets are confidential information that is not generally known or readily accessible to others and that provides a competitive advantage to a business. This can include proprietary formulas, processes, designs, customer lists, and other sensitive business information.

Criteria for Trade Secrets Protection: To qualify for protection as a trade secret, the information must be secret, have commercial value, and be subject to reasonable efforts to maintain its confidentiality. The trade secret must also not be generally known or readily accessible to others.

How to Protect Trade Secrets

i. **Identify and Classify Trade Secrets:** Identify and classify your trade secrets, so you know what information needs protection.

ii. **Confidentiality Agreements:** Use confidentiality agreements or non-disclosure agreements (NDAs) with employees, contractors, and partners to prevent them from sharing your trade secrets.

iii. **Limited Access:** Limit access to trade secrets to those who need to know and ensure that only authorized personnel have access.
iv. **Security Measures:** Implement security measures such as password protection, encryption, and physical security to protect against unauthorized access.
v. **Training and Education:** Train employees and partners on the importance of protecting trade secrets and best practices for maintaining confidentiality.
vi. **Monitoring:** Monitor access to your trade secrets and take action if you suspect unauthorized access or disclosure.
vii. **Contractual Obligations:** Include contractual obligations related to trade secret protection in agreements with third parties, such as suppliers or customers.
viii. **Legal Action:** Take legal action against anyone who misappropriates your trade secrets or breaches confidentiality agreements.

Types of Trade Secrets: Trade secrets can include a wide range of confidential information, including:

i. Proprietary formulas and recipes
ii. Manufacturing processes and techniques
iii. Customer lists and data
iv. Marketing strategies and plans
v. Research and development information
vi. Software and computer code

Measures to Protect Trade Secrets: Businesses can take several measures to protect their trade secrets, including:

i. Implementing policies and procedures to identify, classify, and protect confidential information
ii. Limiting access to confidential information to only those who need it
iii. Requiring employees and contractors to sign non-disclosure agreements (NDAs)
iv. Conducting background checks on employees and contractors
v. Regularly reviewing and updating security measures to protect against unauthorized access or theft

Enforcement of Trade Secret Rights in India

In India, while there is no specific law for the protection of trade secrets, the country is a signatory of the TRIPs Agreement and has the flexibility to frame laws protecting trade secrets. Indian courts and tribunals have upheld protection of trade secrets under other laws such as contract law, copyright law, principles of equity, and the common law action of breach of confidence. Additionally, the Information Technology Act of 2000 provides legal means of protection to confidential information in the form of electronic records. Businesses in India have various legal means to safeguard their trade secrets and maintain a competitive advantage in the marketplace.

Intellectual Property Strategy for Innovators and Entrepreneurs

Intellectual Property (IP) strategy is critical for innovators and entrepreneurs as it can help to protect their ideas, products, and services from infringement, generate revenue, and provide a competitive advantage in the market.

Importance of IP strategy: Developing an IP strategy helps innovators and entrepreneurs to:

i. Protect their valuable ideas and inventions from infringement and misappropriation.

ii. Create revenue streams by licensing, selling or monetizing their IP assets.

iii. Build a strong brand by securing trademark protection.

iv. Secure financing by showing potential investors the company's IP assets.

v. Maintain a competitive edge by preventing competitors from using similar IP.

Key considerations for developing an IP strategy: When developing an IP strategy, innovators and entrepreneurs should consider the following:

i. **Identify and categorize their IP assets:** This includes identifying what IP assets the company has, such as patents, trademarks, copyrights, and trade secrets.

ii. **Prioritize their IP assets:** The company should focus on protecting the most valuable IP assets first.

iii. **Determine how to protect their IP assets:** This involves deciding on the type of protection to seek for each IP asset and the appropriate filing strategy.

iv. **Monitor and enforce their IP rights:** The company should have a plan for monitoring and enforcing their IP rights, such as conducting regular IP audits, filing infringement lawsuits, or sending cease and desist letters.

Aligning IP strategy with business goals: In developing an IP strategy, it is important to align it with the company's overall business goals. The IP strategy should be aligned with the company's product development, branding marketing, and financing objectives.

Several successful companies have implemented effective IP strategies to protect their innovations and build strong brands. For example, Apple Inc. has been successful in protecting its IP through patents, trademarks, and trade secrets, which has helped the company maintain a competitive edge in the market. Coca-Cola has also created a strong brand through trademark protection and strategic marketing, which has helped the company to maintain its position as a market leader in the beverage industry.

The importance of protecting intellectual property (IP) for innovators and entrepreneurs cannot be overstated. Various types of IP, including trademarks, copyrights, trade secrets, industrial designs, and integrated circuits, and discussed the criteria and processes for their protection. Developing an effective IP strategy, which involves aligning IP protection efforts with business goals and considering key factors such as market competition and risk management, is crucial for the long-term success of any business. Looking ahead, emerging trends and challenges such as the increasing importance of data protection and the ongoing evolution of digital technologies will continue to shape the landscape of IP protection. Therefore, it is essential for innovators and entrepreneurs to prioritize IP protection as a critical component of their overall business strategy in order to stay competitive and ensure long-term success in the marketplace.

6

Understanding Geographical Indications and Their Significance in Protecting India's Unique Products

Geographical Indications (GI) are indications or symbols that are affixed to products that possess a specific geographical origin and are known for their unique qualities or reputation. These products can be agricultural, natural, or manufactured goods that are essentially attributed to a particular region, locality, or country where they are produced. According to Article 22.1 of the TRIPS Agreement, GIs identify a good as originating from a specific territory, region, or locality, where a given quality, reputation, or characteristic of the good is essentially attributable to its geographical origin.

GIs are a product of collective tradition and decision-making and continue to evolve. They represent a combination of human effort, culture, land resources, and climatic conditions. GIs are limited to a specific region and are not transferable from one owner or region to another. A product that has a GI tag is associated with a specific geographical area where it is developed, and the GI sign on the product indicates its origin. Products designated with a GI acquire a high reputation and are valuable commercial assets for the geographical region where they originated. However, GIs are also exposed to misappropriation and forgery, and therefore require both national and international protection.

The use of GIs is significant in protecting India's unique products, such as Darjeeling Tea, Alphonso Mangoes, and Kanchipuram Silk. GIs help to safeguard the reputation and distinctiveness of these products and promote their economic value. Producers can differentiate their products from those produced in other regions and enhance consumer trust and loyalty through the use of GIs. In India, GIs have played a vital role in promoting and protecting traditional crafts, agricultural products, and other unique products that are integral to the cultural heritage and identity of the country. By registering a GI, producers can prevent others from misusing the name or falsely claiming the origin of the product, thereby ensuring its quality and authenticity.

The Legal Framework for GIs in India

The legal framework for GIs in India is based on the Geographical Indications of Goods (Registration and Protection) Act, 1999. This Act provides for the registration and protection of GIs in India, and establishes a national registry for GIs. In addition to this Act same other relevant laws and regulations that provide protection for GIs in India.

i. **The Geographical Indications of Goods (Registration and Protection) Act, 1999:** The Geographical Indications of Goods (Registration and Protection) Act, 1999 is the primary law governing GIs in India. The Act defines geographical indication as an indication that identifies goods as originating from a particular geographical location, where a given quality, reputation, or other characteristic of the goods is essentially attributable to its geographical origin. The Act also provides for the registration of GIs, and sets out the procedure for registration, protection, and enforcement of GIs in India. The Act prohibits the use of any misleading or false indication in relation to goods that bear a registered GI, and provides for penalties for the violation of the Act.

ii. **Other Relevant Laws and Regulations:** In addition to the Geographical Indications of Goods (Registration and Protection) Act, 1999, there are other relevant laws and regulations in India that provide protection for GIs. These include the Trade Marks Act, 1999, which prohibits the registration of any trade mark that is likely to deceive or cause confusion with a registered GI, and the Copyright Act, 1957, which provides protection for literary and artistic works, including works related to GIs. Additionally, the Designs Act, 2000, and the Patents Act, 1970, also provide some level of protection for GIs in India.

GI Products in India

i. Description of various GI products in India

India has a diverse range of unique products that have been granted Geographical Indication (GI) status. These products are typically tied to a specific region, and have a reputation for possessing qualities or characteristics that are essentially attributable to their place of origin. Some of the GI products in India include Darjeeling tea, Alphonso mangoes, Basmati rice, Kanchipuram silk sarees, Nagpur oranges, Banarasi silk sarees, Pashmina shawls, Chanderi fabric, and many more.

ii. Famous GI products in India

A total 475 goods have been registered under Sec 2(f) of GIG Act 1999 to 2023 under different categories viz: agricultural goods, food stuff and handicrafts.

Some of the most famous GI products in India not only have a unique identity but also have a significant economic value for their respective regions. Here are few examples of famous GI products in India with brief descriptions:

i. **Darjeeling Tea** - Grown in the Darjeeling district of West Bengal, this tea is known for its unique flavor and aroma.

ii. **Kanchipuram Silk** - A type of silk fabric produced in the Kanchipuram district of Tamil Nadu, known for its durability and intricate designs.

iii. **Alphonso Mangoes** - A popular variety of mangoes grown in the Konkan region of Maharashtra, known for their sweet taste and aroma.

iv. **Banarasi Silk** - A type of silk fabric produced in the city of Varanasi in Uttar Pradesh, known for its intricate designs and gold and silver brocade work.

v. **Chanderi Sarees** - A type of cotton and silk fabric produced in the town of Chanderi in Madhya Pradesh, known for its lightweight texture and delicate embroidery.

vi. **Mysore Sandalwood Oil** - An essential oil extracted from sandalwood trees grown in the Mysore district of Karnataka, used in perfumes, soaps, and cosmetics.

vii. **Nagpur Orange** - A variety of oranges grown in the Nagpur district of Maharashtra, known for their sweet and juicy taste.

viii. **Pashmina Shawls** - A type of woolen shawl produced in the Kashmir region, known for its softness and warmth.

ix. **Ratnagiri Hapus Mangoes** - A variety of mangoes grown in the Ratnagiri district of Maharashtra, known for their sweet and juicy taste.

x. **Salem Fabric** - A type of cotton fabric produced in the city of Salem in Tamil Nadu, known for its durability and color fastness.

xi. **Kolhapuri Chappals** - Handcrafted leather sandals made in the Kolhapur district of Maharashtra, known for their durability and comfort.

xii. **Malabar Pepper** - A type of black pepper grown in the Malabar region of Kerala, known for its strong aroma and flavor.

xiii. **Muga Silk** - A type of silk fabric produced in the Assam state, known for its durability and golden yellow color.

xiv. **Alleppey Green Cardamom** - A type of cardamom grown in the Alleppey district of Kerala, known for its strong aroma and flavor.

xv. **Bangalore Blue Grapes** - A variety of grapes grown in the Bangalore region of Karnataka, known for their sweet taste and juicy texture.

xvi. **Basmati Rice** - A type of long-grain rice grown in the Punjab, Haryana, and Uttar Pradesh states, known for its aromatic fragrance and fluffy texture.

xvii. **Naga Mircha** - A type of chili pepper grown in the Nagaland state, known for its high level of spiciness.

xviii. **Nilgiri Tea** - A type of tea grown in the Nilgiri hills of Tamil Nadu, known for its strong flavor and aroma.

xix. **Salem Mangoes** - A variety of mangoes grown in the Salem district of Tamil Nadu, known for their sweet and juicy taste.

xx. **Tirupati Laddu** - A type of sweet made in the town of Tirupati in Andhra Pradesh, offered as a religious offering in the Sri Venkateswara Temple and known for its unique taste and texture.

Famous GI products at international level

Here are few examples of famous GI products at international level with brief descriptions:

i. **Champagne:** A sparkling wine produced only in the Champagne region of France, according to strict rules and regulations.

ii. **Darjeeling Tea:** A high-quality tea grown only in the Darjeeling district of West Bengal, India, known for its unique flavor and aroma.

iii. **Scotch Whisky:** A distilled alcoholic beverage made in Scotland, with strict regulations governing the production process and ingredients.

iv. **Parmigiano-Reggiano:** A hard cheese produced in the Parma and Reggio Emilia regions of Italy, with strict regulations on production and aging.

v. **Colombian Coffee:** Coffee grown in Colombia is known for its high quality and unique flavor profiles.

vi. **Swiss Watches:** Watches manufactured in Switzerland, with strict regulations on the production process and quality control.

vii. **Kobe Beef:** A high-quality beef produced only in the Hyogo Prefecture of Japan, known for its tenderness and marbling.

viii. **Roquefort Cheese:** A blue-veined cheese produced only in the Roquefort-sur-Soulzon region of France, made from sheep's milk and aged in natural caves.

ix. **Tequila:** A distilled alcoholic beverage made from the blue agave plant, produced only in certain regions of Mexico.

x. **Feta Cheese:** A white cheese made from sheep's milk, produced only in certain regions of Greece, with strict regulations on production and ingredients.

Importance of GIs for India's Unique Products

i. **Benefits of GIs for India's unique products:** Geographical Indications (GIs) offer numerous benefits for India's unique products. One of the main advantages is the protection against imitation and misuse. GIs help prevent unauthorized use of a product's name, branding, or other identifying marks, ensuring that consumers can trust in the authenticity and quality of the product.

ii. **Protection against imitation and misuse:** In addition to protecting against imitation and misuse, GIs also enhance the marketability and economic value of a product. By linking a product to its geographical origin and unique characteristics, GIs provide a means for producers to differentiate their products from those produced in other regions. This differentiation can lead to increased demand and higher prices for the products, which can benefit the local economy and improve the livelihoods of producers.

iii. **Enhancement of marketability and economic value:** GIs also provide a platform for promoting and preserving traditional knowledge and cultural heritage associated with a product. By protecting and promoting the unique qualities and characteristics of a product, GIs contribute to the preservation of traditional crafts and practices, and promote sustainable development. Additionally, GIs can help to promote tourism in the region, attracting visitors interested in experiencing the local culture and unique products.

GI Registration Process in India

a) **Registration process in India:** The registration process for Geographical Indications (GI) in India is governed by the Geographical Indications of Goods (Registration and Protection) Act, 1999. The registration process involves a detailed application process, which includes applying with the appropriate authority, followed by an examination and publication of the application. Once the application is published, interested parties may file an opposition, which is decided by the Registrar of Geographical Indications. If no opposition is filed or if the opposition is unsuccessful, the GI is granted registration and entered in the Register of Geographical Indications.

b) **Eligibility criteria and requirements:** To be eligible for registration as a GI, the product must originate from a specific geographical location and possess a unique quality, reputation, or other characteristic that is essentially attributable to that location. The applicant for registration must be an association of persons or producers of the concerned goods or a competent authority. The application must include details of the product, its origin, and the characteristics that make it unique, as well as evidence to support these claims. The application must also include a draft of the proposed GI tag or logo.

c) **Timeline for the registration process:** The registration process for GIs in India typically takes between 18 to 24 months, depending on the complexity of the application and any opposition filed during the examination and publication stages. However, the actual timeline can vary depending on various factors, including the workload of the authority and any unforeseen delays. It is important to note that the registration of a GI is valid for a period of 10 years and can be renewed for further periods of 10 years each.

Enforcement of GI Rights in India

i. **Enforcement of GI rights in India:** Enforcement of GI rights in India is done through legal proceedings. The legal remedies available for infringement of GI rights include both civil and criminal actions. Civil actions include injunctions, damages, accounts, and other reliefs, while criminal actions include fines and imprisonment. The government of India has established specialized agencies to enforce GI rights, such as the Geographical Indications Registry and the Intellectual Property Appellate Board.

ii. **The Geographical Indications of Goods (Registration & Protection) Act, 1999:** After India joined the TRIPS Agreement as a member state, the Geographical Indications of Goods (Registration and Protection) Act was enacted in 1999. The act came into force on September 15, 2003, along with the Rules of 2002, and the GI registry was established in Chennai. The main objective of the act is to provide registration and protection to goods originating in specific regions linked with specific qualities. The act aims to (i) protect the interests of the goods' producers, (ii) prevent unauthorized misuse of GIs and consumer deception, and (iii) promote Indian GI-tagged goods in the international export market.

iii. **Legal remedies for infringement:** The owner of a registered GI has the right to prevent third parties from using the GI in relation to products that do not conform to the standards set for that GI. Any unauthorized use of a registered GI is considered an infringement, and the owner of the GI can take legal action to stop the infringing activity. In such cases, the owner can seek legal remedies such as injunctions, damages, accounts, and other reliefs. The legal remedies available for infringement of GI rights in India are provided under the Geographical Indications of Goods (Registration and Protection) Act, 1999.

iv. **Role of government agencies in enforcement:** The government of India has established several agencies to enforce GI rights. The Geographical Indications Registry, which operates under the Ministry of Commerce and Industry, is responsible for the registration and administration of GI rights. It maintains a register of GIs and ensures that no unauthorized use of registered GIs takes place. The Intellectual Property Appellate Board hears appeals against decisions made by the Geographical Indications Registry. Additionally, the police and other law enforcement agencies also play a role in enforcing GI rights by conducting raids and seizing infringing goods.

International Protection of Indian GIs

i. **International agreements related to GIs:** Several international agreements and treaties have been established to provide protection and recognition of geographical indications (GIs) at the international level. The most notable of these agreements is the World Trade Organization's Agreement on Trade-Related Aspects of Intellectual Property Rights (TRIPS), which sets out minimum standards for the protection and enforcement of GIs. Other agreements include the Lisbon Agreement for the Protection of Appellations of Origin and their International Registration, the Paris Convention for the Protection of Industrial Property, and the Madrid Agreement Concerning the International Registration of Marks.

ii. **International protection of Indian GIs:** India has taken steps to protect its GIs at the international level by seeking recognition of its products under the TRIPS agreement and through bilateral agreements with other countries. India has also registered several of its GIs under the World Intellectual Property Organization's (WIPO) Lisbon system, which provides international protection for appellations of origin. In addition, India is a member of the International Trademark Association (INTA) and participates in international discussions on the protection of GIs.

iii. **Importance of international protection for Indian GIs:** International protection of GIs is important for Indian products because it allows for wider recognition and awareness of the unique qualities and reputation of these products. It also helps prevent unauthorized use and misuse of the names of these products, which can damage their reputation and economic value. Furthermore, international recognition and protection of GIs can open up new markets and opportunities for Indian producers and enhance the competitiveness of these products in the global marketplace.

Case Studies

i. **Analysis of successful GI registration and enforcement in India:** India has a successful track record in registering and enforcing GIs for its unique products. Some of the successful GI registrations include Darjeeling Tea, Kanchipuram Silk, and Alphonso Mangoes. These products have gained worldwide recognition and protection, leading to increased demand and economic benefits for the producers and the region. The government agencies responsible for GI registration and enforcement have been active in promoting and protecting Indian GIs, and have collaborated with international organizations to strengthen the legal framework and enforcement mechanisms.

ii. **Challenges faced by GIs in India:** Despite the success of GIs in India, there are still challenges faced by these products. One of the main challenges is the lack of awareness among producers and consumers about the importance and benefits of GIs. This can lead to misuse and infringement of GI rights, which can affect the reputation and economic value of the products. Another challenge is the lengthy and complex registration process, which can discourage small producers from seeking GI protection. In addition, the lack of adequate enforcement measures and resources can also hinder the effective protection of GIs.

iii. **Lessons learned from case studies:** Case studies of successful GI registration and enforcement in India provide valuable lessons for other countries and regions seeking to protect their unique products. One lesson is the importance of collaboration between government agencies, producers, and consumers in promoting and protecting GIs. Another lesson is the need for strong legal frameworks and effective enforcement mechanisms to ensure the proper use and protection of GIs. Furthermore, education and awareness campaigns can play a crucial role in promoting and enhancing the reputation and value of GIs. Finally, the success of GIs in India highlights the potential economic benefits for producers and regions, which can lead to sustainable development and the preservation of cultural heritage.

Geographical Indication vs Trademark

Both Geographical Indications (GIs) and trademarks are used to distinguish products from others in the marketplace and convey assurance of quality and information on the origin of the product. However, there are some key differences between the two.

Trademarks identify the company that produced the product and assure consumers of the quality and reputation of the product. Trademarks can be assigned or licensed to other agencies anywhere to produce the goods, but the trademark use is restricted to a single production company.

On the other hand, GIs identify a good originated from a specific locality or region. GIs assure consumers of the product's quality due to its production in a particular geographical area. GIs can be used by many persons within a particular area by maintaining the quality standards. However, the GI cannot be assigned or licensed outside a particular place.

While both GIs and trademarks are used to distinguish products and convey assurance of quality and origin, trademarks identify the company that produced the product and can be assigned or licensed anywhere, while GIs identify a product's origin from a specific locality or region and cannot be assigned or licensed outside that place.

Filing of Geographical Indications Application

Who can be the applicant: Any association of individuals, producers, or any organization or authority established by or under any law that represents the interests of the goods' producers may apply for the registration of a Geographical Indication.

Type of Application

i. **Ordinary/Indian Application:** A simple application filled for registration of goods for GI-tag. The application to be filled in triplicate in Form GI – 1(A) for single class and in GI – 1 (C) for multiple classes.

ii. **Convention Application**: an application filled for GI in convention country along with proof of filling in the home country. The application to be filled in triplicate in Form GI – 1(B) for single class and in GI – 1 (D) for multiple classes

iii. **Single Class Application**: The application is filled for registration of goods in one class.

iv. **Multi Class Application:**The application is filled for registration of goods in multi class.

Application Fee: The prescribed fee Rs. 5000 to be paid in respect of Indian and conventional applications through cash or money order or may be sent by bank draft or cheque.

Application Form and Required Document: The Registrar of Geographical Indications is the Controller-General of Patents, Designs and Trade Marks appointed under section 3(1) of the Trademarks Act, 1999. The application form for GI registration should include details such as the specification and description of the goods, proof of origin, method of production, uniqueness, inspection body, and a statement of case as prescribed in the format. The completed application and related documents should be sent to the Geographical Indications Registry at the Intellectual Property Office Building on G.S.T. Road in Guindy, Chennai – 600032 through post, registered post, speed post, or courier services.

Statement of case (in triplicate):A document of "**Statement of case**" to be enclosed with the application, shall contain the following information.

a. **Specification:** A statement describing the special characteristics and quality parameters of the goods in about 50 to 100 words.

b. **Description of goods giving its uniqueness and geographical linkage:** A brief description for **special characteristics, unique features, linkage to the specific geographical location including human creativity involved**. Environmental factors such as soil, water and climatic condition may be clearly brought out. Fixing such a standard may be necessary for quality control, inspection and enforcement of the GI.

c. **Method of production:** A brief note on method of production, including the process involved, raw materials and tools, packaging specialty if any, etc.

d. **Uniqueness of the product:** The part should explain the uniqueness by comparing with other product not more than in 100 words.

e. **Proof of origin: A historic proof** gazetteers, published documents, news articles, advertisement materials **that prove development of the product in geographical region.**

f. **Inspection body:** An inspection Body set-up by the Applicant to monitor the production in respect of quality, integrity and consistency of the product as well as the genuine use of the GI.

g. **Present scenario of the GI product:** It is detailed information of present market of the product, exports figures, total turnover of the product, and the activities undertaken by the applicant association in development and promotion of the GI.

ii. A certified map of geographical area in triplicate showing the title, name of publisher and date of issue.

iii. A list of members of the association of producers of the good.

iv. Two addition representations are also required at along with application form.

v. An Affidavit, as required under rule 32(6) (a).

Prohibition of registration of certain Geographical Indications: Before application of GI it should confirm that the GI should not fall within the purview of prohibitions as contained in Section 9 for instance that cause confusion, contrary to any law for the time being, religious susceptibilities etc.

Publication in GI Journal: After examination of application details with supportive documents, the communication is done for deficiency if any; otherwise, the Registrar may accept the GI Application absolutely. After acceptance the Registrar publishes all applications in Official GI Journal and make available through the Official Website with key information about goods.

Opposition of GI: Any per can oppose on sound ground with a application along with the fee to Registrar of Geographical Indications at Chennai to GI after three months of publication in GI Journal. In opposition the public interest is more important rather than opponent.

Registrar enter the geographical indication in Part A of the Register on receipt of a request if no opposition is filled against an application after over of three months periods. After entry in the register a certificate of registration of a geographical indication to be releases by registrar on Form-O2. After four months four months if no opposition is filed the application shall be registered and entered in the Part – B of the GI Register.

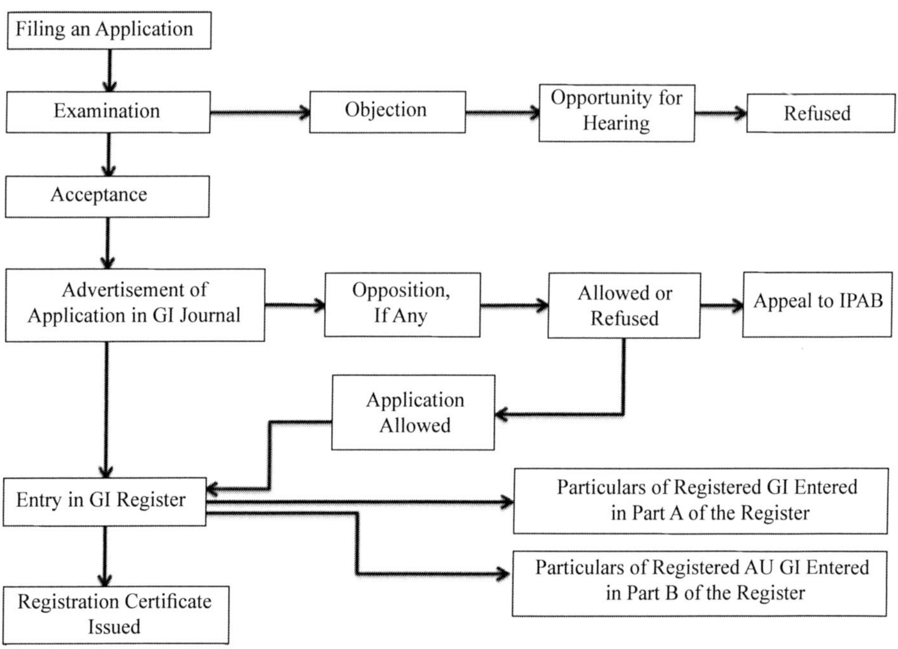

Flow Chart GI Registration Process
Source: http://www.ipindia.nic.in/the-registration-process-gi.htm

What goods need GI Protection: The following steps has to be examined for getting a GI tag.

- Assessment of product quality for potential national and international market
- Formulation of group of producer and other person involved in production of the goods
- Setting up standard quality parameters for production and processing methods
- Use of GI in a limited boundary of the good
- Establishing regulations for continued supply of the product
- Well managed marketing strategies
- Legal protection and enforcement of the GI in the region

Duration of registration Geographical Indication: Initially 10 years are granted to protect the goods under GI. However, it is renewed again for second term.

Benefit of registration of geographical indications

i. GI product are having quality reputation and provides guarantee for originality to customers. If product not protected with GI, the product will be produced and traded without control and its value ultimately lost.

ii. **GI a factor for employment:** The GI products are commercially as premium brand with extra cost. The consumer can pay extra to buy the GI tagged product. GI products also help for employment creations for local residents.

iii. **Quality control for consumer:** The consumer believes on the good with GI that they are buying a genuine product with specific qualities and characteristics.

iv. **Conservation of traditions and goods:** GI protect the knowledge generated through the hundred or thousand years and inherited from generation to generation.

Geographical Indications (GIs) are signs or symbols used on products that have a specific geographical origin and possess qualities, characteristics, or a reputation that is essentially attributable to that origin. In India, GIs have played a vital role in promoting and protecting traditional crafts, agricultural products, and other unique products that are integral to the cultural heritage and identity of the country. The Geographical Indications of Goods (Registration and Protection) Act, 1999, provides a legal framework for the registration and protection of GIs in India. Successful registration and enforcement of GIs in India has led to economic benefits for producers, protection against imitation and misuse, and enhancement of marketability and economic value. However, challenges such as lack of awareness, insufficient resources, and bureaucratic delays still exist. It is important for India to continue to promote and protect its unique products through the registration and enforcement of GIs, both domestically and internationally.

The future outlook for GIs in India is positive, as there is an increasing demand for authentic and unique products in both domestic and international markets. The Indian government has taken steps to promote and protect GIs, such as providing financial assistance for registration and enforcement and creating awareness campaigns. As the Indian economy continues to grow and consumers become more discerning, the protection and promotion of GIs will become even more important for ensuring the economic success and cultural preservation of India's unique products. Additionally, with the increasing importance of international trade, the registration and protection of GIs in foreign countries will be crucial for protecting Indian products from imitation and misuse and increasing their global recognition and economic value.

7

Traditional Knowledge and Cultural Expressions: Understanding their Importance in Preserving Indigenous Heritage

Introduction

Traditional Knowledge (TK) and Traditional Cultural Expressions (TCEs) are crucial intangible heritage passed down through generations in indigenous communities. TK encompasses indigenous knowledge, practices, and innovations related to cultural traditions, natural resources, and ecosystems, while TCEs are heritage artistic and literary expressions that embody traditional culture and knowledge. TK and TCEs are deeply embedded in the cultural heritage of indigenous communities, and play a vital role in preserving their identities, values, and beliefs. However, they are vulnerable to exploitation and misappropriation, which can lead to cultural loss and undermine the rights of indigenous peoples. Thus, it is crucial to involve indigenous communities in decision-making and protect their rights to safeguard and preserve the integrity of TK and TCEs.

TCEs include a wide range of expressions such as handmade textiles, paintings, stories, legends, ceremonies, music, songs, rhythms, and dance, embodying the culture and knowledge of indigenous communities. These expressions are significant for maintaining and transmitting cultural heritage to future generations, enabling communities to retain their cultural identity and self-determination. However, TCEs also face the risk of exploitation and misuse, and their protection and preservation require the recognition of the rights and interests of indigenous communities. It is, therefore, essential to protect TCEs and the rights of indigenous peoples by involving them in the decision-making process and respecting their rights.

Traditional Knowledge (TK)

Traditional Knowledge (TK) is the knowledge, skills, innovations, and practices that are passed down orally or through direct observation and participation from one generation to another within a community. There are different types of TK, including medicinal knowledge, agricultural knowledge, ecological knowledge, and craft and technical skills. TK plays a significant role in the cultural, spiritual, social, and economic life of indigenous communities. It reflects their unique understanding of the natural world, and their relationship with it, as well as their cultural and spiritual beliefs. It also forms the basis of their traditional practices and livelihoods, such as farming, hunting, and fishing. Here are few examples of Traditional Knowledge :

- **Ayurveda:** An ancient Indian system of medicine that uses herbs, minerals, and other natural ingredients to promote health and wellness.
- **Acupuncture:** A Traditional Chinese Medicine technique that involves inserting needles at specific points in the body to alleviate pain and promote healing.
- **Indigenous Agricultural Practices:** Indigenous communities around the world have developed unique agricultural practices that are adapted to their local environments and rely on Traditional Knowledge.
- **Aboriginal Art:** Aboriginal art in Australia is a form of Traditional Knowledge that reflects the cultural and spiritual values of indigenous communities.
- **Traditional Ecological Knowledge:** Indigenous communities have developed Traditional Ecological Knowledge, which involves understanding the relationships between different elements of the natural world and how they interact with each other.

Threats to TK and efforts to preserve it: The threat to TK comes from a range of factors, including modernization, globalization, and climate change, which have led to the loss and erosion of traditional knowledge. Efforts to preserve TK involve documenting and recording it, strengthening traditional education systems, and supporting indigenous initiatives to protect and promote it. Additionally, legal frameworks, such as the Convention on Biological Diversity and the Nagoya Protocol, have been put in place to ensure the protection of TK and to promote fair and equitable sharing of benefits arising from its use.

TK preservation: In India, there are several successful examples of TK preservation. One such case is the revival of the indigenous farming practice of System of Rice Intensification (SRI) in Odisha. SRI is a method of paddy cultivation that was developed by farmers in the area centuries ago. It involves using fewer seeds, spacing them out carefully, and transplanting them at a younger age. Through the efforts of NGOs and government agencies, the practice has been revived, resulting in increased crop yields and income of farmers.

The traditional knowledge of herbal medicine practiced by the Kani tribe in Kerala. The tribe has an extensive knowledge of medicinal plants, which they use to treat a range of ailments. To preserve this knowledge, the tribe has set up a community-based organization called the Kani Medicinal Plant Conservation Centre, which is involved in the conservation and cultivation of medicinal plants and the training of younger generations in traditional medicinal knowledge.

Traditional Cultural Expressions (TCEs)

Traditional Cultural Expressions (TCEs) are artistic and literary expressions that are composed and passed down from generation to generation within indigenous communities. These expressions embody the traditional knowledge and culture of the community and include various forms of art, music, dance, storytelling, and handicrafts. Examples of TCEs include traditional clothing, pottery, jewelry, weaving, painting, music, dance, and other forms of cultural expressions.TCEs are of great significance in indigenous communities as they are a source of cultural identity and pride. They reflect the beliefs, values, and traditions of the community and serve as a means of passing on cultural heritage to future generations. TCEs also have an economic significance as they contribute to the livelihoods of many indigenous communities by providing a source of income through the sale of handicrafts and other cultural products.

Threats to TCEs and efforts to preserve it : TCEs face many threats such as cultural appropriation, misappropriation, and commercialization. Many TCEs have been misused or exploited for profit by non-indigenous individuals or companies, leading to the loss of cultural heritage and economic opportunities for indigenous communities. In efforts to preserve TCEs, indigenous communities have advocated for the protection of their intellectual property rights and for the recognition of their cultural heritage. Governments and international organizations have also made efforts to protect TCEs by establishing legal frameworks for the protection of traditional knowledge and cultural expressions.

TCEs preservation : India has many successful case studies of TCEs preservation. The Pattachitra art form of Odisha, which is a type of traditional painting, has been preserved and promoted by local artists and the government through exhibitions and workshops. The Kullu shawl of Himachal Pradesh, which is a traditional woven shawl, has been protected through a Geographical Indication (GI) registration, which ensures that the product is produced in a specific geographic region and meets certain quality standards. The Warli art of Maharashtra, which is a form of tribal art, has also been promoted and preserved through government initiatives and the efforts of local artists. These successful case studies demonstrate the importance of preserving TCEs for the cultural, economic, and social development of indigenous communities.

The Intersection of TK and TCEs

The relationship between TK and TCEs : TK and TCEs are closely related, as they both form a part of indigenous heritage and are passed down through generations within indigenous communities. TK refers to the knowledge, practices, and innovations of indigenous peoples related to their cultural traditions, natural resources, and ecosystems, while TCEs refer to the expressions of indigenous cultures, such as music, dance, art, storytelling, and handicrafts. Both TK and TCEs are deeply embedded in the identities, values, and beliefs of indigenous communities and play a crucial role in preserving their cultural heritage.

The significance of the intersection in preserving indigenous heritage : The intersection of TK and TCEs is significant in preserving indigenous heritage because they are interdependent and complementary. TK provides the knowledge and understanding of the natural world that is necessary for the creation and transmission of TCEs, while TCEs provide the means for communities to express and transmit their knowledge and culture. Together, they form a holistic system that reflects the interconnectedness of indigenous culture and the natural environment.

TK and TCEs work together in preserving indigenous heritage : Examples of how TK and TCEs work together in preserving indigenous heritage can be seen in various practices within indigenous communities. For instance, in many communities, traditional ecological knowledge is integrated into the creation of TCEs such as handicrafts, textiles, and art, which are used to represent and express the relationship between humans and the natural world. Similarly, traditional dances and songs often reflect the history, stories, and beliefs of indigenous communities, which are based on their knowledge of the environment and their place in it. By using TCEs to express and transmit their TK, indigenous communities are able to preserve and strengthen their cultural heritage.

Legal Framework for the Protection of TK and TCEs

International legal instruments for the protection of TK and TCEs : International legal instruments that protect TK and TCEs include the United Nations Declaration on the Rights of Indigenous Peoples (UNDRIP), the Convention on Biological Diversity (CBD), the World Intellectual Property Organization (WIPO) Intergovernmental Committee on Intellectual Property and Genetic Resources, Traditional Knowledge, and Folklore (IGC), and the UNESCO Convention for the Safeguarding of the Intangible Cultural Heritage.

Indian legal frameworks for the protection of TK and TCEs : In India, the Protection of Plant Varieties and Farmers' Rights Act, 2001, and the Biological Diversity Act, 2002, protect TK related to biodiversity and traditional agricultural practices. The Traditional Knowledge Digital Library (TKDL) was also established in 2001 to document and protect traditional medicinal knowledge. The Indian Copyright Act, 1957, and the Geographical Indications of Goods (Registration and Protection) Act, 1999, protect TCEs related to handicrafts, textiles, and other traditional products.

Protection of Traditional Knowledge in Indian Law : In Indian law, there are several ways to protect traditional knowledge (TK) and traditional cultural expressions (TCEs) from exploitation and misappropriation. One of these ways is through geographical indication (GI) registration. This process helps to protect TK and TCEs from deceptive trading practices and preserves the knowledge for future generations. When a community or social unit develops goods, the GI registration protects the TK and TCEs from misuse, thereby promoting economic growth. This, in turn, benefits the indigenous community through commercial exploitation of their product potential. TK can be protected through various means, such as patent, trademark, geographical indication protection, trade secret, or confidential information. The protection for TK comes in two types : defensive and positive. Defensive protection is meant to stop the acquisition of TK outside the community, while positive protection empowers communities to promote their traditional knowledge for commercial exploitation. Overall, these legal protections are crucial for preserving the rich cultural heritage of indigenous communities and ensuring that their knowledge is not exploited or appropriated without their consent.

Traditional Knowledge Digital Library (TKDL) is an initiative launched by the Indian government in 2001 to safeguard traditional knowledge and prevent its misappropriation at international patent offices. The objective of this initiative is to digitize and translate traditional knowledge into accessible languages for preservation. Since most traditional medical knowledge is documented in languages such as Sanskrit, Hindi, Arabic, Urdu, Tamil, etc., communication technology has been used to translate over 3.6 million formulations into the TKDL database. Thanks to the availability of this database, several incorrect patents have been withdrawn, amended, revoked, corrected, and opposed with accurate information.

India—An overview of wealth : India is a biologically diverse nation, developed four biodiversity hotspots (Himalaya, The Western Ghats, The North-East, and The Andaman and Nicobar Islands) out of total 34 on the globe. Indian cultural development is very historical, evolved many traditional medicinal systems like Ayurveda, Siddha and Unani and some unnotified medicines with plant, animals and natural resources. Many traditional knowledges undergone misappropriation and stealing in the past. Here are some popular examples of TK struggles for protection under IPR.

a) **The Neem case :** W.R. Grace Company patented a stable storage insecticide formulation of azadirachtin in both the United States and the European Union. Our traditional medicine recognizes neem for its antviral, antibacterial, and skin disease -treating properties, as well as its effectiveness as a natural pesticide. However, the patent faced challenges through re examination due to concerns regarding its lack of novelty and inventive step.

b) **The Turmeric case :** The USPTO and EPO patent was granted for "use of turmeric in wound healing" and claimed a method to heal wounds in a patient by administration of an "effective amount" of turmeric. Later on re-examination application against the granted patent with several references from library of Hamdard University resultantly in August 1997, the USPTO ordered revocation of the patent, which lacked novelty.

c) **The Basmati case :** Basmati rice is a traditionally grown aromatic variety of rice, in India and Pakistan. The patent was granted to RiceTec for "Basmati rice lines and grains, a hybrid rice lines that combined desirable grain traits of Basmati rice with desirable plant traits. The patented rice quality was not good as grown in India and Pakistan. A re-examination request was filed, with sound proof, along with several publications on Basmati rice and the research conducted on the rice in India. Resultantly, RiceTec twenty patents reduced to three.

Challenges in implementing legal frameworks for the protection of TK and TCEs : One challenge in implementing legal frameworks for the protection of TK and TCEs is the lack of recognition and respect for indigenous knowledge systems and cultural expressions. Another challenge is the difficulty in defining and identifying TK and TCEs, which can lead to their misappropriation or exploitation. There is also a lack of adequate resources and infrastructure to support the documentation and protection of TK and TCEs.

Case studies : One successful case study is the protection of the GI tag for Darjeeling tea, which recognizes the traditional knowledge and expertise of tea growers in the Darjeeling region of India. The GI tag ensures that only tea grown in this region and meeting specific standards can be marketed as "Darjeeling tea." Another example is the protection of TK related to the medicinal properties of turmeric through the TKDL. The TKDL has documented over 2,000 formulations of traditional medicinal knowledge related to turmeric, which has been used to prevent the misappropriation of this knowledge by patent holders.

Indigenous Community Involvement in Preserving TK and TCEs

Importance of indigenous community : Indigenous community involvement is crucial for the preservation of Traditional Knowledge (TK) and Traditional Cultural Expressions (TCEs) as they are the holders, guardians and transmitters of these intangible heritage forms. Their involvement ensures that TK and TCEs are preserved and maintained in accordance with the cultural practices, beliefs, and values of the community, which is essential for safeguarding the cultural identity and self-determination of indigenous peoples. Furthermore, community involvement ensures that TK and TCEs are not exploited or misused for commercial purposes without their prior and informed consent.

In India, there are several community-led initiatives for the preservation of TK and TCEs. For example, the Kani tribe in Kerala has set up a community radio station to document and transmit their traditional knowledge and cultural expressions, while the Warli tribe in Maharashtra has established a museum to showcase their art and handicrafts. Additionally, the Bishnoi community in Rajasthan has established community-led conservation programs for the protection of their natural resources and ecosystems, which are closely linked to their cultural practices and beliefs.

Challenges and Opportunities for Indigenous Community Involvement: The challenges for indigenous community involvement in TK and TCEs preservation include the lack of recognition and protection of their rights, limited access to resources, and inadequate representation in decision-making processes. However, opportunities for community involvement include the development of community-based initiatives for the documentation, transmission, and protection of TK and TCEs, as well as the use of digital technologies for wider dissemination and access. Furthermore, the increasing recognition and appreciation of TK and TCEs at national and international levels provide opportunities for indigenous communities to assert their rights and promote the value and importance of their cultural heritage.

TK and TCEs are critical components of indigenous heritage, deeply embedded in the identities, values, and beliefs of indigenous communities. They play a crucial role in preserving cultural diversity, promoting sustainable development, and supporting the cultural identity and self-determination of indigenous peoples. However, they are vulnerable to exploitation and threats, both at national and international levels. It is therefore essential to recognize their significance and take action to protect and preserve them for the benefit of present and future generations.The protection and preservation of TK and TCEs require collective action and collaboration between indigenous communities, governments, civil society organizations, and other stakeholders. Governments should develop legal frameworks and policies that recognize and protect TK and TCEs, while indigenous communities should be actively involved in the decision-making processes that affect them. Civil society organizations and other stakeholders can provide support through capacity building, advocacy, and knowledge sharing. It is our collective responsibility to ensure that TK and TCEs are safeguarded and transmitted to future generations as part of our shared cultural heritage.

8

Protection of Plant Varieties Importance and Legal Framework

The plant variety is related to intellectual property that needs to be protected for controlled commercial unitization for a specified period. New plant varieties development requires scientific skill, knowledge, significant investment of land and specialized equipments. These varieties are cultivated by farmers in their adaptive region. In development of new variety 10 to 15 years are required. If developed variety successfully commercialized on farmer's field, then the market business is much higher than input investment. The major issue in developed variety that can be easily reproduced by someone without permission of original breeder. Thus, the original breeder deprived for their financial benefit. It is, therefore, critical to provide an effective system of plant variety protection, which encourages the development of new varieties of plants thereby benefiting the breeder and society at large. The sustainable and long-term breeding effort for development of new varieties needs encouragement and financial reward for investing time and skills. To secure legal protection of the developed variety breeder needs to obtained exclusive right from government.After establishment of WTO in 1995 under TRIP the right to protect plant varieties become mandatory for signatory country. It has also realised that existence of the private seed industry, the protection of plant varieties become necessity.

Why do Farmers and Growers Need New Plant Varieties?

New varieties of agricultural crops with traits such as improved yield, high quality and resistance to insect-pests and diseases are a key element in increasing productivity and product quality in agriculture, horticulture and forestry, while minimizing the pressure on the natural environment. Many inputs need to be combined with such varieties in order to deploy their full potential. The tremendous progress in agricultural productivity in various parts of the world is largely based on improved varieties.

The importance and scale of the contribution of plant breeding can be illustrated by the example of Rapeseed. Originally, only the oil component of rapeseed provides a useful product, as a lubricant for steam engines. It was only when breeders started to work on the crop that it attained major importance for agriculture. Firstly, breeders reduced the glucosinolate content so that the meal could be used for feeding animals. As a following step, breeding was employed to reduce the erucic acid content so that rapeseed could be used as a source of edible oil for human consumption. More recently, efforts are continuing, and breeders are working to develop high oleic and low linoleic acid varieties with nutritional benefits for consumers.

Plant Breeding Product: The crop breeding program aims to develop various types of products, including:

i. **Germplasm Line:** Any genetic material of a plant spp. are described as germplasm line. The germplasm line includes varieties, elite lines, wild relative and wild forms, landrace, cultivars, advanced breeding lines etc.

ii. **Variety:** A variety is a plant grouping within a single botanical taxon with certain expression of characteristics. The variety should be distinguished from any other plant grouping by expression of at least one of the said characteristics and should be considered as a unit with regard to its suitability for being propagated, which remain unchanged after such propagation. The variety are advanced tested product of the elite line that has been released by the concern agency for commercial cultivation in a particular area.

iii. **Cultivar:** The plant material is under cultivation. Plant material doesn't need to be notified or released by any agency. Therefore, a landrace, wild and wild relative farmer selection can be cultivar.

iv. **Landrace:** The traditional bred farmer varieties often referred to as landrace. The landrace are conserved by farmer from long time through saving of seed from generations. These lines are good source of useful diversity or genes for biotic and abiotic stresses.

v. **Test Entry:** It is a pre released material under evaluation in different stages of testing.

vi. **Initial varieties:** Initial variety or novel variety refers to first systematically developed variety by the breeder protected under PBR. These varieties retain the expression of essential character for which the variety were developed. These varieties are not in public domain in India earlier than one year before the date of filing; while outside India is four years. In the case of trees or vines earlier than six years.

vii. **Essentially Derived Variety (EDV):** Essential derived varieties are referred to as varieties which is developed through selection, backcross breeding or transformation by genetic engineering approach. EDV often referred to as dependent varieties.

viii. **Extant variety:** A variety which is notified under section 5 of Seed Act, 1966 or a variety about which there is common knowledge or a farmer's variety or any other variety which is in public domain is considered as an Extant Variety.

ix. **Farmer Variety:** A variety which has been traditionally cultivated and evolved by the farmers in their fields or a variety which is a wild relative or land race of a crop/species about which farmers possess common knowledge.

Why Protection of New Plant Varieties are Important

World population are continuously growing, however natural resources are shrinking day by day. The crop productivity needs to be increase to meet the food requirement despite on limited arable land. On the other hand, nutritional security and low-cost food with better availability are important factor to feed population with nutrient enriched food. Under the climate change scenario outbreak of high input cost, pests and diseases are potential threat to crop production. Considering the above facts Continuous development of new plant varieties is very important activity of agricultural scientist. However, plant breeding is long and expensive activity. On the other hand, Plant varieties can be easily and quickly reproduced from the initial sources. Therefore, an effective and efficient system are required to protect the benefit of researcher, farmers and to promote research and development activities in agriculture. Protection of the plant varieties under the act accelerates agricultural development and stimulates investment for research and development.

Historical Prospects

The protection of agricultural, manufactured, extractive industry and natural product are protected after the Paris convention of 1883. In USA the plant patent act 1930 was enacted to protect plant breeder rights and protection of asexually reproduced cultivars for a period of seventeen years. Several attempts have been made to include sexually propagated spp. in plant patent act. In 1970, The Plant Varieties Protection Act was enacted in USA which provides protection of sexually propagated plants.

In 1961, five European countries (Belgium, France, The Republic of Germany, Italy and Netherland) enacted a sui-generis (Latin ward meaning, its own kind/ genus/ unique in its characteristics) IPR protection system "The International Union for the Protection of New Varieties of Plants" (UPOV) for the Protection of New Varieties of Plants. The headquarters of the UPOV is located in Geneva (Switzerland). It is an intergovernmental organization established by the International Convention in Paris in 1961 and amended in 1972, 1978 and 1991. The OPOV came into force in 1968.

The USA adopted patent system while European Union adopted sui-generis system UPOV to protect plant varieties. The main reason of in appropriateness to protect plant varieties by patent system.

1. The plant material dose not meet the requirement novelty, inventiveness, and industrial application
2. Extensive monopoly is undesirable to protect plant varieties in the public interest
3. The free exchange of protected varieties among researches are desirable for development of new gene combinations. The patents are not allowed to free exchange of protected material.

As per the TRIP agreement, plant varieties must be protected. Indian has enacted a sui-generis system "Protection of Plant Varieties and Farmers' Rights Act 2001" for protection of plant varieties. India is not a of member.

Plant Variety Protection in European Union

The UPOV is derived from the French name, "Union internationale pour la protection des obtentions végétales". UPOV is an independent intergovernmental organization.The International Union for the Protection of New Varieties of Plants (UPOV) is intergovernmental organization was Established in 1961 Geneva, Switzerland. by the International Convention for the Protection of New Varieties of Plants ("UPOV Convention"). The UPOV Convention was adopted on December 2, 1961, by a Diplomatic Conference held in Paris. The UPOV come into force in 1968 and revised in 1972, 1978 and 1991. UPOV, provide intellectual property right to breeder's for new plant varieties and promote an effective system of plant variety protection, with the aim of encouraging the development of new varieties of plants and for the benefit of society.The signatories of WTO are committed to comply with the TRIP requirement to harmonise the protection rights. Accordingly to article 27.3 (b). The WTO member must provide plant variety patent or an effective sui generis system, or both.

UPOV Mission Statement: To provide and promote an effective system of plant variety protection, with the aim of encouraging the development of new varieties of plants, for the benefit of society.

Main Objectives and Role of UPOV

Objectives

i. To provide legal, administrative and technical guidelines for international cooperation in plant variety protection.

ii. To assist in development of legislation and the implementation of an effective plant variety protection system

iii. To enhance public awareness and understanding of the UPOV system of plant variety protection

Role

The UPOV played significant role in protection of plant varieties. The UPOV role are summarized point by point here under:

i. UPOV establishes a framework law which is adopted by countries into their national laws

ii. UPOV provide mechanism for harmonizing national laws and provide standardised definitions and interpretations

Comparison of UPOV 1978 Act and UPOV 1991 Act

The UPOV ensure acknowledgment and achievement of a plant breeder for his/her new plant variety providing them to the exclusive rights. The key differences UPOV 1978 Act and UPOV 1991 Act are below in the table:

Subject	**UPOV 1978 Act**	**UPOV 1991 Act**
Eligibility Requirements	Novelty, distinctness, uniformity and stability.	Novelty, distinctness, uniformity and stability.
Minimum exclusive rights in propagating material	Production for purposes of commercial marketing; offering for sale; marketing; repeated use for the commercial production of another variety.	Production or reproduction; conditioning for the purposes of propagation; offering for sale; selling or other marketing; exporting; importing or stocking for any of these purposes.
Minimum exclusive rights in harvested material	No such obligation, except for ornamental plants used for commercial propagating purposes.	Same acts as above if harvested material obtained through unauthorized use of propagating material and if breeder had no reasonable opportunity to exercise his or her right in relation to the propagating material.
Prohibition on dual protection with patent	Yes, for same botanical genus or species.	No.

Breeders' exemption	Mandatory. Breeders free to use protected variety to develop a new variety.	Permissive, but breeding and exploitation of new variety "essentially derived" from earlier variety require right holder's authorization.
Farmers' privilege	Implicitly allowed under the definition of minimum exclusive rights.	Allowed at the option of the member state within reasonable limits and subject to safeguarding the legitimate interests of the right holder.
Minimum term of protection	18 years for grapevines and trees; 15 years for all other plants.	25 years for grapevines and trees; 20 years for all other plants.

Source: https://www.fao.org/3/y5714e/y5714e03.htm

How PBR work: Like any other system of IP, the PBR works on three requirements:

i. **Identification of protectable subject matter**: A variety is a group of plants having, defined expression of the characteristics, distinguished in at least one character. Certainly, the Plant genetic resource is protected under Convention of Biological Diversity (CBD). In practical, a landrace are heterogeneous in nature and does not satisfy uniformity and stability requirement of PBR protections. Sometimes, it may create some conflict. The protection is granted for planting material, whether it is agricultural and horticultural crop.

ii. **Protection requirement:** Under UPOV 1978 Act and UPOV 1991 Act four essential criteria Novelty (N), Distinctiveness (D), Uniformity (U) and Stability (S) are same for protection of variety. The variety must demonstrate basic requirement for the protection. The following criteria have to be fulfilled by variety for seeking protection under PBR.

   a. Newness (novelty)

   b. Distinctiveness

   c. Uniform

   d. Stable

   a. Novelty, defined that a variety has not been sold for more territory of than one year in country where protections seeking and four years in the tertiary on the member country of the union. In case of tree and wines the variety must not be more than six years under commercial cultivation in the territory of member country of the union.

   b. Distinctiveness: The protected variety should be distinct from any other variety in one or more morphological, physiological characters such as flower colour, plant height, virus resistant, leaf colour, stem length etc.

c. **Uniformity:** The variety reproduces sexually or asexually true to type or plant has to give uniforam character. The variety should be phenotypically homogeneous and variation must be limited.

d. **Stability:** The variety reproduces true to type with essential characteristics across environment (temporal and locations).

Uniformity and stability serve as critical help in identification of propagating material. These two criteria also fulfil commercial need of the variety. The plants of a variety remain unchanged and performed stability over the successive generations.

iii. **Plant Breeder's Right's (PBR):** The right granted to the breeders of a new variety for exclusive control of propagating and harvested material for several years. PBR are intellectual prohibit its rights are exclusive right to a breeder for plant variety to prohibit its unauthorised use, production, marketing and selling of the variety. It is similar to patent right as national legislation system applied to protect only plant variety. The owner of the variety has legal right of protection in country for which the rights are granted.

iv. **Rights of Variety Owner:** The following rights are granted to variety owner and permission must be obtained for:

a. Production (reproduction, multiplication) of protected material for the purpose of commercial marketing.
b. Selling of protected material
c. Exportation
d. Conditioning for propagation
e. Importation
f. Stocking for any of the above purposes

v. **Researcher's Right's:** The protected variety may be used for breeding and experimental purposes are referred to as researcher'right's. The Breeder are exempted under PBR for essential derived varieties (EDV), protected material may be used in breeding programme. The addition of new trait in initial variety developed through backcrossing or transgenic approach. The resultant EDV examined as essential derived material from Initial Variety. The EDV cannot be commercialized without permission of Initial Variety PBR owner. (Section 30)

vi. **Farmer's Right's:** The farmer or grower can retain the crop as seed source for subsequent season, is referred to as Farmer Right or Farmer privilege. The farmer privileges adopted in almost all countries expect European Union where farmer have to pay royalty on saved seed (small farmers exempted).

Examination method of variety: A testing is required to grant right, compliance with requirement. To test distinctive trait, examination has to be conducted with the existing variety which is in common knowledge at the time. Thus, a national examination system must be exits with large data base of descriptor. The field trial of applicant variety is conducted with reference variety to examine the NDUS.

Duration of protection under PBR: Under UPOV 1991 the duration of protection of grapevine and tree including root stock is 25 years. The duration for all other plant spp. is 20 years.

Who can protect a plant variety? Only the breeder of a new plant variety can protect that new plant variety. It is not permitted for someone other than the breeder to obtain protection of a variety. There are no A restrictions on who can be considered to be a breeder under the UPOV system: A breeder might be an individual, a farmer, a researcher, a public institute, a private company etc.

Members of UPOV upto November, 2021

S.No.	State/Organization	Date on which State/Organization became member of UPOV
1	African Intellectual Property Organization	July 10, 2014
2	Albania	October 15, 2005
3	Argentina	December 25, 1994
4	Australia	March 1, 1989
5	Austria	July 14, 1994
6	Azerbaijan	December 9, 2004
7	Belarus	January 5, 2003
8	Belgium3	December 5, 1976
9	Bolivia (Plurinational State of)	May 21, 1999
10	Bosnia and Herzegovina	November 10, 2017
11	Brazil	May 23, 1999
12	Bulgaria	April 24, 1998
13	Canada	March 4, 1991
14	Chile	January 5, 1996
15	China	April 23, 1999
16	Colombia	September 13, 1996
17	Costa Rica	January 12, 2009
18	Croatia	September 1, 2001
19	Czech Republic	January 1, 1993
20	Denmark5	October 6, 1968
21	Dominican Republic	June 16, 2007

22	Ecuador	August 8, 1997
23	Egypt	December 1, 2019
24	Estonia	September 24, 2000
25	European Union6	July 29, 2005
26	Finland	April 16, 1993
27	France	October 3, 1971
28	Georgia	November 29, 2008
29	Germany	August 10, 1968
30	Ghana	December 3, 2021
31	Hungary	April 16, 1983
32	Iceland	May 3, 2006
33	Ireland	November 8, 1981
34	Israel	December 12, 1979
35	Italy	July 1, 1977
36	Japan	September 3, 1982
37	Jordan	October 24, 2004
38	Kenya	May 13, 1999
39	Kyrgyzstan	June 26, 2000
40	Latvia	August 30, 2002
41	Lithuania	December 10, 2003
42	Mexico	August 9, 1997
43	Montenegro	September 24, 2015
44	Morocco	October 8, 2006
45	Netherlands	August 10, 1968
46	New Zealand	November 8, 1981
47	Nicaragua	September 6, 2001
48	North Macedonia	May 4, 2011
49	Norway	September 13, 1993
50	Oman	November 22, 2009
51	Panama	May 23, 1999
52	Paraguay	February 8, 1997
53	Peru	August 8, 2011
54	Poland	November 11, 1989
55	Portugal	October 14, 1995
56	Republic of Korea	January 7, 2002
57	Republic of Moldova	October 28, 1998
58	Romania	March 16, 2001
59	Russian Federation	April 24, 1998
60	Saint Vincent and the Grenadines	March 22, 2021
61	Serbia	January 5, 2013

62	Singapore	July 30, 2004
63	Slovakia	January 1, 1993
64	Slovenia	July 29, 1999
65	South Africa	November 6, 1977
66	Spain	May 18, 1980
67	Sweden	December 17, 1971
68	Switzerland	July 10, 1977
69	Trinidad and Tobago	January 30, 1998
70	Tunisia	August 31, 2003
71	Turkey	November 18, 2007
72	Ukraine	November 3, 1995
73	United Kingdom.	August 10, 1968
74	United Republic of Tanzania	November 22, 2015
75	United States of America.	November 8, 1981
76	Uruguay	November 13, 1994
77	Uzbekistan	November 14, 2004
78	Viet Nam	December 24, 2006

Note: India is not a member of UPOV. India has developed its own sui generis system PPV & FRA 2001 to protect new plant varieties and farmer rights.

PBR Vs Patent

There are more similarities between patent and PBR, both control protected variety to be available in the public domain, freely.

a) **Protection requirement:** The novelty, inventive step and utility or industrial application are required for protection under patent right. The Novelty are required in both PBR and patent. Inventive step in patent is similar to distinctiveness in PBR. For utility under patent, some use of the product or process must be identified. The varieties under PBR, intended for sale and display some practical merit. Uniformity and stability have no comparable requirement under patent law however it is a criterial under PBR.

b) **Protectable subject matter:** The protectable subject matter is quite different under patent and PBR. Under patent, saving of seed is comes under infringement. The farmer privilege and researcher right are the weaker section of PBR than patent. Therefore, commercial breeder selects, patent, or patent plus PBR, over PBR alone.

c) **Patent** are costly to file and maintain. Generally an attorney assistance required for filing. The PBR application generally filled by the breeder himself/herself.

Exceptions and limitation of PBR

i. **Breeder Exemptions:** Under UPOV 1978, the researcher can use any protected variety (initial variety) for development of a new variety. The PBR will be to the breeder who developed the new variety not to initial variety PBR holder. This provision is called Breeder exemptions. Under UPOV 1991, the researcher of the EDV has to take permission for insertion of the gene or trait to the initial variety.

ii. **Farmer's Privilege:** Under UPOV 1978, the farmers are allowed to use the protected material produced on the farm without any obligation. This is called Farmer privilege. Under UPOV 1991, the Farmer privilege is optional to adopt by the member state to allow and disallow the use of protected material.

Plant Variety Protection in USA

In the United States, new plant varieties can be protected through three types of intellectual property (IP):

i. **Plant Patents:** The new and distinct asexually-reproduced distinct plants are protected by Plant Patents. A Plant patent is granted to inventor by United States Patent and Trademark Office (USPTO) for 20 years from date of filling the application.

ii. **Utility Patents:** A Utility Patents is granted to new inventions created by man related to genes, traits, methods, and plant parts are protected for 20 years from date of filling the application. The United States Patent and Trademark Office (USPTO) grant utility patents for protection of plant material.

iii. **Plant Variety Protection (PVP):** The Plant Variety Protection Office (PVPO) of the U.S. Department of Agriculture (USDA) provides intellectual property protection in the form of plant variety protection certificates to breeders for new varieties of seeds, tubers, and asexually propagated plants. The protection duration of varieties is 20 years and for vines and trees it is 25 years.

Additionally, plant breeders can also protect information of value as trade secrets.

International Treaty on Plant Genetic Resources for Food and Agriculture

The International Treaty on Plant Genetic Resources for Food and Agriculture (ITPGRFA) was adopted in the Conference of the Food and Agriculture

Organization of the United Nations on 3 November 2001. Under treaty, the members may freely share plant seeds subject to the treaty's standard material transfer agreement (SMTA). The treaty formulated an access and benefit sharing system for certain crops among the member country.

Treaty aims

i. Recognizing the enormous contribution of farmers to the diversity of crops that feed the world;

ii. Establishing a global system to provide farmers, plant breeders and scientists with access to plant genetic materials;

iii. Ensuring that recipients share benefits they derive from the use of these genetic materials with the countries where they have been originated.

Why Protection for Plants is Required in India?

The breeding activities and exploitation of new varieties are the decisive factors for improving rural income and their overall economic development. Plant variety development process of crop breeding programme is long term and expensive task. Therefore, legal rights are required to provide an effective system of plant variety protection with an aim to encourage the development of new varieties of plants for the benefit of the society. Protection of the plant varieties under the act accelerates agricultural development and stimulates investment for research and development.

Plant Genetic Resources (PGRs) are the prerequisite for the development of nutritionally rich high yielding crop varieties. PGRs are treated as 'heritage of mankind' and were shared freely among the nations till the concerns for conservation of biological diversity were raised by the Conservation on Biological Diversity (CBD) which came into force in 1993. The conservation and sustainable utilization and access to biological diversity were considered as national sovereignty by CBD. Consequently, many issues regarding the rights of the conservers, users, breeders, farmers and intellectual property have emerged.The idea of Plant Breeders' Rights arises to protect plant varieties to provide financial benefit to plant breeder sustains agricultural progress. The protection of plant varieties has been in focus worldwide after the implementation of WTO rules and regulations. After India become signatory to TRIPs in 1994 a legislation was required to be formulated. TRIP Article 27.3 (B) agreement it requires that the member countries to provide for protection of plant varieties either by patent or by an effective sui generis system or a combination of two.Sui generis is a Latin term meaning "a special kind" or "Unique". In Intellectual Property Rights (IPRs) discourse the term

'Sui generis' refers to a special form of protection regime outside the known framework.It can also be viewed as a regime especially tailored to meet a certain need.

Protection of Plant Varieties and Farmer's Rights Act 2001

A sui generis system is an attempt by the Indian government to recognize and protect the rights of both commercial plant breeders and farmers in respect of their contribution in conserving, improving and making available the plant genetic resources for the development of new plant varieties. After a signatory member of WTO, and as per TRIP Article 27.3 (B), India has adopted sui-generis system and brought "Protection of Plant Varieties and Farmers' Rights Act 2001" and Rules 2003. The PPV&FRA 2001 enacted by India is undoubtedly the trend setter as it is the only one that covers both Plant Breeder's and Farmer's Rights. The following aim are set up through PPV & FR Act.

India developed PPV& FRA 2001 a 'Sui generis' system to protect plant varieties by merging of streams (i) Researcher's Right's (ii) Breeder's Right's (iii) Farmer's Right's. The role of PPV& FRA are:

i. Establishment of effective system for protection of plant varieties
ii. Encourage scientists, farmers & communities for the development of new plant varieties having quality and production potential
iii. Safeguard the interest of plant breeders
iv. Recognize and protect farmer's right for their contribution in conserving, improving, and making available plant genetic resources for the development of new plant varieties
v. Stimulate investment for research and development and to facilitate growth of seed industry
vi. Ensure availability of high-quality seeds and planting material of improved varieties to farmers

Registration of Varieties: Four types of varieties can be registered under the PPV & FRA act.

i. **Novel Variety** which is not in public domain in India earlier than one year before the date of filing. While outside India, in the case of trees or vines earlier than six years and in any other case earlier than four years.

ii. **Extant Variety** which is notified under section 5 of Seed Act, 1966 or a variety about which there is common knowledge or a farmer's variety or any other variety which is in public domain is considered as an Extant Variety.

iii. **Farmer's Variety** which has been traditionally cultivated and evolved by the farmers in their fields or a variety which is a wild relative or landrace of a crop/species about which farmers possess common knowledge.

iv. **Essentially Derived Varieties (EDV):** These varieties are predominantly derived from initial variety while retaining the expression of essential characteristics. EDV developed from combination of genotypes of such initial variety and is clearly distinguishable from such initial variety. EDV derived through backcross method and genetic enginerring.

Any variety that contains any technology which is injurious to the life or health of human beings or animals or plants including "Terminator Technology" cannot be registered.

Criteria for the Protection of a Variety

For a variety to be eligible for registration, it mu… rm the criteria of NDUS:

i. **Novel:** Variety should not have been commercially exploited for more than one year before the grant of protection

ii. **Distinct:** Clearly distinguishable by at least one essential characteristics from any other variety whose existence is a matter of common knowledge

iii. **Uniform:** A variety must be sufficiently uniform in its essential characteristics (all plants should be uniform)

iv. **Stability:** Essential characteristics of a variety must be stable after repeated propagation or in the case of a particular cycle of propagation at the end of each cycle

Rights under PPV& FRA 2001

i. **Breeder's Right:** Certificate of registration for a variety issued under this act shall confer an exclusive right to the breeder or his successor or his agent or licensee to:

   i. Produce

   ii. Sell

   iii. Market

   iv. Distribute

   v. Export or Import seeds of the protected variety

   However, if the breeders' variety is essentially derived (EDV) from a farmers' variety, the breeder cannot give any authorisation without the consent of the farmers or communities from whose varieties the protected variety is derived.

ii. **Farmer's Right:** The Act recognizes the farmer not just as a cultivator but also as a conserver of the agricultural gene pool and a breeder who has bred several successful varieties.

iii. **Rights as a conserver and cultivator:** Right to save, use, sow, re-sow, exchange, share or sell farm produced seed, including the seed of registered variety (Seeds for sale should not be branded).

iv. **Rights as a conserver:** Right to share benefit when farmers' variety is used for breeding new commercial variety and right to farmers and communities to receive reward and recognition for conserving plant genetic resources.

v. **Rights as a Breeder:** Right to register farmer's variety and in the event of an essentially derived variety (EDV) is developed from farmer's variety, the commercialization of such EDV cannot be done without prior consent from the concerned farmer.

vi. **Protection against unintentional use:** Farmers are protected from the act of innocent infringement. If the farmers who is not aware of the existence of this act and unknowingly breaks the law of protection, then the farmer will be considered innocent but he has to prove his innocence before the court.

vii. **Researcher's Right:** The researchers have been provided access to protected varieties for bonafide research purposes (section 30). This section states, nothing contained in this section shall prevent:

- Use of any variety registered under this act by any person for conducting experiments or research
- Use of variety as an initial source of variety for the purpose of creating other varieties provided that the authorization of the breeder of a registered variety is required where the repeated use of such variety as a parental line is necessary (for EDV)

Compulsory Licensing System

Seeds of a protected variety are not available after 3 years of registration in adequate quantity, and at reasonable price, the authority may grant license to a third party to undertake, production, distribution and sale of the seed on reasonable royalty.

Applicant for Protection of a Plant Variety

1. Any person claiming to be the breeder of the variety
2. Any successor of the breeder of the variety
3. Any person being the assignee of the breeder of the variety

4. A farmer or group of farmers or community of farmers claiming to be the breeder of the variety
5. Any person authorized in the prescribed manner by the persons mentioned above
6. Any university or publicly funded agricultural institution claiming to be the breeder of the variety

PPV & FRA headquarter is located at the NASC complex, DPS Marg, Opp-Todapur, New Delhi-110 012

Period of Protection of a Variety

The following are the duration of plant material protection under different categories.

i. **Field crops (New variety):** 15 years total protection
ii. **Trees and vines (New variety):** 18 years total protection
iii. **Extant variety (Notified):** Fifteen years from date of notification

Benefits of Plant Variety Protection

Protection of Breeder's Rights

1. By granting Certificates of Registration
2. By granting exclusive rights to produce, sale, market, distribute, import or export the seeds of protected variety

Protection of Researcher's Rights

1. Can use any 'Protected Variety' for experimental purpose
2. Can use it as an initial source to develop other varieties

Protection of Farmer's Rights

1. Rights to use seed of protected variety for sowing, re-sowing, exchange and sale of its produce except in brand name
2. Right to register traditional varieties as breeder
3. Right to get rewards and recognition from National Gene Fund
4. Right to get compensation
5. Right to not pay any fee for registration
6. Right for protection for any innocent infringements

Plant Genome Saviour Awards

The PPV&FRA, has instituted Plant Genome Savior Community Award for rewarding and recognizing farmers contribution in plant genetic resource conservation, and crop improvement. The award is given from gene fund, since 2007. The gene fund receives from Act, national and international organizations. The award targets for community of farmers/farming community for conservation, improvement and preservation of genetic resources of economic plants and their wild relatives particularly in the areas identified as agro bio-diversity hotspots.

i. **Plant Genome Saviour Community Awards:** There are maximum of five awards in a year, consisting of a cash of Rs. 10 lakh each, a citation and a memento.

ii. **Plant Genome Saviour Farmer Awards:** The Authority has setup ten Plant Genome Saviour Farmer Reward of Rs 1.50 Lakh each with citation & memento.

iii. **Plant Genome Saviour Farmer Recognition:** The Authority has setup twenty Plant Genome Saviour Farmer Recognition annually from 2012-13 to the farmers engaged in the conservation of the genetic resources of the landraces and wild relatives of economic plants and their improvement through selection and preservation.

9

Intellectual Property Rights in Biotechnology

Biotechnology is a field of science that applies technology to living organisms, cells, and their molecular components, to develop products and services that improve human life. Biotechnology covers a wide range of applications, from agriculture and food production to medicine and healthcare, environmental management, and industrial biotechnology. Intellectual property rights (IPRs) are legal rights that protect the creations of the human mind, such as inventions, literary and artistic works, symbols, and designs. In the context of biotechnology, IPRs are critical to incentivize research and development, protect the fruits of scientific discoveries, and ensure that the inventors and investors of biotechnological inventions can benefit financially from their investment.

The importance of protecting biotechnological inventions lies in the fact that these inventions can be expensive and time-consuming to develop, and may require significant financial resources bring to market. Without appropriate legal protection, biotechnological inventions may be vulnerable to unauthorized use, duplication, and exploitation by third parties, leading to reduced incentives for research and development and potential loss of economic benefits for inventors and investors.

For instance, in the case of genetic engineering, which involves modifying the DNA of living organisms to create new products or processes, Biotechnology inventions may include genetically modified organisms (GMOs), engineered enzymes, new medicines, or diagnostic tools. Patent protection can provide inventors and investors with a monopoly on the use of their inventions for a certain period of time, allowing them to recoup their investment and gain a competitive advantage in the market.

Moreover, IPRs can also contribute to the dissemination of knowledge and technology transfer, by encouraging licensing agreements, collaborations, and other forms of partnerships between research institutions, companies, and public organizations. Through the protection of biotechnological inventions, IPRs can foster innovation, create new jobs, and promote economic growth, while ensuring that the ethical and safety concerns of biotechnology are addressed.

Need to Protect Biotechnological Inventions

Economic Benefits of Protecting Biotechnology

1. **Promotes investment:** Biotechnology is a highly capital-intensive field that requires significant investment in research and development. Intellectual property protection provides a way for investors to recoup their investment and earn a return on their capital.
2. **Boosts economic growth:** Biotechnology has potential to create new products and services, improve existing ones, and provide solutions to some of the world's biggest challenges, such as food security, health, and the environment. By protecting biotechnological inventions, intellectual property rights can stimulate economic growth and job creation.
3. **Increases competition:** Intellectual property protection encourages competition among inventors and companies, driving them to innovate and improve their products and services to stay ahead of the competition. This can lead to better quality products, lower prices, and increased consumer choice.
4. **Facilitates technology transfer:** By allowing inventors and companies to license their technologies to others, intellectual property protection can facilitate the transfer of technology and knowledge to other parts of the world, potentially leading to new collaborations and partnerships.

Encouraging Research and Development

1. **Provides incentives:** Intellectual property protection provides inventors and companies with a legal framework that encourages them to invest in research and development by providing a period of exclusivity over their inventions.
2. **Attracts investment:** Intellectual property protection can make it easier for inventors and companies to attract investment, as investors are more likely to invest in ventures that have strong intellectual property portfolios.
3. **Enables partnerships:** Intellectual property protection can encourage partnerships between universities, research institutions, and private companies, leading to collaborations that can drive innovation and create new products and services.

Promoting Innovation and Creativity

1. **Fosters competition:** Intellectual property protection can create a competitive environment that encourages inventors and companies to develop new and innovative products and services.

2. **Protects creativity:** Intellectual property protection provides a framework for protecting the creativity and ingenuity of inventors and companies, ensuring that their ideas are not stolen or copied by others.
3. **Encourages collaboration:** Intellectual property protection can encourage inventors and companies to collaborate with others, leading to new and innovative ideas that may not have been possible otherwise.

Protectable Subject Matters of Biotechnology

a) **Gene Patents:** Gene patents refer to patents on isolated and purified DNA sequences that are associated with specific traits or conditions. These patents allow the patent holder exclusive rights to the use, production, and sale of the gene sequence. Gene patents include patents on the BRCA1 and BRCA2 genes associated with an increased risk of breast and ovarian cancer, as well as patents on the CFTR gene associated with cystic fibrosis.

   Legal Requirements for Gene Patenting: In the United States, gene patents are subjected to the same legal requirements as any other type of patent. To be patentable, the gene sequence must be novel, non-obvious, and have a specific and useful application.

   Controversies and Ethical Issues: Gene patents have been the subject of significant controversy and ethical debate. Critics argue that gene patents stifle innovation, limit patient access to medical treatments, and may result in increased healthcare costs. They also argue that gene patents raise ethical concerns by allowing for the commercialization of genetic information and may lead to discrimination based on genetic predispositions.

b) **Microorganism Patents:** Microorganism patents refer to patents on living organisms such as bacteria, fungi, and viruses. These patents allow the patent holder exclusive rights to the use, production, and sale of the microorganism. Microorganism patents include patents on genetically modified bacteria used in industrial processes, such as the production of insulin, and patents on viruses used in vaccine development.

   Legal Requirements for Microorganism Patenting: In the United States, microorganism patents are subjected to the same legal requirements as any other type of patent. The microorganism must be novel, non-obvious, and have a specific and useful application.

   Controversies and Ethical Issues: Microorganism patents have also been the subject of controversy and ethical debate. Critics argue that the commercialization of living organisms raises ethical concerns and may lead to environmental damage or public health risks. They also argue that the exclusive rights granted by these patents can stifle innovation and limit access to essential technologies.

c) **DNA Sequence Patents:** DNA sequence patents refer to patents on specific DNA sequences, including both gene and non-gene sequences. These patents allow the patent holder exclusive rights to the use, production, and sale of the DNA sequence.DNA sequence patents include patents on the HIV virus genome, used in the development of HIV diagnostic tests, and patents on synthetic DNA sequences used in gene therapy.

   Legal Requirements for DNA Sequence Patenting: DNA sequence patents are subject to the same legal requirements as any other type of patent in the United States. The sequence must be novel, non-obvious, and have a specific and useful application.

   Controversies and Ethical Issues: Like gene and microorganism patents, DNA sequence patents have been the subject of controversy and ethical debate. Critics argue that DNA sequence patents raise ethical concerns by allowing for the commercialization of genetic information, and may limit patient access to medical treatments. They also argue that the exclusive rights granted by these patents may lead to a monopoly on essential technologies and stifle innovation.

Mechanisms to Protect Biotechnology

Overview of Intellectual Property Protection Mechanisms to Protect Biotechnology: Biotechnology is a field that involves the use of living organisms, cells, and biomolecules to develop products and technologies for various industries. Biotechnology inventions can be protected by a range of intellectual property (IP) protection mechanisms, including patents, trademarks, copyrights, and trade secrets. Patents are the most commonly used form of IP protection in biotechnology. Patents can be granted for biotechnology inventions such as genetically engineered organisms, biologically active molecules, and methods of diagnosing or treating diseases. In addition to patents, biotechnology companies may also use trademarks to protect their brand names and logos, and copyright to protect their software and other creative works.

National and International Legal Frameworks to Protect Biotechnology: The national legal frameworks for biotechnology IP protection vary from country to country. However, most countries have laws that provide for the protection of biotechnology inventions. In the United States, for example, biotechnology inventions are protected under the Patent Act of 1952, while in Europe, biotechnology inventions are protected under the European Patent Convention.

At the international level, there are several treaties and agreements that provide protection of biotechnology inventions. These include the Patent Cooperation Treaty, the Budapest Treaty on the International Recognition of the Deposit of Microorganisms for the Purposes of Patent Procedure, and the Convention on Biological Diversity. The World Trade Organization's Agreement on Trade-Related Aspects of Intellectual Property Rights (TRIPS) also provides for the protection of biotechnology inventions.

Roles of Patent Offices, Courts, and Other Regulatory Bodies to Protect Biotechnology: Biotechnology is a rapidly growing field that involves the use of living organisms or their derivatives for various purposes such as healthcare, agriculture, and environmental protection. Given the potential for biotechnology to create significant economic value, various regulatory bodies are responsible for protecting biotech innovations and their associated intellectual property rights. Patent offices play a crucial role in protecting biotechnology by granting patents to new inventions that meet certain criteria such as novelty, non-obviousness, and industrial applicability. Patents give inventors the exclusive right to use, make, and sell their inventions for a certain period, typically 20 years from the date of filing. This protection encourages innovation by providing inventors with a legal framework to invest in and profit from their creations. Courts also play a critical role in protecting biotech innovations. In cases of patent infringement, courts can issue injunctions that prevent infringing parties from using, making, or selling the patented invention. They may also award damages to the patent owner, which can serve as a deterrent to others who might consider infringing on biotech patents. Other regulatory bodies, such as the Food and Drug Administration (FDA) in the United States, play a crucial role in regulating the use of biotech innovations in the public interest. The FDA regulates the use of biotech products in food, drugs, and medical devices, ensuring that they are safe and effective for their intended uses.

Protection Through Patents

Advantages

1. **Exclusive Rights:** Patents provide inventors with exclusive rights to use, make, and sell their inventions, preventing competitors from copying or using their ideas without permission.

2. **Economic Benefits:** Patents can provide inventors with economic benefits through licensing, sale, or commercialization of their inventions.
3. **Incentives for Innovation:** Patents encourage innovation by providing inventors with the legal framework necessary to invest in and profit from their creations.
4. **Public Disclosure:** Patents require inventors to disclose their inventions, providing valuable information to the public and encouraging further research and development.

Disadvantages

1. **Cost:** Obtaining a patent can be expensive due to the fees associated with filing and prosecuting a patent application.
2. **Time:** The patent application process can take several years, delaying the commercialization of an invention.
3. **Limited Duration:** Patents have a limited duration, typically 20 years from the date of filing, after which the invention falls into the public domain and can be used freely by others.
4. **Public Disclosure:** While public disclosure can be an advantage, it can also provide competitors with valuable information about an inventor's technology.

Patentability requirements and criteria: To be patentable, an invention must meet several criteria, including:

1. **Novelty:** The invention must be new and not disclosed in any prior art.
2. **Non-Obviousness:** The invention must not be obvious to a person having ordinary skill in the relevant field.
3. **Industrial Applicability:** The invention must be capable of industrial application and have a practical use.
4. **Enablement:** The patent specification must provide enough detail for a person skilled in the relevant field to reproduce the invention.

Patent prosecution process and strategies: The patent prosecution process involves filing a patent application with the relevant patent office, which then examines the application to determine whether it meets the patentability criteria. Strategies for successful prosecution include:

1. Conducting a thorough prior art search to identify potential obstacles to patentability.

2. Drafting a clear and detailed patent specification that fully describes the invention and meets the enablement criteria.
3. Responding effectively to office actions, which are objections or rejections by the patent office, by providing arguments and amendments to address the examiner's concerns.

Patent enforcement and litigation: If a patent holder believes that their patent rights are being infringed, they can enforce their rights through litigation. Strategies for successful patent litigation include:

1. Conducting a thorough analysis of the patent and potential infringing products or services to identify potential arguments and defenses.
2. Hiring experienced patent attorneys and expert witnesses to assist with the litigation.
3. Seeking an injunction to prevent further infringement.
4. Pursuing damages for lost profits or reasonable royalties caused by the infringement.

Patent protection provides many advantages for biotech inventions, including exclusive rights, economic benefits, and incentives for innovation. However, it also has some disadvantages, such as cost, time, and limited duration. To obtain a patent, inventors must meet the patentability criteria, including novelty, non-obviousness, industrial applicability, and enablement. Patent prosecution requires effective strategies to overcome office actions and obtain a granted patent. If infringement occurs, patent enforcement and litigation can be pursued to protect the patent holder's rights.

Types of Protection in Biotechnology

Trademarks and branding: Trademarks and branding are essential components of any successful business strategy. A trademark is a symbol, word, or phrase used to distinguish the goods or services of one company from those of another. Trademarks can be registered with the relevant national or regional trademark office and provide the owner with exclusive rights to use the mark in connection with their goods or services. Effective branding helps businesses to build customer loyalty, distinguish themselves from their competitors, and promote their products or services in the marketplace.

Copyrights and data protection: Copyrights protect original works of authorship, such as books, music, and artwork, by providing the owner with exclusive rights to reproduce, distribute, and perform the work. Copyrights can be registered with the relevant national or regional copyright office, but registration is not required for protection. Copyrights also protect computer software, which can be a valuable asset in the biotechnology industry.Data protection is another important consideration in the biotech industry, where confidential and proprietary information is often at risk of theft or misuse. Data protection laws, such as the General Data Protection Regulation (GDPR) in the European Union, require companies to take measures to protect personal data and prevent unauthorized access.

Trade secrets and confidentiality agreements: Trade secrets are confidential information that provides a company with a competitive advantage. Examples of trade secrets in the biotech industry include formulas, manufacturing processes, and research data. Trade secrets can be protected by confidentiality agreements, which are legal contracts that require employees, contractors, or other parties to keep certain information confidential.

Confidentiality agreements can be one of the most effective ways to protect trade secrets, but they must be carefully drafted to ensure that they are enforceable. The agreement should clearly define the information that is confidential, the parties to the agreement, and the consequences of breach. It is also important to ensure that all parties to the agreement understand the obligations and restrictions imposed by the agreement.

Protection of Other Biological Materials

Plant varieties and plant patents: Plant varieties can be protected by plant breeders' rights (PBR) or plant patents, depending on the country. PBR provides the breeder with exclusive rights to produce, sell, and distribute the protected variety for a certain period of time. In contrast, a plant patent is a form of intellectual property that provides the owner with exclusive rights to the asexually reproduced plant. To be eligible for a plant patent, the plant must be novel, non-obvious, and distinct from other known plants.

Animal patents and welfare issues: Animal patents are a controversial topic, as they raise ethical and animal welfare concerns. In some countries, animal patents are granted for genetically modified animals or other animals that have been artificially bred or engineered. Animal welfare activists argue that animal patents encourage the commodification of animals and can lead to animal cruelty.In addition to animal patents, there are also animal welfare issues related to the use of animals in biotechnology research. Animal testing is often necessary to ensure the safety and efficacy of new drugs and treatments, but it can also be controversial. Many organizations advocate for the use of alternative methods, such as in vitro testing or computer modeling, to reduce the number of animals used in research.

Traditional knowledge and indigenous rights: Traditional knowledge refers to the knowledge, innovations, and practices of indigenous and local communities that have been developed over generations. In the biotech industry, traditional knowledge can be a valuable source of information for the development of new drugs, agricultural products, and other innovations. However, there are concerns about the exploitation of traditional knowledge without proper recognition or compensation for the communities that hold it.

Indigenous rights are also an important consideration in the biotech industry, as many indigenous communities have a deep connection to the natural resources and biodiversity that are the focus of biotechnology research. There have been efforts to recognize and protect indigenous rights in biotech research, such as the Nagoya Protocol on Access to Genetic Resources and the Fair and Equitable Sharing of Benefits Arising from their Utilization, which aims to ensure that indigenous communities receive fair compensation for the use of their genetic resources and traditional knowledge.

Future Directions and Challenges in Biotechnology IP Rights

Biotechnology is a rapidly evolving field, and the intellectual property landscape is constantly changing. Some of the key challenges facing biotechnology IP rights in the future include:

1. **Emerging technologies:** As new technologies, such as gene editing and synthetic biology, continue to develop, there will be new questions about the patentability and ownership of these technologies.
2. **Access to genetic resources:** The Nagoya Protocol and other international agreements aim to ensure that indigenous communities receive fair compensation for the use of their genetic resources, but there are still challenges in implementing these agreements and ensuring that all parties are treated fairly.
3. **Patent eligibility:** The criteria for patent eligibility in the biotech industry are constantly evolving, and there are ongoing debates about the patentability of certain types of inventions, such as diagnostic tests and gene sequences.
4. **Data protection:** As the amount of data generated by biotech research continues to grow, there will be new challenges in protecting this data and ensuring that it is not misused or accessed by unauthorized parties.

Importance of Balancing Innovation and Ethical Concerns: While intellectual property rights are important for promoting innovation and investment in the biotech industry, it is also important to balance these concerns with ethical considerations. Some of the key ethical concerns in biotechnology include:

1. **Access to healthcare:** Biotech innovations can provide life-saving treatments and cures, but these treatments are often expensive and may not be accessible to everyone who needs them.
2. **Animal welfare:** As discussed earlier, there are ethical concerns about the use of animals in biotech research and the granting of animal patents.
3. **Human rights:** Biotech research also raises important questions about human rights, such as the right to privacy and the right to control one's own genetic information.
4. **Environmental concerns:** Biotech research can have significant environmental impacts, such as the release of genetically modified organisms into the environment.

Balancing these ethical concerns with the need for innovation and investment in the biotech industry requires careful consideration and dialogue between stakeholders. It is important to ensure that the benefits of biotech innovations are accessible to everyone, while also protecting the rights and welfare of animals, indigenous communities, and other stakeholders.

10

Protecting Biodiversity: The Importance of Conservation and Sustainable Use

Introduction

Biodiversity conservation and sustainable use are critical for maintaining the health and well-being of the Earth and all its inhabitants. Biodiversity provides essential ecosystem services, resources, and cultural values, and is under threat from various human activities. Conservation and sustainable use are necessary for preserving biodiversity and ensuring a sustainable future, and require the participation and collaboration of all stakeholders.

The protection and restoration of biodiversity are complex challenges that require a multidisciplinary approach and a commitment to sustainability. This includes addressing the underlying drivers of biodiversity loss, promoting sustainable land-use practices, protecting ecosystems and wildlife, and engaging in education and awareness-raising activities. By working together to conserve and sustainably use of biodiversity, we can ensure a healthy planet for future generations.

Understanding Biodiversity

Biodiversity refers to the variety of living organisms present on Earth, including species diversity, genetic diversity, and ecosystem diversity. It is important because it supports ecosystem functioning, provides essential goods and services to humans, and contributes to cultural, spiritual, and aesthetic values.

Types of biodiversity include

1. **Species diversity:** This refers to the variety of species present in an ecosystem, ranging from microorganisms to plants and animals.
2. **Genetic diversity:** This refers to the variation of genes within and among populations of the same species.
3. **Ecosystem diversity:** This refers to the variety of ecosystems, including terrestrial, freshwater, and marine ecosystems, and the interactions among them.

The current state of global biodiversity loss is alarming. According to the Intergovernmental Science-Policy Platform on Biodiversity and Ecosystem Services (IPBES), one million species are at risk of extinction due to human activities such as habitat destruction, overexploitation, pollution, and climate change. This loss of biodiversity has severe implications for human well-being, including food security, health, and livelihoods, and threatens the stability of ecosystems and the services they provide. Conservation efforts and sustainable management practices are crucial to mitigate biodiversity loss and ensure a healthy and resilient planet for future generations.

Threats to Biodiversity

1. **Habitat loss and fragmentation:** Habitat loss refers to the destruction or conversion of natural habitats, such as forests, grasslands, and wetlands, into human-made landscapes for agriculture, urbanization, and infrastructure development. Fragmentation occurs when habitats are divided into smaller and isolated patches, resulting in the loss of connectivity and reducing the size of viable populations. Habitat loss and fragmentation are major threats to biodiversity, as they can lead to species extinctions, loss of genetic diversity, and changes in ecosystem functioning.
2. **Climate change:** Climate change refers to the long-term changes in the Earth's climate, such as rising temperatures, changing precipitation patterns, and increasing frequency and intensity of extreme weather events. Climate change affects biodiversity by altering habitats, disrupting species interactions, and causing shifts in the timing of ecological events. For example, rising temperatures can shift the distribution of species, while changing precipitation patterns can affect the availability of water resources. Climate change also exacerbates other threats to biodiversity, such as habitat loss and fragmentation.
3. **Overexploitation and illegal wildlife trade:** Overexploitation refers to the unsustainable harvesting of wildlife, including hunting, fishing, and logging, for commercial or subsistence purposes. Overexploitation can lead to population declines, loss of genetic diversity, and changes in ecosystem structure and function. Illegal wildlife trade involves the unlawful harvest and trade of wildlife, including endangered species, for use in traditional medicines, luxury goods, and the pet trade. Overexploitation and illegal wildlife trade are major threats to biodiversity, particularly in developing countries where weak governance, poverty, and limited law enforcement make it difficult to control.
4. **Invasive species:** Invasive species refer to non-native species that are introduced to new environments, either intentionally or unintentionally,

and that cause harm to native species, ecosystems, and human health and economies. Invasive species can outcompete native species for resources, alter ecosystems through changes in nutrient cycling and fire regimes, and transmit diseases to native species. Invasive species are a major threat to biodiversity worldwide, and their impacts are projected to increase with global trade, travel, and climate change.

Conservation Strategies

1. **Protected areas and their role in biodiversity conservation:** Protected areas are designated areas that are managed to conserve biodiversity and ecosystem services. They include national parks, wildlife reserves, and other conservation areas. Protected areas play a critical role in conserving biodiversity by providing habitat for wildlife, protecting key ecological processes, and ensuring the provision of ecosystem services. Protected areas also contribute to sustainable development by supporting tourism, recreation, and cultural activities. However, the effectiveness of protected areas depends on adequate funding, management, and enforcement of regulations.

2. **Restoration and rehabilitation of degraded ecosystems:** Restoration and rehabilitation refer to the process of returning degraded ecosystems to their original state or improving their ecological function. Restoration can include activities such as reforestation, wetland restoration, and soil conservation. Restoring degraded ecosystems can help to enhance biodiversity, increase the provision of ecosystem services, and reduce the impacts of climate change. Successful restoration requires careful planning, monitoring, and adaptive management.

3. **Species conservation and management:** Species conservation and management involve actions aimed at conserving endangered or threatened species. This can include habitat protection, captive breeding and reintroduction, and control of invasive species. Species conservation and management are critical for maintaining biodiversity and preventing extinctions. It is essential to involve local communities and stakeholders in species conservation efforts to ensure their long-term success.

4. **Community-based conservation:** Community-based conservation involves the participation of local communities in conservation efforts. It recognizes the importance of traditional knowledge and local practices in biodiversity conservation and seeks to involve communities in decision-making and management. Community-based conservation can lead to more sustainable and effective conservation outcomes by promoting local

stewardship, reducing conflicts, and improving livelihoods. However, successful community-based conservation requires strong partnerships, clear communication, and equitable distribution of benefits.

Sustainable Use of Biodiversity

1. **Sustainable use principles and practices:** Sustainable use principles and practices refer to the responsible use of natural resources to ensure their long-term availability and to maintain ecosystem integrity. Sustainable use includes practices such as sustainable agriculture, forestry, and fishing, which ensure that natural resources are used in a way that supports ecological and social sustainability. Sustainable use principles also promote the conservation of biodiversity by minimizing negative impacts on ecosystems and species.
2. **Ecotourism as a tool for conservation and sustainable use:** Ecotourism refers to responsible travel to natural areas that conserves the environment and improves the well-being of local people. Ecotourism can provide economic incentives for conservation and sustainable use by creating jobs, generating revenue for local communities, and promoting the conservation of biodiversity. Ecotourism also raises awareness about the value of natural resources and supports the development of sustainable practices.
3. **Sustainable harvesting of natural resources:** Sustainable harvesting of natural resources involves the extraction of resources in a way that ensures their long-term availability and maintains ecosystem integrity. Sustainable harvesting can include practices such as selective logging, sustainable fishing, and responsible hunting, which aim to minimize negative impacts on ecosystems and species. Sustainable harvesting also promotes the conservation of biodiversity by ensuring that resources are used in a way that supports ecological and social sustainability.
4. **Bioprospecting and benefit sharing:** Bioprospecting refers to the exploration of biodiversity for commercially valuable genetic and biochemical resources. Benefit sharing involves the fair and equitable distribution of benefits arising from bioprospecting to local communities and countries that provide the genetic resources. Bioprospecting and benefit sharing can provide economic incentives for biodiversity conservation and support the development of sustainable practices. However, it is essential to ensure that benefit sharing is transparent, equitable, and culturally appropriate to avoid exploitation and to promote the conservation of biodiversity.

Biodiversity Governance

Efforts at the international level to conserve biodiversity are critical given the global nature of biodiversity and the challenges facing its conservation. Two important international frameworks for biodiversity conservation are:

a) **Convention on Biological Diversity (CBD):** The CBD is a multilateral treaty signed by over 190 countries, which aims to promote the conservation and sustainable use of biodiversity, as well as the equitable sharing of benefits derived from genetic resources. The CBD has three main objectives: the conservation of biodiversity, the sustainable use of its components, and the fair and equitable sharing of benefits arising from the use of genetic resources. The CBD provides a framework for the development of national biodiversity strategies and action plans and for the implementation of measures to conserve biodiversity.

b) **International Treaty on Plant Genetic Resources for Food and Agriculture (ITPGRFA):** The ITPGRFA is a global agreement that aims to ensure the conservation and sustainable use of plant genetic resources for food and agriculture (PGRFA), as well as the fair and equitable sharing of benefits arising from their use. The ITPGRFA recognizes the role of farmers in the conservation and development of PGRFA and encourages the sustainable use of these resources to support food security and sustainable agriculture.

These international frameworks provide a basis for global cooperation on biodiversity conservation and offer guidance on the development of national policies and strategies. They also emphasize the importance of equitable sharing of benefits and recognize the role of local communities and indigenous peoples in the conservation and sustainable use of biodiversity.

Efforts at the national level are critical for biodiversity conservation, as each country has unique biodiversity and conservation challenges that require tailored approaches. Important efforts in India are:

1. **Indian Biological Diversity Act, 2002:** The Indian Biological Diversity Act, 2002 provides a legal framework for the conservation and sustainable use of India's biological resources. The Act aims to regulate access to biological resources and associated traditional knowledge, as well as ensure the fair and equitable sharing of benefits arising from their use. The Act also establishes the National Biodiversity Authority, State Biodiversity Boards, and Biodiversity Management Committees at the local level to oversee the implementation of the Act.
2. Indian Biological Diversity Salient features and functioning:

i. The Indian Biological Diversity Act, 2002 aims to conserve biological diversity, promote sustainable use, and ensure fair and equitable sharing of benefits from the use of biological resources.
ii. The Act establishes the National Biodiversity Authority, State Biodiversity Boards, and Biodiversity Management Committees to oversee biodiversity conservation and management.
iii. The Act includes provisions for the registration of access to biological resources and associated knowledge, as well as the establishment of a national registry for such access.
iv. The Act recognizes the rights of local communities and indigenous peoples in the conservation and sustainable use of biological resources and associated traditional knowledge.

Legal and Contractual Mechanisms for Benefit Sharing and Technology Transfer

a) **Legislation for benefit sharing:** Benefit sharing is an important aspect of biodiversity conservation and sustainable use, as it ensures that the benefits arising from the use of genetic resources are fairly and equitably shared with the providers of those resources. Many countries have enacted legislation that requires benefit sharing in cases where genetic resources are accessed for commercial purposes. The Convention on Biological Diversity (CBD) also includes provisions on benefit sharing.

b) **Licensing of technologies:** Licensing of technologies is a mechanism for the transfer of technology from one party to another. In the context of biodiversity conservation and sustainable use, licensing can be used to facilitate the transfer of technologies that support conservation and sustainable use practices, such as those related to habitat restoration or sustainable harvesting of natural resources.

c) **Material transfer agreements:** Material transfer agreements (MTAs) are legal agreements that govern the transfer of biological materials, such as genetic resources, from one party to another. MTAs can include provisions related to benefit sharing, intellectual property rights, and confidentiality.

d) **Research collaboration agreement:** Research collaboration agreements are legal agreements that govern the collaboration between two or more parties in research activities. In the context of biodiversity conservation and sustainable use, research collaboration agreements can facilitate collaborative research on conservation and sustainable use practices, such as those related to ecosystem restoration or sustainable management of natural resources.

National policies and strategies play a crucial role in promoting biodiversity conservation by providing a framework for coordinated and effective action. These policies and strategies can help to align efforts among different sectors and stakeholders, promote sustainable use of natural resources, and protect important habitats and species.

Role of Local Communities in Biodiversity Governance

a) Local communities are key stakeholders in biodiversity governance as they rely on natural resources for their livelihoods and have traditional knowledge and practices that can inform conservation efforts.

b) Involving local communities in decision-making and management can lead to more sustainable and effective conservation outcomes by promoting local stewardship, reducing conflicts, and improving livelihoods.

c) It is important to ensure that local communities have equitable access to benefits arising from biodiversity conservation and that their rights and interests are protected.

d) Participatory approaches such as community-based natural resource management (CBNRM) and co-management can facilitate the involvement of local communities in biodiversity governance.

e) In some cases, the recognition and protection of customary land and resource rights may be necessary to ensure effective local participation in biodiversity governance.

f) The United Nations Declaration on the Rights of Indigenous Peoples recognizes to maintain and develop their traditional knowledge, practices, and innovations, and to have their lands and territories protected and respected.

g) Effective engagement with local communities requires building trust, promoting transparency & establishing effective communication channels.

h) Private sector engagement in biodiversity conservation and sustainable use: The private sector can play an important role in biodiversity conservation and sustainable use by adopting sustainable practices and investing in conservation efforts. Private sector engagement can include initiatives such as corporate social responsibility programs, sustainable sourcing, and biodiversity offsets. It is essential to ensure that private sector engagement is transparent, accountable, and aligned with biodiversity conservation goals. Public-private partnerships can also help to leverage private sector resources and expertise to promote effective biodiversity conservation.

Monitoring and Assessment

1. **Biodiversity monitoring and assessment methodologies:** Biodiversity monitoring and assessment methodologies are used to measure changes in biodiversity over time and to identify conservation and management priorities. These methodologies can include techniques such as species inventories, habitat mapping, and population monitoring. Biodiversity assessments can provide valuable information on the status and trends of biodiversity and can help to inform decision-making and management actions.
2. **Use of indicators to measure progress towards conservation and sustainable use goals:** Indicators are used to measure progress towards conservation and sustainable use goals and to assess the effectiveness of management actions. Indicators can include measures of biodiversity, ecosystem services, and human well-being. Developing indicators that are scientifically robust, easy to measure, and relevant to stakeholders can help to promote effective monitoring and evaluation of conservation and sustainable use efforts.
3. **Importance of data sharing and collaboration:** Data sharing and collaboration are essential for effective biodiversity conservation and management. Sharing data and information among stakeholders can help to identify knowledge gaps, inform decision-making, and promote transparency and accountability. Collaboration can also help to leverage resources and expertise to promote more effective conservation outcomes. It is essential to ensure that data sharing and collaboration are conducted in a way that respects intellectual property rights, protects sensitive information, and promotes equitable access to benefits arising from biodiversity conservation.

Challenges and Opportunities

1. **Key challenges to biodiversity conservation and sustainable use:** Biodiversity conservation and sustainable use face several challenges, including habitat loss and fragmentation, climate change, overexploitation and illegal wildlife trade, invasive species, pollution, and unsustainable use of natural resources. These challenges are often interlinked and require integrated and cross-sectoral approaches to address them effectively.
2. **Opportunities for innovation and collaboration:** Innovative solutions and collaboration among different sectors and stakeholders can provide opportunities to address the challenges facing biodiversity conservation and sustainable use. Examples of innovative solutions include the use of

technology to monitor and manage biodiversity, the development of new sustainable practices, and the promotion of ecosystem-based approaches to management. Collaboration can help to leverage resources, expertise, and knowledge to promote more effective conservation outcomes.

3. **Role of education and awareness-raising in promoting biodiversity conservation and sustainable use:** Education and awareness-raising are critical to promoting biodiversity conservation and sustainable use. By raising public awareness of the importance of biodiversity and the threats it faces, individuals and communities can become more engaged in conservation efforts. Education can also help to build the capacity of stakeholders to participate in decision-making and management actions. Education and awareness-raising efforts can include formal and informal education, outreach programs, and public engagement initiatives.

Conclusion

Biodiversity is the variety of life on Earth and is essential for human well-being and the functioning of ecosystems. However, biodiversity is currently under threat from a range of human activities. To promote biodiversity conservation and sustainable use, a range of approaches are required, including the protection of habitats and ecosystems, the restoration of degraded areas, the sustainable use of natural resources, and the promotion of education and awareness-raising. Effective biodiversity conservation and sustainable use require collaboration among different sectors and stakeholders, as well as the use of innovative solutions and the sharing of data and knowledge.

i. **Call to action for biodiversity conservation and sustainable use:** Given the urgent need to address the challenges facing biodiversity, there is a need for action from individuals, communities, governments, and the private sector. This includes supporting initiatives that promote biodiversity conservation and sustainable use, reducing our ecological footprint, advocating for policy changes, and engaging in education and awareness-raising efforts.

ii. **Future directions for research and policy:** Future research and policy efforts should focus on developing innovative solutions to address the challenges facing biodiversity conservation and sustainable use. This includes exploring the potential of new technologies, promoting the use of ecosystem-based approaches to management, and enhancing collaboration among stakeholders. Policy efforts should aim to promote integrated and cross-sectoral approaches to conservation and management, as well as the equitable sharing of benefits arising from biodiversity conservation.

11

Biosafety Regulations and Risk Assessment: Promoting Safe and Responsible Use of Biotechnology

Biotechnology has the potential to revolutionize many aspects in our lives, from healthcare to agriculture. However, along with its benefits, it also poses significant risks to human health, environment, and ecology. Therefore, it is essential to establish biosafety regulations and risk assessment frameworks to promote safe and responsible use of biotechnology. This chapter compiled an overview of biotechnology, the importance of biosafety regulations and risk assessment, IPR protection, contaminants, risk analysis, risk management in laboratories and fields, biosafety concerns to human health, environment and ecology, biosafety at international and national levels, bioethics, and general principles.

Overview of Biotechnology and its Potential Benefits and Risks

Biotechnology refers to the use of living organisms or their components to develop products and processes for various fields such as healthcare, agriculture, and environmental remediation. Biotechnology applications include genetic engineering, bioremediation, biopharmaceuticals, and biobased products.

Potential Benefits: Biotechnology has the potential to provide numerous benefits, including:

i. Improved crop yields through genetically modified crops

ii. More effective disease treatments through gene therapy and recombinant proteins

iii. New sources of energy through biofuels and biopower

iv. Bioremediation of contaminated environments using microorganisms

v. Improved food quality and safety through genetic engineering and other biotechnology applications

vi. Increased efficiency in industrial processes through biocatalysts and bioreactors

Potential Risks: Biotechnology also poses significant risks, including:

i. Potential for unintended consequences and unintended harm to human health, the environment, and ecology
ii. Potential for the development of antibiotic-resistant bacteria
iii. Potential for genetic contamination of non-target organisms and ecosystems
iv. Potential for negative impacts on traditional farming practices and indigenous communities
v. Concerns about the long-term effects of genetically modified organisms (GMOs) on human health and the environment

Ethical Concerns: Biotechnology raises ethical concerns related to the manipulation of living organisms and the potential for unintended consequences. These concerns include issues related to human dignity, social justice, and environmental stewardship.

Regulatory Frameworks: Governments around the world have established regulatory frameworks to manage the risks associated with biotechnology. These frameworks include risk assessments, risk management strategies, and regulations governing the use and release of genetically modified organisms.

Importance of Biosafety Regulations and Risk Assessment

1. **Biosafety regulations prevent harm**: Regulations ensure safe use, transport, and disposal of biotech products.
2. **Risk assessment evaluates hazards:** Frameworks identify risks and guide management decisions.
3. **Risk management is guided by assessment:** Decisions use assessment results to mitigate risks.
4. **Regulations minimize risks:** They control product use, transport, and disposal to ensure safety.
5. **Risk assessment improves products:** Identifying hazards early improves safety and risk management.
6. **Biosafety and risk assessment build public confidence:** Public trust is gained with effective biosafety and risk assessment.
7. **Biosafety and risk assessment support sustainability:** They promote safe use, minimizing potential harm to people and the environment.

Biosafety regulations and risk assessment are crucial to promote safe and responsible use of biotechnology. These measures help to ensure that biotechnology products are safe for humans, environment, and ecology, and they support sustainable development.

IPR Protection, Biosafety, and Bioethics

i. **IPR protection encourages innovation:** Legal rights provided by IPR protection encourage innovation and investment in biotechnology while ensuring fair access to its benefits.

ii. **Biosafety minimizes risks:** Biosafety measures minimize potential risks associated with biotechnology products to ensure their safety for human health, environment, and ecology.

iii. **Bioethics addresses ethical concerns:** Bioethics provides a framework for addressing ethical concerns surrounding the use of biotechnology products, including issues of equity and justice.

iv. **IPR and biosafety are interdependent:** IPR protection and biosafety regulations are interdependent, promoting investment and innovation while ensuring safe and environmentally responsible products.

v. **Ethical considerations are crucial:** Ethical considerations ensure that the use of biotechnology products aligns with societal values and principles.

Integration of IPR, biosafety, and bioethics promotes responsible development: The integration of IPR protection, biosafety, and bioethics promotes safe, ethical, and sustainable biotechnology development, benefiting society as a whole.

Contaminants, Risk Analysis, Risk assessment and Risk Management in Laboratories and Fields

i. **Contaminants**, such as chemicals or biological agents, can pose a significant risk to human health and the environment if not properly handled in laboratories and fields. Risk analysis is a crucial step in identifying and evaluating potential hazards associated with biotechnology products and contaminants.

ii. **Risk Analysis,** Biosafety risk analysis is the process of identifying, assessing, and controlling potential risks associated with biological materials, to ensure safe and responsible handling and use. This involves identifying hazards, evaluating the likelihood and severity of harm, implementing strategies to mitigate or eliminate risks, and communicating with stakeholders about potential risks and safety measures. Biosafety risk analysis is crucial for minimizing potential risks to humans, animals, plants, and the environment in biological research.

iii. **Risk assessment,** biosafety risk assessment is a crucial step in biosafety risk analysis that evaluates the likelihood and severity of harm from exposure to biological materials. The assessment considers various factors such as the virulence of the organism, the route and duration of exposure, and the susceptibility of the population. The results of the assessment helps in identifying appropriate control measures to mitigate or eliminate potential risks, such as personal protective equipment, engineering or administrative controls. The biosafety risk assessment helps to ensure the safe handling and use of biological materials in research and other activities.

iv. Risk management measures, including the use of personal protective equipment, proper storage and handling procedures, and safe disposal methods, are essential to prevent or mitigate the risks associated with biotechnology products and contaminants. In laboratories, strict biosafety measures and procedures must be followed to prevent contamination and accidental exposure to hazardous materials. Similarly, in agricultural fields, proper handling and disposal of pesticides and fertilizers are crucial to prevent environmental contamination and health risks. Effective risk management in laboratories and fields helps to ensure the safe and responsible use of biotechnology products and protects human health and the environment from potential harm. It also helps in maintaining public confidence in biotechnology and its applications.

Good laboratory and field practices: Good laboratory and field practices are essential to ensure biosafety in research and other activities involving biological materials. Some of the best practices for biosafety include:

i. **Use of personal protective equipment (PPE):** Gloves, laboratory coats, and other protective clothing are essential for minimizing the risk of exposure to biological materials.

ii. **Implementation of engineering controls:** This includes measures such as biosafety cabinets, fume hoods, and other engineering controls to minimize the risk of exposure to biological materials.

iii. **Proper handling and storage of biological materials:** This includes the proper labeling, packaging, and storage of biological materials to minimize the risk of exposure and cross-contamination.

iv. **Implementation of standard operating procedures (SOPs):** This involves developing and implementing written procedures for handling biological materials, including emergency procedures.

v. **Training of personnel:** Training should be provided to all personnel working with biological materials to ensure that they are aware of the potential risks and know how to handle the materials safely.

vi. **Regular inspection and maintenance of equipment:** This ensures that equipment is functioning correctly and reduces the risk of accidents and contamination.

vii. **Waste management:** Proper disposal of biological waste is essential to minimize the risk of exposure to biological materials.

viii. **Environmental monitoring:** Regular monitoring of the laboratory or field environment is important to identify potential hazards and ensure that biosafety measures are effective.

Biosafety Concerns to Human Health, Environment, and Ecology: Biotechnology products can have a range of potential impacts on human health, the environment, and ecology. Some of these risks include:

1. **Human Health:** Biotechnology products can pose risks to human health, such as through exposure to harmful chemicals or allergens. In addition, biotechnology products used in medicine can have unknown long-term effects on human health.
2. **Environment:** Biotechnology products can also have a negative impact on the environment. For example, genetically modified crops may harm non-target organisms, such as insects or birds, or lead to unintended changes in the ecosystem.
3. **Ecology:** Biotechnology products can also affect the ecology of an area, such as through the spread of invasive species or changes to the natural balance of an ecosystem.

To address these risks, biosafety regulations and risk assessment frameworks are put in place. These regulations require appropriate labeling, testing, and monitoring of biotechnology products to ensure they are safe for humans, environment, and ecology. Risk assessment frameworks help to identify potential hazards and guide risk management decisions to prevent or mitigate these risks. Overall, biosafety concerns to human health, environment, and ecology must be carefully considered to promote the responsible and sustainable use of biotechnology.

Biosafety at National and International Levels

Biosafety regulations and risk assessment frameworks exist at both international and national levels. International frameworks include the Cartagena Protocol on Biosafety, which regulates the transboundary movement of biotechnology products, and the Convention on Biological Diversity, which aims to conserve biodiversity and promote the sustainable use of natural resources. National biosafety regulations and risk assessment frameworks vary by country but aim to promote safe and responsible use of biotechnology products.

Here are some detailed points on biosafety at international and national levels:

1. **International biosafety frameworks:** The Cartagena Protocol on Biosafety is a legally binding international agreement governing the transboundary movement of living modified organisms (LMOs), such as genetically modified organisms (GMOs). It aims to protect biological diversity and human health by regulating the import, export, and transit of LMOs. The Convention on Biological Diversity also addresses biosafety issues, including the conservation and sustainable use of biodiversity, the fair and equitable sharing of benefits arising from the use of genetic resources, and the safe transfer, handling, and use of living organisms.
2. **National biosafety regulations:** India has its own set of biosafety regulations and risk assessment frameworks. The Ministry of Environment, Forest and Climate Change (MoEFCC) has established the National Biodiversity Authority (NBA) and the Genetic Engineering Appraisal Committee (GEAC) to regulate the use of biotechnology products. The NBA oversees the conservation and sustainable use of India's biological resources, while the GEAC evaluates the safety and environmental impact of genetically engineered organisms and products before their release into the environment. In addition, the Department of Biotechnology (DBT) and the Department of Science and Technology (DST) also play a role in regulating biotechnology research and development in India. These regulatory bodies work to ensure that biotechnology products are safe for human health, the environment, and ecology, while promoting innovation, investment, and fair access to its benefits.
3. **The importance of international cooperation:** Biosafety regulations and risk assessment frameworks require international cooperation to be effective. Countries must work together to share information, establish common standards, and monitor compliance with biosafety regulations. International organizations such as the World Health Organization (WHO) and the World Trade Organization (WTO) play important roles in facilitating this cooperation.
4. **Biosafety training and capacity-building:** Biosafety regulations and risk assessment frameworks require specialized knowledge and expertise. Many countries, especially those with limited resources, require training and capacity-building programs to develop their biosafety infrastructure. International organizations such as the International Atomic Energy Agency (IAEA) and the United Nations Environment Programme (UNEP) provide assistance to developing countries in this area.
5. **Emerging biosafety challenges:** As biotechnology continues to evolve, biosafety challenges also evolve. New technologies such as gene editing

and synthetic biology raise new biosafety concerns that must be addressed. International and national biosafety frameworks must adapt to these emerging challenges to ensure the continued safe and responsible use of biotechnology products.

Bioethics and General Principles: Bioethics concerns the ethical issues surrounding the use of biotechnology products, including issues related to human dignity, social justice, and environmental sustainability. General principles for promoting safe and responsible use of biotechnology include transparency, stakeholder involvement, risk communication, and continuous monitoring and evaluation.

Biosafety regulations and risk assessment frameworks play a crucial role in ensuring the safe and responsible use of biotechnology products. Such frameworks should prioritize the protection of human health, the environment, and ecology, while also promoting innovation, investment, and fair access to the benefits of biotechnology. By implementing effective biosafety regulations and risk assessment frameworks at both international and national levels, we can ensure that biotechnology is developed and used in a safe, ethical, and sustainable manner, benefiting society as a whole.

Subject Index